T0180991

Environmental Chemistry for a Sustainable World

Volume 34

Series Editors
Eric Lichtfouse, Aix Marseille University, CNRS, IRD, INRA, Coll France, CEREGE, Aix-en-Provence, France
Jan Schwarzbauer, RWTH Aachen University, Aachen, Germany
Didier Robert, CNRS, European Laboratory for Catalysis and Surface Sciences, Saint-Avold, France

Other Publications by the Editors

Books
Environmental Chemistry
http://www.springer.com/978-3-540-22860-8

Organic Contaminants in Riverine and Groundwater Systems
http://www.springer.com/978-3-540-31169-0

Sustainable Agriculture
Volume 1: http://www.springer.com/978-90-481-2665-1
Volume 2: http://www.springer.com/978-94-007-0393-3

Book series
Environmental Chemistry for a Sustainable World
http://www.springer.com/series/11480

Sustainable Agriculture Reviews
http://www.springer.com/series/8380

Journals
Environmental Chemistry Letters
http://www.springer.com/10311

More information about this series at http://www.springer.com/series/11480

Mu. Naushad • Saravanan Rajendran
Eric Lichtfouse

Editors

Green Photocatalysts

 Springer

Editors
Mu. Naushad
Department of Chemistry,
College of Science
King Saud University
Riyadh, Saudi Arabia

Saravanan Rajendran
Faculty of Engineering,
Department of Mechanical Engineering
University of Tarapacá
Arica, Chile

Eric Lichtfouse
Aix Marseille University
CNRS, IRD, INRA, Coll France,
CEREGE
Aix-en-Provence, France

ISSN 2213-7114   ISSN 2213-7122   (electronic)
Environmental Chemistry for a Sustainable World
ISBN 978-3-030-15610-7   ISBN 978-3-030-15608-4   (eBook)
https://doi.org/10.1007/978-3-030-15608-4

This Springer imprint is published by the registered company Springer Nature Switzerland AG.
The registered company address is: Gewerbestrasse 11, 6330 Cham, Switzerland

Preface

Water is one of the most important substances on earth. All plants and animals must have water to survive. If there was no water there would be no life on earth. (WHO report)

The industrial and population growth is affecting our water body system. Especially, the wastage and hazardous and toxic chemical from the leather, textile, and pharmaceutical industries are spoiling our water resource. In this connection, "photocatalyst" is one of the finest green technologies to remove the contaminations from the water bodies. This method has numerous advantages including safe, clean, cost-effective, suitable, and green method for effective degradation of water contaminations. This book contains ten chapters; first two chapters described the general principle, definition, synthesis of green catalysts, description, chemical reaction, and mechanism of photocatalyst. The remaining chapters of this book deal with the depth analysis of photocatalyst technology using different catalysts such as:

  (i) Nanostructured catalysts
 (ii) Semiconductors and metal- and nonmetal-doped catalyst
(iii) Surface plasmon materials
 (iv) Graphene oxide-based materials
  (v) Polymer-based composite materials
 (vi) Heterogenous type I and type II catalysts

Riyadh, Saudi Arabia   Mu. Naushad
Arica, Chile   Saravanan Rajendran
Aix-en-Provence, France   Eric Lichtfouse

Acknowledgments

We would like to first and foremost thank God for giving us good health to complete this book successfully.

We acknowledge our sincere gratitude to the Springer for accommodating this book as part of the series "Environmental Chemistry for a Sustainable World." Further, we extend our heartfelt thanks to series editor and advisory board for accepting our book as a part of this series. Without reviewers and contributing authors, we could not complete this task. We are very grateful to reviewers and contributing authors for their valuable involvement throughout this book. We would like to express our sincere thanks to the researchers and publisher for permitting us the copyright to use their figures and tables. We would still like to offer our deep apologies to any copyright holder if unknowingly their right is being overstepped.

R. Saravanan would like to express his sincere thanks to Prof. Francisco Gracia (DIQBT, University of Chile), Prof. Lorena Cornejo Ponce (EUDIM, Universidad de Tarapacá), and Prof. Rodrigo Palma (Director, SERC) for their constant encouragement and valuable support that helped him to complete the task. He further extends his thanks to the Government of Chile (CONICYT-FONDECYT-Project No.: 11170414), SERC (CONICYT/FONDAP/15110019), and School of Mechanical Engineering (EUDIM), Universidad de Tarapacá, Arica, Chile, for their financial support.

Dr. Mu. Naushad expresses his deep gratitude to the chairman, Department of Chemistry, College of Science, King Saud University, Saudi Arabia, and extends his appreciation to the Deanship of Scientific Research at King Saud University for the support.

Mu. Naushad
Saravanan Rajendran
Eric Lichtfouse

Contents

About the Editors

Dr. Mu. Naushad is presently working as an associate professor in the Department of Chemistry, College of Science, King Saud University (KSU), Riyadh, Kingdom of Saudi Arabia. He obtained his M.Sc. and Ph.D. in Analytical Chemistry from Aligarh Muslim University, Aligarh, India, in 2002 and 2007, respectively. He has a vast research experience in the fields of Analytical Chemistry, Materials Chemistry, and Environmental Science. He holds several US patents, over 250 publications in the international journals of repute, 20 book chapters, and several books published by renowned international publishers. He has >7200 citations with a Google Scholar h-index of >52. He has successfully run several research projects funded by the National Plan for Science and Technology (NPST) and King Abdulaziz City for Science and Technology (KACST), Kingdom of Saudi Arabia. He is the editor/editorial member of several reputed journals like *Scientific Reports* (Nature), *Process Safety and Environmental Protection* (Elsevier), *Journal of Water Process Engineering* (Elsevier), and *International Journal of Environmental Research and Public Health* (MDPI). He is also the associate editor for *Environmental Chemistry Letters* (Springer) and *Desalination and Water Treatment* (Taylor and Francis). He has been awarded the Scientist of the Year Award 2015 by the National Environmental Science Academy, New Delhi, India, Scientific Research Quality Award 2019, King Saud University and Almarai Award 2017, Saudi Arabia.

Dr. Saravanan Rajendran has received his Ph.D. in Physics-Material Science in 2013 from the Department of Nuclear Physics, University of Madras, Chennai, India. He was awarded the University Research Fellowship (URF) during the year 2009–2011 by the University of Madras. After working as an assistant professor in Dhanalakshmi College of Engineering, Chennai, India, during the year 2013–2014, he was awarded SERC and CONICYT-FONDECYT Postdoctoral Fellowship, University of Chile, Santiago, Chile, in the year 2014–2017. He has worked (2017–2018) in the research group of Professor John Irvine, School of Chemistry, University of St Andrews, UK, as a postdoctoral research fellow within the framework of an EPSRC Global Challenges Research Fund for the removal of blue-green algae and their toxins. Currently, he is working as a research scientist in the School of Mechanical Engineering (EUDIM), University of Tarapacá, Arica, Chile, as well as a research associate in SERC, University of Chile, Santiago, Chile. He is associate editor for *International Journal of Environmental Science and Technology* (Springer). His research interests focuses in the area of nanostructured functional materials, hotophysics, surface chemistry, and nanocatalysts for renewable energy and waste water purification. He has published several international peer-reviewed journals, five book chapters, and three books published by renowned international publishers.

Dr. Eric Lichtfouse (Ph.D.) born in 1960, is an environmental chemist working at the University of Aix-Marseille, France. He has invented carbon-13 dating, a method allowing to measure the relative age and turnover of molecular organic compounds occurring in different temporal pools of any complex media. He is teaching scientific writing and communication and has published the book *Scientific Writing for Impact Factor Journals*, which includes a new tool – the micro-article – to identify the novelty of research results. He is founder and chief editor of scientific journals and series in environmental chemistry and agriculture. He got the Analytical Chemistry Prize by the French Chemical Society, the Grand Prize of the Universities of Nancy and Metz, and a Journal Citation Award by the Essential Indicators.

About the Authors

Rosalin Beura Centre for Nanoscience and Technology, Pondicherry University, Puducherry, India

Rabah Boukherroub University Lille, CNRS, Centrale Lille, ISEN, University Valenciennes, UMR 8520–IEMN, Lille, France

Moo Hwan Cho School of Chemical Engineering, Yeungnam University, Gyeongsan-si, Gyeongbuk, South Korea

David Contreras Facultad de Ciencias Químicas, Centro de Biotecnología, Universidad de Concepción, Concepción, Chile

Anitha Devadoss Centre for Nano Health, College of Engineering, Swansea University, Swansea, Wales, UK

C. Ravi Dhas PG & Research Department of Physics, Bishop Heber College, Trichy, Tamil Nadu, India

D. Duraibabu The Key Laboratory of Low-Carbon Chemistry & Energy Conservation of Guangdong Province/State Key Laboratory of Optoelectronic Materials and Technologies, School of Materials Science and Engineering, Sun Yat-Sen University, Guangzhou, People's Republic of China

D. Durgalakshmi Department of Medical Physics, Anna University, Chennai, Tamil Nadu, India

Pravin K. Dwivedi Physical & Material's Chemistry Division, CSIR-National Chemical Laboratory, Pune, India

Lisdelys González Centro de Biotecnología, Universidad de Concepción, Concepción, Chile

Adolfo Henríquez Centro de Biotecnología, Universidad de Concepción, Concepción, Chile

Sagar M. Jain Multi-functional Photocatalyst & Coatings Group, SPECIFIC, College of Engineering, Swansea University (Bay Campus), Swansea, Wales, UK

J. Nimita Jebaranjitham P.G. Department of Chemistry, Women's Christian College, Chennai, TN, India

Mohammad Ehtisham Khan Department of Chemical Engineering and Technology, College of Applied Industrial Technology (CAIT), Jazan University, Jazan, Kingdom of Saudi Arabia

School of Chemical Engineering, Yeungnam University, Gyeongsan-si, Gyeongbuk, South Korea

D. David Kirubakaran PG & Research Department of Physics, Bishop Heber College, Trichy, Tamil Nadu, India

Baskaran Ganesh Kumar Department of Electrical and Electronics Engineering, Koc University, Istanbul, Turkey

Department of Chemistry, PSR Arts and Science College, Sivakasi, TN, India

J. Santhosh Kumar Centre for Nanoscience and Technology, Pondicherry University, Puducherry, India

R. V. Mangalaraja Department of Materials Engineering, Faculty of Engineering, University of Concepción, Concepción, Chile

Technological Development Unit (UDT), University of Concepcion, Coronel Industrial Park, Coronel, Chile

Héctor D. Mansilla Department of Organic Chemistry, Faculty of Chemical Sciences, University of Concepción, Concepción, Chile

Luiz H. C. Mattoso National Nanotechnology Laboratory for Agribusiness, Embrapa Instrumentação, São Carlos, Brazil

Victoria Melin Facultad de Ciencias Químicas, Centro de Biotecnología, Universidad de Concepción, Concepción, Chile

Akbar Mohammad School of Chemical Engineering, Yeungnam University, Gyeongsan-si, Gyeongbuk, South Korea

Esther Santhoshi Monica PG & Research Department of Physics, Bishop Heber College, Trichy, Tamil Nadu, India

Mu. Naushad Department of Chemistry, College of Science, King Saud University, Riyadh, Saudi Arabia

Gabriel Pérez-González Facultad de Ciencias Químicas, Centro de Biotecnología, Universidad de Concepción, Concepción, Chile

Sudhagar Pitchaimuthu Multi-functional Photocatalyst & Coatings Group, SPE-CIFIC, College of Engineering, Swansea University (Bay Campus), Swansea, Wales, UK

Sebastian Raja National Nanotechnology Laboratory for Agribusiness, Embrapa Instrumentação, São Carlos, Brazil

Saravanan Rajendran Escuela Universitaria de Ingeniería Mecánica (EUDIM), Universidad de Tarapacá, Arica, Chile

R. Ajay Rakkesh Centre for Advanced Sciences in Crystallography and Biophysics, University of Madras, Chennai, Tamil Nadu, India

Paola Santander Center of Biotechnology, University of Concepción, Concepción, Chile

Millenium Nuclei on Catalytic Processes towards Sustainable Chemistry (CSC), Concepción, Chile

Y. Sasikumar Laboratory of Experimental and Applied Physics (LAFEA), Centro Federal de Educação Tecnológica (CEFET/RJ), Celso Suckow da Fonseca, Maracanã Campus, Rio de Janeiro, Brazil

Manjusha V. Shelke Physical & Material's Chemistry Division, CSIR-National Chemical Laboratory, Pune, India

Kavitha Shivaji Department of Biotechnology, K.S.R. College of Technology, Tiruchengode, Tamil Nadu, India

R. Suresh Department of Analytical and Inorganic Chemistry, Faculty of Chemical Sciences, University of Concepción, Concepción, Chile

P. Thangadurai Centre for Nanoscience and Technology, Pondicherry University, Puducherry, India

Surendar Tonda Department of Environmental Engineering, Kyungpook National University, Daegu, South Korea

Poonam Yadav Physical & Material's Chemistry Division, CSIR-National Chemical Laboratory, Pune, India

Jorge Yáñez Department of Analytical and Inorganic Chemistry, Faculty of Chemical Sciences, University of Concepción, Concepción, Chile

Chapter 1
Principles and Mechanisms of Green Photocatalysis

D. Durgalakshmi, R. Ajay Rakkesh, Saravanan Rajendran, and Mu. Naushad

Contents

Abstract The accessibility of clean and fresh water is a major requirement for the survival and societal development of humans. Photocatalysis is an artificial photosynthesis technique that uses green, sustainable chemistry to solve energy and environmental issues. There is now great demand for a low-cost, high-performance device to treat wastewater. The role of nanotechnology in field water purification has significantly advanced by altering the shape, size, and properties of nanomaterials. This chapter outlines recent developments in green photo-active nanostructures, which includes metals, metal oxides, metal-doped metal-oxides, and plasmonic

D. Durgalakshmi (✉)
Department of Medical Physics, Anna University, Chennai, Tamil Nadu, India

R. Ajay Rakkesh
Centre for Advanced Sciences in Crystallography and Biophysics, University of Madras, Chennai, Tamil Nadu, India

S. Rajendran
Escuela Universitaria de Ingeniería Mecánica (EUDIM), Universidad de Tarapacá, Arica, Chile

M. Naushad
Department of Chemistry, College of Science, King Saud University, Riyadh, Saudi Arabia

© Springer Nature Switzerland AG 2020
M. Naushad et al. (eds.), *Green Photocatalysts*, Environmental Chemistry for a Sustainable World 34, https://doi.org/10.1007/978-3-030-15608-4_1

photocatalytic materials in water splitting. The differences in photosynthesis and photocatalysis energy concepts, as well as an emerging heterogeneous photocatalyst-based z-scheme approach, are also discussed in detail.

Keywords Green materials · Metal oxides · Photocatalyst · Fuel cells · Solar disinfectants · Activated carbon

1.1 Introduction

Water pollution is one of the worst environmental problems that the world faces today (Alothman et al. 2012; Sharma et al. 2017). Most water bodies have been polluted by humans due to modernization, the release of industrial waste, oil spills in the ocean, and plastic waste accumulation in the water resources (Ma et al. 2009; Bhateria and Jain 2016). To obtain clean (or even unclean) water for household purposes, people may need to walk very long distances every day. Thus, there is currently a real water crisis affecting people worldwide. In other cases, the consumption of polluted or untreated water causes malnutrition or disease; this is mainly due to drought, floods, or inadequate sanitation. Still, there is a lack of knowledge, funds, and awareness to solve the problems related to water allocation and usage (Organization 2001; Programme 2015).

Sustainable development can be divided into three important concepts: economic, social, and environmental. It is the key to reducing water-related problems, as well as a focus of Rio 2020: Green Economy and institutional frameworks for sustainable development (Biermann 2013; Bina 2013). The ideas and the objectives of a "green economy" and "sustainable development" only make sense if the ocean is completely integrated. Sustainable development is considered to be "growth that helps the desires of the present without compromising the capability of next generations to convene their own needs" (Dittmar 2014). Sustainable development is a process of dealing with the economic, social, and environmental problems that the world has confronted in the past decades. It has been accepted and maintained by the international community. However, the problems still continue and have even become worse in the ocean regions. Coastal communities are unable to manage current problems, and government bodies are incapable of addressing these current changes in institutional issues.

The major sources of water pollutions are industrialization, human activities such as agriculture and deforestation, and other environmental and global changes. The nature of water resources is worsening day by day because of the incessant accumulation of undesirable contaminants and chemicals in the water bodies. These chemicals include both inorganic and organic pollutants, which contaminate water resources. Among inorganic water pollutants, heavy metal ions such as mercury (Hg^{2+}), lead (Pb^{2+}), arsenic, cadmium (Cd^{2+}), chromium (Cr^{6+}), and nickel (Ni^{2+}) are the most hazardous pollutants found in industrial effluents (Ghasemi et al. 2014a, b; Naushad et al. 2016; Pandey and Ramontja 2016). Various types of hazardous organic contaminants, including organic dyes, coloring agents, pesticides, oils, phenols, fertilizers, pharmaceutical waste, hydrocarbons, phenols, and detergents, have also been found in various water bodies (Gupta et al. 2012; Javadian et al. 2014; Rashed 2013; Sharma et al. 2015).

The World Bank has estimated that 1.3 billion tons of solid waste were produced by cities in 2013; with the present urbanization rates, this is expected to grow to 2.2 billion tons per year by 2025, which is an increase of 70% (Daniel and Perinaz 2012). Thus, managing waste will become more expensive. The best way to resolve the problem is to change the way we use and reuse the waste. Recently, many initiatives have attempted to reutilize solid waste. One such initiative converts solid waste into a value-added product. The sorbent is a type of product made from solid waste; the successful use of these sorbents for gas separation processes has been reported. Moreover, the development of inexpensive and high-capacity sorbents has been receiving extensive interest from various researchers for the removal of toxic substances (Crini 2006; Yagub et al. 2014). The effectiveness of adsorbents in removing pollutants has been recognized in many other applications because of their high adsorption capacity, ease of operation, and simplicity of design (Alqadami et al. 2017; Hayashi et al. 2005). Furthermore, there is a need to produce adsorbents from alternative materials that are renewable, cheap, and more readily available than commercial coal-based adsorbents, which are relatively expensive (Attia et al. 2008). For this reason, research interest in the production of adsorbents from cheaper and renewable precursors is strong (Mittal et al. 2016).

1.2 Green Energy

The sun is the ultimate energy source for human beings. Solar energy is transmitted onto the earth's surface through a constant flow of electromagnetic radiation waves (Fig. 1.1). Among the total radiated energy in the solar system, only a very small portion is captured by the earth. Solar energy utilization by photosynthesis has been

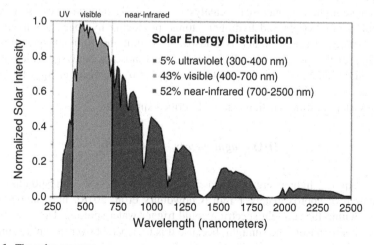

Fig. 1.1 The solar spectrum

the dominant approach for solar energy conversion, which is essential to the life of human beings via agriculture and forestry. The total power of solar radiation is 384.6 yotta-watts (3.846 × 1026 W) (Hussey 2014). This large amount of energy flow is uniformly radiated into all directions from the sun. At a distance of 150 million kilometers away, exposed areas of the earth receive approximately 1368 Wm^2 (Foster et al. 2009). Because 30% of the irradiated solar energy is reflected by the atmosphere, approximately 1000 Wm^2 of solar energy is received everywhere on the surface of the earth. Although this amount of total energy provided by the sun is far more than sufficient, the real challenge is how to efficiently collect and use solar energy.

1.3   Photosynthesis and Photocatalysts

A light-induced system that excites electrons and holes and promotes a chemical reaction is an excitonic chemical conversion system. The term *photocatalysis* refers to the acceleration of a chemical reaction from the presence of light and a catalyst. A refined definition based on Nozik (Nozik 1977) and Bard (1979) differentiated between photosynthetic and photocatalytic processes, based on the thermodynamics of the coupled reaction: "Photo electrolytic cells can be classified as photosynthetic or photocatalytic. In the former case, radiant energy provides Gibbs energy to drive a reaction such as $H_2O+H_2+1/2O_2$, and electrical or thermal energy may be later recovered by allowing the reverse, spontaneous reaction to proceed. In a photocatalytic cell, the photon absorption promotes a reaction with $\Delta G < 0$ so there is no net storage of chemical energy, but the radiant energy speeds up a slow reaction." (Osterloh 2017).

Photocatalysis was first introduced by Fujishima and Honda in the 1970s and was later referred to as the Honda-Fujishima Effect (Hashimoto et al. 2005). One typical example of photocatalysis is water splitting (Fig. 1.2), in which the interactions between sunlight and nanoscale catalysts (e.g., TiO_2 nanoparticles) split water molecules into oxygen and hydrogen atoms, producing hydrogen gas as a clean energy source. Because water is transparent within the sunlight spectrum, the efficiency of this process crucially relies on the nanoscale catalysts, which absorb the sunlight and convert the photon energy into chemical energy for water molecules.

The water-splitting reaction can be described simply as

$$H_2O + \textit{light energy} \rightarrow \frac{1}{2}O_2 + H_2$$

The required light energy for water splitting is 1.23 eV, corresponding to a wavelength of approximately 1 mm (Chowdhury et al. 2018). In principle, most photons within the sunlight spectrum could trigger water splitting. However, water cannot directly absorb the sunlight, because it is transparent over the entire sunlight

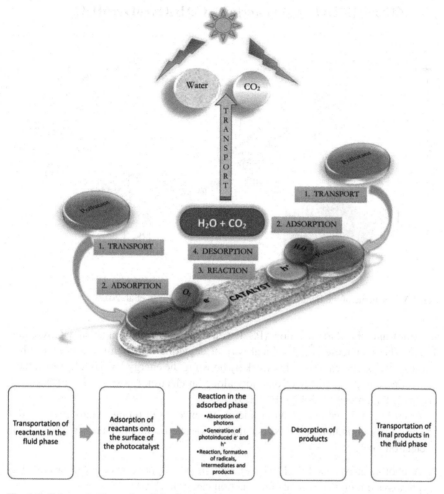

Fig. 1.2 Schematic illustration of the the various steps involved in a heterogeneous photocatalytic reaction. (Bora and Mewada 2017)

spectrum. To introduce the photon energy to the water molecules, a catalyst is needed, which first absorbs the sunlight and then transfers the energy to the water.

Photosynthesis is a process that converts the energy from sunlight into chemical energy that is stored in certain types of chemical materials. Plants use the photosynthesis processes to synthesize carbohydrates ($C_6H_{12}O_6$) and oxygen (O_2) via the degradation of carbon dioxide (CO_2) and water (H_2O) (Lems et al. 2010):

$$6CO_2 + 12H_2O + \textit{light energy} \rightarrow C_6H_{12}O_6 + 6O_2 + 6H_2O$$

Photosynthesis is a method of transforming light energy into chemical energy. It is separated into two broad levels: the light reaction (photosystem [PS] II; Hill

$$6CO_2 + 12H_2O + \textit{light energy} \rightarrow C_6H_{12}O_6 + 6O_2 + 6H_2O$$

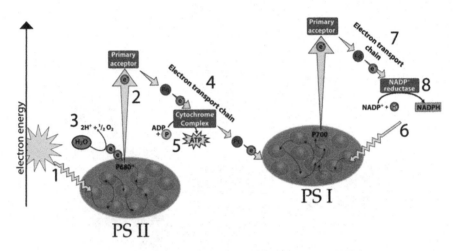

Fig. 1.3 Z-scheme of the light reaction of photosynthesis. (Dayan et al. 2017)

reaction) and the dark reaction (PS I; Calvin cycle) (Whitmarsh and Govindjee 1999). The **Z-scheme** is a graphical way of depicting how chloroplasts (and other photosynthesizers) carry out this work by boosting the energy level of electrons, then harvesting energy as the electrons flow along an electron transport chain (Dayan et al. 2017; Sayama et al. 2002) (Fig. 1.3).

From Fig. 1.3, we can create a linear sequence, even though many parts of the process occur simultaneously. Notice that the Z-scheme diagram has a Y-axis, which is electron energy:

1. A photon arrives at PS II. The energy from this photon bounces around the photosystem until it reaches the reaction center at P680.
2. An electron in a chlorophyll molecule gets an energy boost that sends it to PS II's primary electron acceptor. Note that in this diagram, the higher position of the electron at PS II's primary acceptor means that it has more energy than the original electrons in the unenergized chlorophyll in PS II.
3. As PS II's primary electron acceptor accepts chlorophyll's electron (oxidizing it), an oxygen-evolving complex in PS II rips electrons away from a water molecule and permits them to p680. The water dissociates into protons and oxygen.
4. PS II's primary electron acceptor passes this energized electron to an electron transport chain. As the electron moves from one embedded electron carrier in the chain to the next, it releases energy.
5. This energy release shown in step 4 is coupled with cellular work—in this case, the work of creating adenosine triphosphate from adenosine diphosphate and P_i.
6. The electron from PS II arrives at PS I in a low-energy condition. However, as it has been travelling, photons have been shining on PS I and energizing its

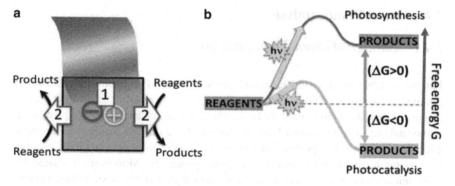

Fig. 1.4 (a) Basic procedures occurring during photocatalysis and an artificial photosynthesis reaction. (b) Energy of photocatalysis and photosynthesis. (Osterloh 2017)

electrons. The energy bounces to P700, and its electrons are boosted in energy and then snatched away by PS I's primary electron acceptor. These lost electrons are replaced by the electrons arriving from PS II (which can ultimately be traced to water). Notice that the electron boost from PS I boosts these electrons to their highest energy level.

7. PS I's primary electron acceptor passes its energized electrons to PS I's electron transport chain. This electron transport chain, however, does not pump protons. Instead, it channels PS I's high-energy electrons to the enzyme $NADP^+$ reductase.

8. $NADP^+$ reductase uses the incoming electron's energy to convert oxidized, low-energy $NADP^+$ into a high-energy, reduced NADPH.

The electron-hole pair energy conversion, photocatalysis, and photosynthetic systems play vital roles in the conversion of light energy into chemical energy by means of photogenerated charge carriers, thus promoting the radical reactive species at the solid-gas or solid-liquid interfaces (Fig. 1.4a) (Osterloh 2017). Accordingly, both photocatalytic and photosynthetic systems should follow two main criteria: (i) light absorption and the creation of electrons and hole pairs and (ii) photogenerated charge carriers promoting redox reactions at the surface of the chemical species.

The major variation in the concepts of photocatalysis and photosynthesis devices relies on the thermodynamics of surface chemical reactions (Fig. 1.4b). Photocatalysts use light energy to convert chemical energy, which is thermodynamically acceptable ($\Delta G < 0$). The chemical reaction is promoted by the photosynthesis systems, which are thermodynamically forbidden ($\Delta G > 0$), and it requires more input energy from the light source for the photochemical reaction and conversion to proceed. Hence, the products in the photosynthesis reaction have greater free energy than the reagents, and the reverse photosynthetic reaction is thermodynamically favored ($-\Delta G < 0$). Therefore, in addition to the basic functions (i) and (ii) mentioned previously, photosynthesis systems also must be able to (iii) suppress this reverse photosynthesis reaction.

1.4 Green Photocatalysis

1.4.1 Sunlight-Driven Photocatalysts

Finding suitable metal oxide semiconductors to be used as photocatalysts for the removal or treatment of contaminants/pollutants in water bodies using solar energy is one of the most important tasks in the field of material science. An ideal photocatalytic material should have properties of the desired bandgap to absorb a wide range of the solar spectrum, dissociate the water molecules, and remain stable in a water environment during the reaction processes. Moreover, it should be economically viable, easy to process, easily available, and non-toxic to the environment. For the past few decades, various metal oxide semiconducting-based nanoassemblies have been designed and demonstrated as catalytic materials for water remediation applications under sunlight. Transition or d block metal ions have shown excellent efficiency in highly explored semiconductors for efficient photocatalytic materials (Wang et al. 2008).

The following factors affect the photocatalyst process:

1. **Dye concentration:** The concentration of the dye used in the photocatalytic reaction is an important factor. The catalyst should be able to degrade an average quantity of the dye. A quantity of the dye is adsorbed on the surface of the catalyst, which involves a photocatalytic reaction process under stimulated light conditions. The adsorption of dye on the surface of the photocatalyst is directly proportional to the original concentration of dye. The original concentration of dye is a significant feature that should be monitored carefully. Usually, the degradation percentage decreases with an increasing quantity of dye concentration, although the required quantity of photocatalyst should be maintained (Reza et al. 2017).

2. **Catalyst amount:** The amount of catalyst in the photocatalytic reaction also affects the degradation of the dye. In a heterogeneous photocatalytic process, increasing the amount of photocatalyst in the reaction process also increases the percentage of photodegradation of the dye. More active sites can be produced in the photocatalytic reaction by increasing the number of catalysts, which promotes the creation of a greater number of reactive radicals in the photodegradation process. (Akpan and Hameed 2009).

3. **pH:** The solution pH also plays a vital role in the degradation process. It can either induce or suppress the photocatalytic reaction, depending upon the nature of the material and pollutant properties. By changing the pH of the solution, the surface potential of the catalyst (metal oxide nanoparticles) may vary. As a result, the pollutant adsorption on the surface of the photocatalyst could be altered, thus initiating a change in the photodegradation rate (Davis et al. 1994).

4. **Surface morphology of the photocatalyst:** Surface morphology includes significant features to be measured for the photodegradation activity, such as particle size and shape. Each morphology is a direct relationship between the surface of the catalyst and organic pollutant(Kormann et al. 1988). The rate of

photocatalytic activity can be controlled by the number of photons striking on the surface of the photocatalyst. The reaction occurs faster if the photocatalysts have a variety of morphologies (Zhu et al. 2006).

5. **Surface area:** For better photocatalytic performance, materials with greater surface areas should be used. These materials are able to create a number of active sites on the photocatalyst surface, thus leading to the creation of more radical reactive species for efficient photodegradation activity (Ameen et al. 2012).

6. **Temperature-dependent reaction:** To achieve efficient photocatalytic performance, the reaction temperature should be in the range of 0–80 °C. When the temperature exceeds 80 °C, the catalyst will promote the recombination of the electron-hole pair and suppress the photocatalytic activity. Hence, the reaction temperature plays a vital role in photocatalytic activity (Kazuhito et al. 2005; Mamba et al. 2014).

7. **Nature of the pollutants and their concentrations:** The degree of photodegradation can be determined by the concentration and nature of certain pollutants in a water matrix. Some photocatalysts, such as TiO_2, are not able to disinfect the pollutant when the concentration of the pollutant is higher. It saturates the surface of the photocatalyst and does not allow the creation of active radicals, thus reducing the photocatalytic efficiency (Mills et al. 1993).

8. **Irradiation period and intensity of the light:** The intensity of the incoming light and irradiation period are major factors affecting the photodegradation of pollutants. At high light intensities, the photodegradation percentage is inversely proportional to the intensity of light, because the creation of excitons is predominant at low light intensities and also hinders the recombination of electron-hole pairs. Alternatively, while increasing the intensity of the irradiation light, the recombination of electron-hole pairs occurs at the photocatalytic surfaces and thereby reduces the catalytic activity in the reaction medium (Asahi et al. 2001).

9. **Dopants on dye degradation:** Various methods are available to fabricate TiO_2 nanomaterials that can absorb photons at very low energy. These methods consist of band gap engineering by modifying and continuously altering valence and conduction bands by introducing metals and non-metals in the photocatalytic materials. Surface modification can be achieved by coupling with organic materials and semiconductors (Rajeshwar et al. 2008).

1.4.2 Metal Oxides

Semiconducting metal oxide–based nanostructures have been used for photocatalytic applications to treat wastewater and produce hydrogen fuel by splitting oxygen and hydrogen, among others. The most important criteria for an efficient photocatalytic material are the required band gap, desired band edge position, large surface area, perfect morphology, chemical stability, and reusability capability. Among various metal oxide semiconductors, TiO_2, ZnO, SnO_2, Cu_2O, and WO_3

with these characteristics have identical photocatalytic properties, such as absorption of light. This excites photogenerated charge carriers with the creation of holes, which are able to oxidize organic substances (Maeda 2011).

In this reaction process, semiconducting metal oxide nanostructures are activated with direct sunlight, visible light, ultraviolet (UV) light or a combination of both visible and UV light irradiation. The photogenerated charge carriers are excited from the valence band to the conduction band, creating electron/hole pairs. The photogenerated electron-hole pairs are involved in the oxidation and reduction reaction to break down the molecular chains of the organic pollutants. The photocatalytic activity of the semiconducting metal oxide comes from two broad aspects (Abe 2010): (i) the creation of hydroxyl radicals by the oxidation of OH^- anions, (ii) the production of superoxide radicals by the reduction of O_2. These radical reactive species can counter with organic pollutants to disinfect or mineralize to nontoxic byproducts. Hence, it is of great scientific significance in the environmental, hydrogen fuel production, and energy industries. The applications of these photocatalytic materials are generally used for treating wastewater by the elimination of pathogens and other harmful pollutants.

Figure 1.5 portrays the band gap and band edge position of some commonly used photocatalytic semiconducting materials (Ola and Maroto-Valer 2015). Even though some of these semiconducting materials, such as ferric oxide (Fe_2O_3), have suitable band gap energies that are able to act in the visible light region, they have some drawbacks to their performance as efficient photocatalysts. To overcome these issues, researchers are searching for alternate materials. For example, metal chalcogenide-based semiconductors (e.g., PbS, CdS) have demonstrated photocatalytic activity. However, those materials have very low stability, toxic effects, and are

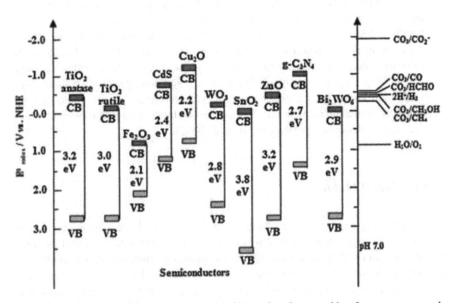

Fig. 1.5 Band edge position of certain metal oxide semiconductors with reference to a normal hydrogen electrode. (Ola and Maroto-Valer 2015)

prone to photocorrosion. However, metal oxide semiconductors such as SnO_2, WO_3, and Fe_2O_3 have conduction band edges below the Normal hydrogen electrode (NHE) potential and have great stability and photocorrosive properties in aqueous solutions. Gupta et al. reported the use of external voltage to trigger the charge transfer property for hydrogen evolution during water splitting (Gupta and Tripathi 2011). Fox et al. demonstrated the lower photoactivity of Fe_2O_3 owing to corrosion or the creation of short-lived metal-to-ligand or ligand-to-metal charge transfer states compared to TiO_2 and ZnO materials (Fox and Dulay 1993). Bahnemann et al. developed the formation of ZnO and observed that the material starts dissolution in water by causing instability over time (Bahnemann et al. 1987). In contrast, TiO_2 nanomaterial is corrosion resistant and possesses great stability in an aqueous medium.

1.4.3 Metal-Doped Metal Oxides

The method of introducing impurity atoms into the lattice system of any semiconducting material is called doping, which was described by Neamen (Neamen 1997). The dopant atoms in semiconductor lattices influence and engineer the properties of the host material. Figure. 1.6 provides a pictorial representation of the defects in a lattice structure. Any impurity or foreign atom is located at the lattice sites of the host atom in the form of a substitution; hence, it is termed a substitution or substitutional doping. This type of doping can be achieved with the following conditions: (i) The host and dopant metals have the same crystal structure, electronegativity, and solubility state; and (ii) the atomic radii of the dopant atoms do not exceed 15% of its difference.

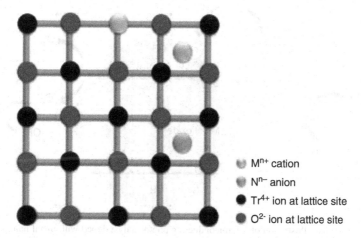

M^{n+} cation

N^{n-} anion

Tr^{4+} ion at lattice site

O^{2-} ion at lattice site

Fig. 1.6 Schematic illustration of a lattice structure of TiO_2 crystal depicting two types of doping: substitutional and interstitial. (Ola and Maroto-Valer 2015)

When the foreign atoms are sandwiched between the usual lattice position, this kind of doping is called interstitial doping. The atoms are removed from the lattice position and create voids between the host atoms. The probabilities of atoms getting into the interstitial site have been identified by measuring the interstitial and host atom radius. The variations in the atomic radius show the exact location of the dopant atoms in the interstitial site. The cations residing in the definite interstitial sites can be determined by the ratio of the cation/anion (r^+/r^-) ionic radius values. The coordination numbers of cations in the interstitial sites are based on the ratio of the ionic radius, such as 6 (octahedral), 4 (tetrahedral), etc. To increase the ratio of the ionic radius, the quantity of anions filled around the cations increases as well. When the ratio of the ionic radius lies between 0.225 and 0.414, the cations are sandwiched between the anion planes in the tetrahedral holes of the close-packed structures. If they are filled with octahedral holes, then it lies within 0.414 and 0.732 (Neamen 1997).

To improve the absorption capability and engineer the electronic properties of a semiconducting photocatalyst, doping is the best strategy, whether it be metal, non-metal, or the doping of molecules with suitable semiconductors (Fig. 1.7). The oxidation state, surface chemistry and ionic radii of the dopant can powerfully induce the efficiency of the doping process. In recent years, metal and non-metal doping have been highly investigated. Particularly, doping with non-metal ions (e.g., carbon, nitrogen, boron, sulfur) have been examined systematically to alter the bandgap and the band edge position of a semiconductor to make it a visible-light active photocatalyst.

Some semiconducting photocatalysts (e.g., TiO_2, SnO_2, ZnO) are generally active under a UV light range because of its wider bandgap energy. To drag the absorption range into the visible light region of the semiconductors by doping with suitable metal or non-metal ions, it helps to alter the electronic structure of the semiconductor

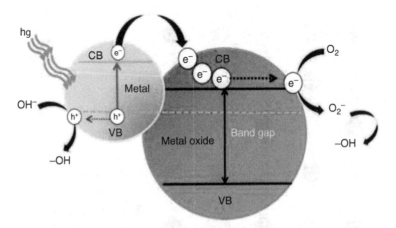

Fig. 1.7 Graphical illustration of a semiconducting photocatalyst doped with metal nanoparticles. (Malwal and Gopinath 2016)

and modify the band edge positions. That doped semiconductor exhibits unusual photocatalytic performance under visible light irradiation (Zhang et al. 2017).

Over recent decades, metal-doped TiO_2 nanomaterials have been studied extensively for the visible light active photodegradation of harmful organic pollutants, pharmaceuticals waste, and toxic dyes. This kind of structural variation could alter the absorption of TiO_2 into the visible range. Moreover, TiO_2 nanomaterials doped with nonmetals such as nitrogen, carbon, fluorine, and sulfur have been examined widely snice the early 1990s. Those photocatalysts have shown extraordinary activity against various pollutants under sunlight irradiation.

1.4.4   Plasmonic Photocatalysis

Plasmonic photocatalysts have attracted great interest among researchers due to their enhanced photodegradation efficiency under visible light irradiation, wide range of sunlight absorption, and excellent charge transport properties (Xuming et al. 2013; Zhou et al. 2016; Liang et al. 2018; Dong et al. 2014; Yang et al. 2014). This kind of material architecture can be created by dispersing noble metal nanoparticles onto the semiconductors. In that way, two distinct features are achieved: localized surface plasmon resonance (LSPR) and a Schottky barrier (Xuming et al. 2013; Wang and Astruc 2014). These properties will help to separate and transfer the charge carriers effectively under visible light irradiation.

The LSPR is the most important feature of plasmonic photocatalysis, signifying strong oscillation on the surface of the metal nanoparticles and the semiconductor photocatalysts. Figure 1.8 lists the effects that occur during the photocatalytic activity of the plasmonic metal nanoparticles. The metal-semiconductor junction in

Fig. 1.8 Flowchart of plasmonic photocatalysis and its major effects. (Xuming et al. 2013)

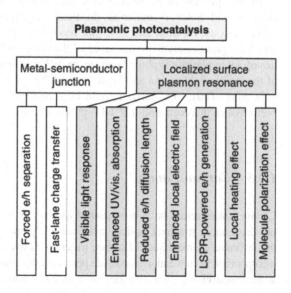

plasmonic photocatalysts helps to separate the electron-hole efficiently and is able to transfer the charge carriers rapidly due to its short diffusion length and interfacial charge transfer effect at the heterojunction.

Plasmonic metal nanoparticles (e.g., Ag, Au, Pt) have resonance oscillation at certain wavelengths that could modify the absorption range of a UV light active photocatalyst (TiO_2) into a visible range, depending upon the noble metal nanoparticle morphology, shape, and size. The surface plasmon resonance effect in the metal nanoparticle can drastically increase the visible-light absorption capability of a low-bandgap semiconducting photocatalyst, such as Fe_2O_3. It is able to enhance the electron transport properties of a poor electron transport semiconductor by the strong absorbing capacity of the total incoming light in a very thin layer of metal nanoparticles. The distance between the photogenerated electron/hole and the surface of noble metal nanoparticles is very short; similarly, the diffusion length is also short. Thus, the transport properties of the charge carriers are enhanced under the excitation of photons (Xuming et al. 2013; Thomann et al. 2011).

Figure 1.9 shows the most important mechanisms involved in plasmonic photocatalysis. Au nanomaterials are surrounded partially on the surface of the TiO_2 photocatalyst. Generally, TiO_2 nanoparticles have n-type characteristics due to native oxygen defects and subsequent surplus electrons in the material feature (Morgan and Watson 2010). To determine the photocatalytic efficacy of the Au–TiO_2 nanostructures, conventional TiO_2 photocatalysts have been compared and examined extensively. Figure 1.9a shows that a plasmonic Au nanoparticle on the surface of TiO_2 can collect the whole visible range spectrum in the electromagnetic radiation due to the LSPR effect. This effect could adsorb the collective oscillation of electrons and holes in TiO_2. It involves an interfacial charge transfer process; thus, it facilitates a reduction and oxidation reaction for the photodegradation of harmful pollutants in an aqueous medium.

In this process of an interfacial charge transfer effect, the recombination of electron and holes is inhibited drastically compared with conventional TiO_2 photocatalysts. Electron and hole pair recombination is an important factor could affect the photocatalytic performance of the semiconductor. The prime characteristics of Au in plasmonic photocatalysts are its facilitation of the absorption capability of the visible light region and hinderance of the recombination rate of the electron-hole pairs. Thus, this material could be capable of more efficient photocatalytic performance than conventional TiO_2 photocatalysts.

1.4.5   Carbon Family

Carbon nanomaterials have been studied extensively in recent times for their unusual physiochemical, structural, optical, and electronic properties (Zhu 2017). While keeping the favorable characteristics of carbon materials, nanocomposites can be developed for incorporation in conventional and stale photocatalytic materials for efficient light-derived water remediation applications (Hebbar et al. 2017; Jiang et al. 2016).

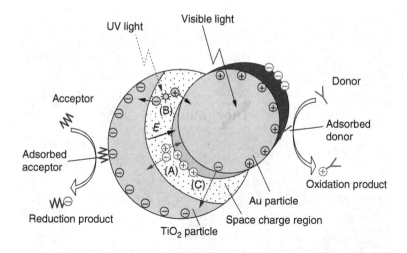

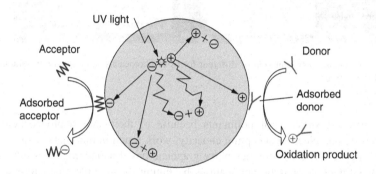

Fig. 1.9 Pictorial illustration of the charge transfer mechanism in plasmonic photocatalysts compared with semiconductor photocatalysts. (Xuming et al. 2013; (ed.) 1993)

Different forms of nanomaterials have been investigated for water disinfection, oil adsorbent, and the removal of pathogens from water bodies. Among them, carbon-based nanostructures possess excellent physiochemical properties, which helps them to remove organic, inorganic, and other heavy metal materials from water bodies. Some carbon-based adsorbents are commercially available in the market. Fullerenes, graphene, and carbon nanotubes are currently used as a co-catalysts in conventional photocatalytic materials such as TiO_2, ZnO, SnO_2, etc., for their rapid photodegradation and charge transport properties. Above all, carbon-based nanomaterials could provide strong reduction proficiency to break down complex and harmful pollutants in water bodies.

Kamat et al. reported that graphene-decorated TiO_2 nanocomposites showed excellent photocatalytic activity under UV light illumination (Williams et al. 2008). The breakthrough of this nanocomposite was demonstrated deeply with various material combinations for better activity. Among the various carbon nanostructures, graphene and its derivatives show excellent catalytic performance

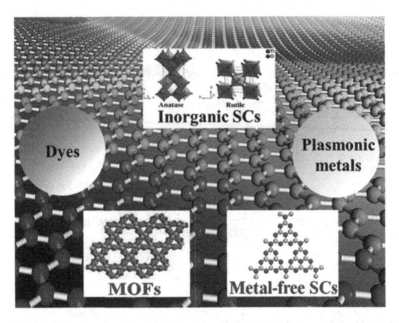

Fig. 1.10 Schematic representation of different forms of photocatalysts coupled with graphene-based nanostructures

toward different varieties of pollutants because of their superb physiochemical properties, strong electron accepting capacity, work function modulation, and electronic properties. These properties make graphene-based nanocomposites excellent candidates for photocatalytic applications by tuning the suitable characteristics to influence the photocatalytic activity of semiconductors.

Various types of photocatalysts have been used to fabricate diverse morphologies and combine with graphene, either by electrostatic interaction or chemical bonds for photocatalytic applications (Suárez-Iglesias et al. 2017). Different forms of photocatalysts are being used, such as organic, inorganic, a combination of metal-organic frameworks, semiconductors, plasmonic metals, non-metal plasmonic materials, and dyes (Fig. 1.10).

Carbon nanotubes (CNTs) have been widely used as efficient catalytic materials with enhanced properties compared to other standard catalyst supports, such as graphite, activated carbon, and soot (Hebbar et al. 2017). It was recently found that CNTs are able to absorb some toxic materials effectively because CNTs exhibit more active sites and also have a very high surface area. The incorporation of TiO_2 in the CNT matrix was found to be an excellent photocatalytic material with enhanced activity compared with pristine CNT and pristine TiO_2. These nanocomposites could drastically inhibit the rate of electron-hole pair recombination by trapping the electrons in the valence band and capturing a wide range of visible light absorption. Visible light absorption of the nanocomposite could be attained by reducing the energy gap of the composite nanostructures.

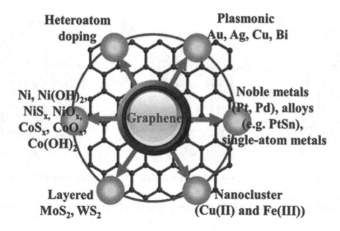

Fig. 1.11 The various arrangements of catalysts and co-catalysts in graphene architecture. (Li et al. 2016)

The applicability of organic material in the field of energy and environmental sectors has added benefits, such as being chemically inert, inexpensive, and proficient in addressing many environmental-related issues. Based on this, metal organic frameworks, graphitic carbon nitride, and other organic dyes have been used as photocatalysts by coupling with graphene or CNT nanostructures (Fig. 1.11) (Li et al. 2016). The formation of a Schottky-type heterojunction between the organic semiconductor and graphene interfaces during the coupling strategy promotes charge transport effectively throughout the junction. This kind of layered architecture with metal-free photocatalysts facilitates the structural, electronic, and physiochemical properties of a graphene-organic semiconductor for outstanding photocatalytic performances. The combination of chemical reduction and impregnation strategies were demonstrated by Xiang et al. using graphene/g-C_3N_4nanocomposites. The authors reported that this material could generate hydrogen by splitting into hydrogen and oxygen from water.

1.4.6 Z-Scheme in Photocatalysis

The design and development of properly assembled metal oxide-based semiconducting photocatalysts are encouraging the use of nanomaterials to address environmental issues because of their ability to capture the visible range of the electromagnetic spectrum to cause numerous photodegradation reactions. Particularly, the development of the Z-scheme in photocatalysis has many advantageous processes, such as excellent harvesting capability of sunlight, rapid charge separation, the creation of active species for oxidation and reduction reactions, and very good redox proficiency, which makes for better photocatalytic activity (Fig. 1.12).

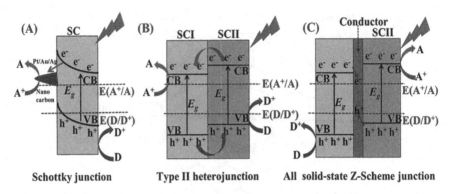

Fig. 1.12 Band edge diagram for three different models of photocatalysis systems. (Li et al. 2016)

In this Z-scheme, a photocatalysis system consists of two different semiconducting photocatalyst materials coupled in a suitable redox mediator. The utilization or absorption of sunlight in this system more efficient than in the conventional photocatalysis system. The amount of energy required to active the Z-scheme is also lower. It can be used as a dual-purpose system for water oxidation or reduction potential to any one of the photocatalysts in this system.

Tada et al. reported a preparation strategy for CdS-Au-TiO$_2$ nanocomposites and photocatalytic performance influenced by the charge transfer processes. In detail, the Au nanoparticles can be sandwiched between the nanoparticles of CdS and TiO$_2$, which influence very strong photoinduced electron transfer by TiO$_2$ and photoinduced hole transfer by CdS nanoparticles. Then, subsequent photoinduced electrons are left with intense reduction capability on CdS and the photoexcited holes are left with strong oxidation ability on the TiO$_2$ nanoparticles. This electron transfer mechanism is also known as a shuttle redox mediator system. This type of charge transfer mechanism only occurs with Z-scheme photocatalysts. The path of the charge transfer processes in these two semiconducting systems resembles the shape of the letter Z; hence, they are known as Z-scheme photocatalysts (Qin et al. 2017; Gnanasekaran et al. 2016; Romero Sáez et al. 2017; Saravanan et al. 2011, 2018).

1.5 Green Materials for Green Photocatalysts

A number of technologies are currently available for the treatment of contaminated water, such as reverse osmosis, ion exchange, biological treatment, and adsorption. Reverse osmosis and ion exchange require high capital investments and operational costs, whereas adsorption and biological treatment are considered to be low-value processes. The adsorption process is a versatile technique for the treatment of wastewater and the removal of hazardous pollutants. The biggest obstacle to this process is the price of the adsorbent. Currently, most industrial adsorption processes for water and wastewater treatment processes use activated carbons as the adsorbents (Sen 2017). The application of activated carbon is used for the effective removal of organic pollutants in the environment. Commercially available activated carbons are very expensive; thus, reducing the cost of the production process is necessary.

Alternative adsorbent materials or substitute activated carbons should meet the following criteria: have high adsorption capacity, be low cost, be available in large quantities in one location, and be able to be easily regenerated (Febrianto et al. 2009). One of the potential candidates for alternative adsorbents is agricultural wastes (Sen 2017.). Hundreds or even thousands of studies on the utilization of agricultural wastes as alternative adsorbents for water and wastewater treatment have been conducted in the last three decades. A massive amount of literature is available on biomass materials for the removal of hazardous substances (e.g., Abdolali et al. 2014; Bhatnagar and Sillanpää 2017; Levchuk et al. 2018; Mohan et al. 2015; Silva et al. 2015). Various aspects of the utilization of low-cost adsorbents, including agricultural wastes, have been discussed in those review papers. However, with the growing numbers of articles published on this subject, new proposed ideas, and the complexity of the adsorption process, it is necessary to provide up-to-date information and comprehensive discussions on the utilization of agricultural wastes as alternative adsorbents.

1.6 Conclusion

Nanotechnology has great potential in the design of artificial photosynthesis systems to store solar energy and to reduce organic contaminants in the environment because of the unique properties of nanomaterials. The advent of green, accessible, and safe methods of producing these nanomaterials is necessary to address modern environmental concerns. Important aspects of green or biosynthesis solutions using biomass materials include a rapid reaction process, low temperature, and minimal usage of toxic substances. By mimicking photoactive green nanomaterials found in nature, we can create light-harvesting assemblies, new methods for synthesizing fuels, and tools to synthesize novel functional materials for solar cells, water-splitting units, pollution control devices, and more.

Acknowledgement D. Durgalakshmi gratefully acknowledges the DST-INSPIRE faculty fellowship under the sanction DST/INSPIRE/04/2016/000845 for its funding. R. Saravanan gratefully acknowledges financial support from the SERC (CONICYT/FONDAP/15110019), FONDECYT, Government of Chile (Project. No: 11170414), and School of Mechanical Engineering (EUDIM), Universidad de Tarapaca, Arica, Chile.

References

Abdolali A, Guo WS, Ngo HH, Chen SS, Nguyen NC, Tung KL (2014) Typical lignocellulosic wastes and by-products for biosorption process in water and wastewater treatment: a critical review. Bioresour Technol 160:57–66. https://doi.org/10.1016/j.biortech.2013.12.037

Abe R (2010) Recent progress on photocatalytic and photoelectrochemical water splitting under visible light irradiation. J Photochem Photobiol C: Photochem Rev 11(4):179–209. https://doi.org/10.1016/j.jphotochemrev.2011.02.003

Akpan UG, Hameed BH (2009) Parameters affecting the photocatalytic degradation of dyes using TiO2-based photocatalysts: a review. J Hazard Mater 170(2):520–529. https://doi.org/10.1016/j.jhazmat.2009.05.039

AL-Othman ZA, Ali R, Naushad M (2012) Hexavalent chromium removal from aqueous medium by activated carbon prepared from peanut shell: adsorption kinetics, equilibrium and thermodynamic studies. Chem Eng J 184:238–247. https://doi.org/10.1016/j.cej.2012.01.048

Alqadami AA, Naushad M, Alothman ZA, Ghfar AA (2017) Novel Metal–Organic Framework (MOF) based composite material for the sequestration of U(VI) and Th(IV) metal ions from aqueous environment. ACS Appl Mater Interfaces 9:36026–36037. https://doi.org/10.1021/acsami.7b10768

Ameen S, Seo H-K, Shaheer Akhtar M, Shin HS (2012) Novel graphene/polyaniline nanocomposites and its photocatalytic activity toward the degradation of rose Bengal dye. Chem Eng J 210:220–228. https://doi.org/10.1016/j.cej.2012.08.035

Asahi R, Morikawa T, Ohwaki T, Aoki K, Taga Y (2001) Visible-light Photocatalysis in nitrogen-doped titanium oxides. Science 293(5528):269–271. https://doi.org/10.1126/science.1061051

Attia AA, Girgis BS, Fathy NA (2008) Removal of methylene blue by carbons derived from peach stones by H3PO4 activation: batch and column studies. Dyes Pigments 76(1):282–289

Bahnemann DW, Kormann C, Hoffmann MR (1987) Preparation and characterization of quantum size zinc oxide: a detailed spectroscopic study. J Phys Chem 91(14):3789–3798. https://doi.org/10.1021/j100298a015

Bard AJ (1979) Photoelectrochemistry and heterogeneous photo-catalysis at semiconductors. J Photochem 10(1):59–75

Bhateria R, Jain D (2016) Water quality assessment of lake water: a review. Sustain Water Resour Manag 2(2):161–173

Bhatnagar A, Sillanpää M (2017) Removal of natural organic matter (NOM) and its constituents from water by adsorption – a review. Chemosphere 166:497–510. https://doi.org/10.1016/j.chemosphere.2016.09.098

Biermann F (2013) Curtain down and nothing settled: global sustainability governance after the 'Rio+ 20' earth summit. Environ Plann C Gov Policy 31(6):1099–1114

Bina O (2013) The green economy and sustainable development: an uneasy balance? Environ Plan C Gov Policy 31(6):1023–1047

Bora LV, Mewada RK (2017) Visible/solar light active photocatalysts for organic effluent treatment: fundamentals, mechanisms and parametric review. Renew Sust Energ Rev 76:1393–1421

Chowdhury FA, Trudeau ML, Guo H, Mi Z (2018) A photochemical diode artificial photosynthesis system for unassisted high efficiency overall pure water splitting. Nat Commun 9:1707

Crini G (2006) Non-conventional low-cost adsorbents for dye removal: a review. Bioresour Technol 97(9):1061–1085

Daniel H, Perinaz B (2012) What a waste: a global review of solid waste management, Urban development series knowledge paper. World Bank, Washington, DC

Davis RJ, Gainer JL, O'Neal G, Wu IW (1994) Photocatalytic decolorization of wastewater dyes. Water Environ Res 66(1):50–53

Dayan FE, Duke SO, Grossmann K (2017) Herbicides as probes in plant biology. Weed Sci 58(3):340–350. https://doi.org/10.1614/WS-09-092.1

Dittmar M (2014) Development towards sustainability: how to judge past and proposed policies? Sci Total Environ 472:282–288

Dong F, Xiong T, Sun Y, Zhao Z, Zhou Y, Feng X, Wu Z (2014) A semimetal bismuth element as a direct plasmonic photocatalyst. Chem Commun 50(72):10386–10389. https://doi.org/10.1039/C4CC02724H

Febrianto J, Kosasih AN, Sunarso J, Ju Y-H, Indraswati N, Ismadji S (2009) Equilibrium and kinetic studies in adsorption of heavy metals using biosorbent: a summary of recent studies. J Hazard Mater 162(2):616–645. https://doi.org/10.1016/j.jhazmat.2008.06.042

Foster R, Ghassemi M, Cota A (2009) Solar energy: renewable energy and the environment. CRC Press, Boca Raton

Fox MA, Dulay MT (1993) Heterogeneous photocatalysis. Chem Rev 93(1):341–357. https://doi.org/10.1021/cr00017a016

Ghasemi M, Naushad M, Ghasemi N, Khosravi-fard Y (2014a) Adsorption of Pb(II) from aqueous solution using new adsorbents prepared from agricultural waste: adsorption isotherm and kinetic studies. J Ind Eng Chem 20:2193–2199. https://doi.org/10.1016/j.jiec.2013.09.050

Ghasemi M, Naushad M, Ghasemi N, Khosravi-fard Y (2014b) A novel agricultural waste based adsorbent for the removal of Pb(II) from aqueous solution: kinetics, equilibrium and thermodynamic studies. J Ind Eng Chem 20:454–461. https://doi.org/10.1016/j.jiec.2013.05.002

Gnanasekaran S, Saravanan N, Ilangkumaran M (2016) Influence of injection timing on performance, emission and combustion characteristics of a DI diesel engine running on fish oil biodiesel. Energy 116:1218–1229

Gupta SM, Tripathi M (2011) A review of TiO2 nanoparticles. Chin Sci Bull 56(16):1639. https://doi.org/10.1007/s11434-011-4476-1

Gupta VK, Ali I, Saleh TA, Nayak A, Agarwal S (2012) Chemical treatment technologies for wastewater recycling—an overview. RSC Adv 2(16):6380–6388

Hashimoto K, Irie H, Fujishima A (2005) TiO2 photocatalysis: a historical overview and future prospects. Jpn J Appl Phys 44(12R):8269

Hayashi J, Yamamoto N, Horikawa T, Muroyama K, Gomes V (2005) Preparation and characterization of high-specific-surface-area activated carbons from K2CO3-treated waste polyurethane. J Colloid Interface Sci 281(2):437–443

Hebbar RS, Isloor AM, Inamuddin AAM (2017) Carbon nanotube- and graphene-based advanced membrane materials for desalination. Environ Chem Lett 15:643–671

Hussey J (2014) Bang to eternity and betwixt: cosmos. John Hussey, New York

Javadian H, Angaji MT, Naushad M (2014) Synthesis and characterization of polyaniline/γ-alumina nanocomposite: a comparative study for the adsorption of three different anionic dyes. J Ind Eng Chem 20:3890–3900. https://doi.org/10.1016/j.jiec.2013.12.095

Jiang Y, Biswas P, Fortner JD (2016) A review of recent developments in graphene-enabled membranes for water treatment. Environ Sci Water Res Technol 2(6):915–922

Kazuhito H, Hiroshi I, Akira F (2005) TiO 2 Photocatalysis: a historical overview and future prospects. Jpn J Appl Phys 44(12R):8269

Kormann C, Bahnemann DW, Hoffmann MR (1988) Photocatalytic production of hydrogen peroxides and organic peroxides in aqueous suspensions of titanium dioxide, zinc oxide, and desert sand. Environ Sci Technol 22(7):798–806. https://doi.org/10.1021/es00172a009

Lems S, Van Der Kooi H, De Swaan Arons J (2010) Exergy analyses of the biochemical processes of photosynthesis. Int J Exergy 7(3):333–351

Levchuk I, Rueda Márquez JJ, Sillanpää M (2018) Removal of natural organic matter (NOM) from water by ion exchange – a review. Chemosphere 192:90–104. https://doi.org/10.1016/j.chemosphere.2017.10.101

Li X, Yu J, Wageh S, Al-Ghamdi AA, Xie J (2016) Graphene in Photocatalysis: a review. Small 12(48):6640–6696. https://doi.org/10.1002/smll.201600382

Liang X, Wang P, Li M, Zhang Q, Wang Z, Dai Y, Zhang X, Liu Y, Whangbo M-H, Huang B (2018) Adsorption of gaseous ethylene via induced polarization on plasmonic photocatalyst Ag/AgCl/TiO2 and subsequent photodegradation. Appl Catal B Environ 220:356–361. https://doi.org/10.1016/j.apcatb.2017.07.075

Ma J, Ding Z, Wei G, Zhao H, Huang T (2009) Sources of water pollution and evolution of water quality in the Wuwei basin of Shiyang river, Northwest China. J Environ Manag 90(2):1168–1177

Maeda K (2011) Photocatalytic water splitting using semiconductor particles: history and recent developments. J Photochem Photobiol C: Photochem Rev 12(4):237–268. https://doi.org/10. 1016/j.jphotochemrev.2011.07.001

Malwal D, Gopinath P (2016) Fabrication and applications of ceramic nanofibers in water remediation: a review. Crit Rev Environ Sci Technol 46(5):500–534. https://doi.org/10.1080/ 10643389.2015.1109913

Mamba G, Mamo MA, Mbianda XY, Mishra AK (2014) Nd,N,S-TiO2 decorated on reduced graphene oxide for a visible light active Photocatalyst for dye degradation: comparison to its MWCNT/Nd,N,S-TiO2 analogue. Ind Eng Chem Res 53(37):14329–14338. https://doi.org/10. 1021/ie502610y

Mills A, Davies RH, Worsley D (1993) Water purification by semiconductor photocatalysis. Chem Soc Rev 22(6):417–425. https://doi.org/10.1039/CS9932200417

Mittal A, Naushad M, Sharma G et al (2016) Fabrication of MWCNTs/ThO2 nanocomposite and its adsorption behavior for the removal of Pb(II) metal from aqueous medium. Desalin Water Treat 57:21863–21869. https://doi.org/10.1080/19443994.2015.1125805

Mohan M, Banerjee T, Goud VV (2015) Hydrolysis of bamboo biomass by subcritical water treatment. Bioresour Technol 191:244–252. https://doi.org/10.1016/j.biortech.2015.05.010

Morgan BJ, Watson GW (2010) Intrinsic n-type defect formation in TiO2: a comparison of rutile and Anatase from GGA+U calculations. J Phys Chem C 114(5):2321–2328. https://doi.org/10. 1021/jp9088047

Naushad M, Vasudevan S, Sharma G et al (2016) Adsorption kinetics, isotherms, and thermodynamic studies for Hg2+ adsorption from aqueous medium using alizarin red-S-loaded amberlite IRA-400 resin. Desalin Water Treat 57:18551–18559. https://doi.org/10.1080/19443994.2015. 1090914

Neamen DA (1997) Semiconductor physics and devices, vol 3. McGraw-Hill, New York

Nozik A (1977) Photochemical diodes. Appl Phys Lett 30(11):567–569

ODFaA-EH (ed) (1993) Photocatalytic purification and treatment of water and air. Elsevier, Amsterdam

Ola O, Maroto-Valer MM (2015) Review of material design and reactor engineering on TiO2 photocatalysis for CO2 reduction. J Photochem Photobiol C: Photochem Rev 24:16–42. https:// doi.org/10.1016/j.jphotochemrev.2015.06.001

Organization WH (2001) Water for health: taking charge. World Health Organization (WHO), Geneva

Osterloh FE (2017) Photocatalysis versus photosynthesis: a sensitivity analysis of devices for solar energy conversion and chemical transformations. ACS Energy Lett 2(2):445–453

Pandey S, Ramontja J (2016) Turning to nanotechnology for water pollution control: applications of nanocomposites. Focus Sci 2(2):1–10

Programme UNWWA (2015) World water development report 2015: water for a sustainable world. Unesco Paris

Qin J, Yang C, Cao M, Zhang X, Saravanan R, Limpanart S, Mab M, Liu R (2017) Two-dimensional porous sheet-like carbon-doped ZnO/g-C3N4nanocomposite with high visible-light photocatalytic performance. Mater Lett 189:156–159. https://doi.org/10.1016/j.matlet. 2016.12.007

Rajeshwar K, Osugi ME, Chanmanee W, Chenthamarakshan CR, Zanoni MVB, Kajitvichyanukul P, Krishnan-Ayer R (2008) Heterogeneous photocatalytic treatment of organic dyes in air and aqueous media. J Photochem Photobiol C: Photochem Rev 9(4):171–192. https://doi.org/10. 1016/j.jphotochemrev.2008.09.001

Rashed MN (2013) Adsorption technique for the removal of organic pollutants from water and wastewater. In: Organic pollutants-monitoring, risk and treatment. InTech, New York

Reza KM, Kurny ASW, Gulshan F (2017) Parameters affecting the photocatalytic degradation of dyes using TiO2: a review. Appl Water Sci 7(4):1569–1578. https://doi.org/10.1007/s13201-015-0367-y

Romero-Sáez M, Jaramillo LY, Saravanan R, Benito N, Pabón E, Mosquera E, Gracia F (2017) Notable photocatalytic activity of TiO_2-polyethylene nanocomposites for visible light degradation of organic pollutants. Express Polym Lett 11(11):899–909. https://doi.org/10.3144/expresspolymlett.2017.86

Saravanan R, Shankar H, Prakash T, Narayanan V, Stephen A (2011) ZnO/CdO composite nanorods for photocatalytic degradation of methylene blue under visible light. Mater Chem Phys 125(1–2):277–280. https://doi.org/10.1016/j.matchemphys.2010.09.030

Saravanan R, Manoj D, Qin J, Naushad M, Gracia F, Lee AF, MansoobKhan MM, Gracia-Pinilla MA (2018) Mechanothermal synthesis of Ag/TiO_2 for photocatalytic methyl orange degradation and hydrogen production. Process Saf Environ Prot 120:339–347. https://doi.org/10.1016/j.psep.2018.09.015

Sayama K, Mukasa K, Abe R, Abe Y, Arakawa H (2002) A new photocatalytic water splitting system under visible light irradiation mimicking a Z-scheme mechanism in photosynthesis. J Photochem Photobiol A Chem 148(1):71–77. https://doi.org/10.1016/S1010-6030(02)00070-9

Sen TK (2017) Air, gas, and water pollution control using industrial and agricultural solid wastes adsorbents. CRC Press, Boca Raton

Sharma G, Naushad M, Pathania D et al (2015) Modification of *Hibiscus cannabinus* fiber by graft copolymerization: application for dye removal. Desalin Water Treat 54:3114–3121. https://doi.org/10.1080/19443994.2014.904822

Sharma G, Naushad M, Kumar A et al (2017) Efficient removal of coomassie brilliant blue R-250 dye using starch/poly (alginic acid-cl-acrylamide) nanohydrogel. Process Saf Environ Prot. https://doi.org/10.1016/j.psep.2017.04.011

Silva L, Gasca-Leyva E, Escalante E, Fitzsimmons K, Lozano D (2015) Evaluation of biomass yield and water treatment in two Aquaponic systems using the dynamic root floating technique (DRF). Sustainability 7(11):15384

Suárez-Iglesias O, Collado S, Oulego P, Díaz M (2017) Graphene-family nanomaterials in wastewater treatment plants. Chem Eng J 313:121–135

Thomann I, Pinaud BA, Chen Z, Clemens BM, Jaramillo TF, Brongersma ML (2011) Plasmon enhanced solar-to-fuel energy conversion. Nano Lett 11(8):3440–3446. https://doi.org/10.1021/nl201908s

Wang C, Astruc D (2014) Nanogold plasmonic photocatalysis for organic synthesis and clean energy conversion. Chem Soc Rev 43(20):7188–7216. https://doi.org/10.1039/C4CS00145A

Wang X, Maeda K, Thomas A, Takanabe K, Xin G, Carlsson JM, Domen K, Antonietti M (2008) A metal-free polymeric photocatalyst for hydrogen production from water under visible light. Nat Mater 8:76. https://doi.org/10.1038/nmat2317. https://www.nature.com/articles/nmat 2317#supplementary-information

Whitmarsh J, Govindjee (1999) The photosynthetic process. In: Singhal GS, Renger G, Sopory SK, Irrgang KD, Govindjee (eds) Concepts in photobiology: photosynthesis and photomorphogenesis. Springer, Dordrecht, pp 11–51. https://doi.org/10.1007/978-94-011-4832-0_2

Williams G, Seger B, Kamat PV (2008) TiO2-graphene Nanocomposites. UV-assisted Photocatalytic reduction of graphene oxide. ACS Nano 2(7):1487–1491. https://doi.org/10.1021/nn800251f

Xuming Z, Yu Lim C, Ru-Shi L, Din Ping T (2013) Plasmonic photocatalysis. Rep Prog Phys 76(4):046401

Yagub MT, Sen TK, Afroze S, Ang HM (2014) Dye and its removal from aqueous solution by adsorption: a review. Adv Colloid Interf Sci 209:172–184

Yang Y, Guo W, Guo Y, Zhao Y, Yuan X, Guo Y (2014) Fabrication of Z-scheme plasmonic photocatalyst Ag@AgBr/g-C3N4 with enhanced visible-light photocatalytic activity. J Hazard Mater 271:150–159. https://doi.org/10.1016/j.jhazmat.2014.02.023

Zhang Q, Thrithamarassery Gangadharan D, Liu Y, Xu Z, Chaker M, Ma D (2017) Recent advancements in plasmon-enhanced visible light-driven water splitting. J Mater 3(1):33–50. https://doi.org/10.1016/j.jmat.2016.11.005

Zhou L, Zhang C, McClain MJ, Manjavacas A, Krauter CM, Tian S, Berg F, Everitt HO, Carter EA, Nordlander P, Halas NJ (2016) Aluminum Nanocrystals as a Plasmonic Photocatalyst for hydrogen dissociation. Nano Lett 16(2):1478–1484. https://doi.org/10.1021/acs.nanolett.5b05149

Zhu Z (2017) An overview of carbon nanotubes and graphene for biosensing applications. Nano-Micro Lett 9(3):25

Zhu J, Deng Z, Chen F, Zhang J, Chen H, Anpo M, Huang J, Zhang L (2006) Hydrothermal doping method for preparation of Cr3+-TiO2 photocatalysts with concentration gradient distribution of Cr3+. Appl Catal B Environ 62(3):329–335. https://doi.org/10.1016/j.apcatb.2005.08.013

Chapter 2
Synthesizing Green Photocatalyst Using Plant Leaf Extract for Water Pollutant Treatment

Kavitha Shivaji, Esther Santhoshi Monica, Anitha Devadoss,
D. David Kirubakaran, C. Ravi Dhas, Sagar M. Jain,
and Sudhagar Pitchaimuthu

Contents

K. Shivaji
Department of Biotechnology, K.S.R. College of Technology, Tiruchengode, Tamil Nadu, India

E. S. Monica · D. D. Kirubakaran · C. R. Dhas
PG & Research Department of Physics, Bishop Heber College, Trichy, Tamil Nadu, India

A. Devadoss
Centre for Nano Health, College of Engineering, Swansea University, Swansea, Wales, UK

S. M. Jain · S. Pitchaimuthu (✉)
Multi-functional Photocatalyst & Coatings Group, SPECIFIC, College of Engineering, Swansea
University (Bay Campus), Swansea, Wales, UK
e-mail: S.Pitchaimuthu@swansea.ac.uk

© Springer Nature Switzerland AG 2020   25
M. Naushad et al. (eds.), *Green Photocatalysts*, Environmental Chemistry
for a Sustainable World 34, https://doi.org/10.1007/978-3-030-15608-4_2

Abstract Developing toxic-free material synthesis route shows great demand for sustainable environment. In particular, green-synthesized catalyst material perceived great attention in environmental- and biomedical-based applications. Mostly conventional chemical synthesis route contains particle stabilizer agent for obtaining low-dimensional nanoparticle or avoiding particle aggregation. Unfortunately, these chemicals are highly toxic to the environment. In this context, plant extract, biomass, and algae can be utilized as green bio-surfactant without polluting the environment. This chapter explores the recent development of green synthesis routes for photocatalyst synthesis using plant extract, and their capability in organic pollutant removal from the water will be discussed in detail.

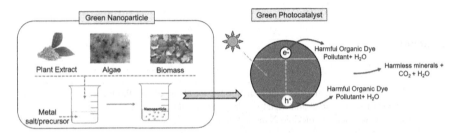

Keywords Photocatalyst · Green chemistry · Bio-surfactant · Pollutant degradation · Solar energy · Water treatment

2.1 Introduction

Rapid industrial growth has cozied up human beings; however, its adverse effects came out in the form of environmental deterioration. Tons of organic wastes are being dumped into the aquatic system in the form of chemical spills and effluents each year (Naushad et al. 2016; Gnanasekaran et al. 2018). Abundant diseases, health issues, and even fatalities have been reported with water pollution. The pollution not only threatens the environment, but human beings and animals as well. Several industries such as leather, plastic, food, tanneries, cosmetic, pharmaceutical, and textile industries discharge toxic dye effluents into the aquatic system, land, etc., after various chemical processes, thereby disturbing the ecosystem ecology. In addition, the organic pollutants also induce skin irritation and blood disorder, and they also carry venomous effect to the central nervous system in humans and animals (Alshehri et al. 2014; McCann and Ames 1976).

In particular, textile industries devour large volumes of water in the processing operations that include pre-treatment, dyeing, printing, and finishing. It is estimated that out of approximately 10 million kg of dyestuff per year, about 1–2 million kg of active dye stuff passes into the biosphere every year exhibiting severe wallop on the water bodies and the land surrounding them (Robinson et al. 2001). Dyes have a complex aromatic molecular structure, making them inert and troublesome to bio-degrade when expelled into the environment. As dyes are engineered to be stable both chemically and photolytically, the effect is visually more unpleasant rather than hazardous.

Powerful regulations that are concerning the quality of water accentuate the need to establish new methods for the removal of organic pollutants from industrial wastewater. Hence, it is most pivotal to develop ecologically clean and safe tech-nologies to address the global issue on water pollution. Degradation through photocatalysis is ecologically clean, safe, and promising key for addressing water pollution which is a major cause of concern in many countries such as India and other developing nations. Photocatalysis is simply the process of accelerating chem-ical reaction for the eradication of pollutants (phenols, pesticides, alkenes, alkanes, and aromatics) and complete mineralization of the organic compounds through photocatalysts that absorb light energy of wavelength equivalent to its band gap.

Nanoparticles are playing a vital role in building up future sustainable technolo-gies for humanity. At present, research interest is being focused toward the devel-opment of nanostructured semiconductor photocatalysts due to its effective decolorization of dyes (Nishio et al. 2006). In a photocatalytic reaction, the activity is dependent on the surface area, size distribution, porosity, crystal structure, band gap, and surface hydroxyl group density of the catalyst (Pandimurugan and Thambidurai 2016). Nanostructured semiconductors exhibit better efficiency than the bulk materials due to its special properties including energy mass transfer, large surface area to volume ratio, and quantum confinement effects of charge carriers for the removal of contaminants from water (Chakrabarti and Dutta 2004; He et al. 2012; Sinha and Jana 2012). In the photocatalytic pathway, nanoparticles activated by sunlight form a redox atmosphere in aqueous solution and act as a sensitizer for a light-induced redox mechanism which is involved in complete elimination of pol-lutants from the environment (Beydoun et al. 1999). Innumerable physical and chemical processes have been developed for the nanoparticle synthesis. However, the production cost and the usage of toxic chemicals remain a limitation for such synthesis methods. This necessitates the alternate eco-friendly approach for the synthesis of nanoparticles.

The biosynthesis approach of using biological agents such as plants and micro-organisms for the synthesis of metal and metal oxide nanoparticles is being consid-ered as a highly potential alternative (Ahmed 2014). It bridges nanotechnology and plant biotechnology for the metal ion reduction using plant extracts without involv-ing high pressure, energy, temperature, or toxic chemicals, and also the method restricts the usage of additional capping and stabilizing agent, thereby enhancing the efficacy of biocompatible treatment. This chapter exploits the mechanism behind the green synthesis of metal and metal oxide nanoparticle catalysts using different biological sources and their role in photocatalytic degradation of pollutants in the environment.

2.2 Theory

2.2.1 Why Green Synthesis Approach Needed for Nanoparticle Production

In general, nanoparticles are synthesized using any of the two approaches, namely, top-down and bottom-up approach. The top-down approach involves milling of large-scale macroscopic particles to nanoscale-level particles through plastic deformation. However, this process is slow and expensive (Ahmed et al. 2016). In bottom-up approach, nanoparticles are synthesized by chemical (chemical reduction) and biological (plants and microorganisms) methods. For decades, physical and chemical methods have been utilized for the synthesis of nanoparticles (Huy et al. 2013). Physical method involves physical force, costly equipment, and high temperature and pressure for the synthesis of nanoparticles with well-defined structures (Chandrasekaran et al. 2016). On the other hand, chemical methods utilize toxic chemicals and lead to toxic by-product evolution during the synthesis process. Being low-cost and eco-friendly with minimal by-product evolution procedure, green synthesis technique is of utmost importance in nanotechnological era (Abdul Salam et al. 2014).

Green synthesis of nanoparticles is an alternative strategy for conventional method that can be realized through the usage of plant tissues like leaf, stem, root, fruit, peel, and flower. For the green synthesis of nanoparticles, plant extracts were prepared by washing plant samples with distilled water to remove organic moieties and then boiling it in a suitable solvent. The prepared extracts were centrifuged for the removal of undesired impurities and then subjected to salt solution (silver, gold, copper, etc.) in varying concentrations. The biosynthesis will be confirmed through color change which differs from plant to plant. The presence of phytochemicals (proteins, enzymes, polysaccharides, vitamins, carbohydrates, amino acids, and organic acids) in the prepared plant extracts restricts the usage of external stabilizing agent since the phytochemicals themselves act as stabilizing agents sufficient to produce nanoparticles. Green synthesis of nanoparticles can be achieved through single-step one-pot method which is biocompatible and highly suitable for large-scale production (Fig. 2.1). This environmental benign synthesis technology is emerging as recent advancement in the field of biotechnology and nanotechnology (Abdul Salam et al. 2014).

2.3 Green Synthesis of Metal and Metal Oxide Nanoparticle

Feasibility and low adverse impact on the environment of the green synthesis techniques for the synthesis of metal and metal oxide nanoparticles have attracted significant attention in recent research areas. It is interesting to note the biogenic reduction of metal ions to the base metal. This chapter summarizes the green

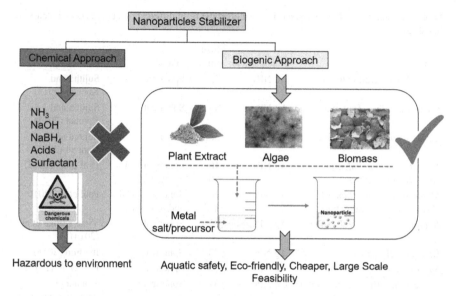

Fig. 2.1 Schematic illustration of conventional and green-synthesized nanoparticle

approach for metal and metal oxide nanoparticle production. Table 2.1 displays various attempts for the green synthesis of metal and metal oxide nanoparticles as elaborated in this chapter.

2.3.1   Green Synthesis of Gold Nanoparticles

Gold nanoparticles have been widely exploited for its applications such as drug delivery, cosmetics, catalysts, biosensors, and various other biomedical fields due to its chemical stability and unique properties. Sujitha et al. (Sujitha and Kannan 2013) reported the synthesis of gold nanoparticles (Au NPs) using citrus fruit extracts (*Citrus limon, Citrus reticulata,* and *Citrus sinensis*). The citrus fruit extract-mediated Au NPs exhibited distinct size of the particles on varying the concentration of the extract as confirmed through the UV spectrum and transmission electron microscope (TEM) analysis. Varying morphologies were observed for varying concentrations of the fruit extract. Among the three mentioned citrus fruit extracts, Au NPs synthesized through *Citrus limon* extract showed better efficiency followed by *C. reticulata* and *C. sinensis.* In another study, Au NPs were synthesized from *Pogostemon benghalensis (B) O.Ktz.* leaf extract. These plant-mediated Au NPs were well dispersed and predominantly triangular and spherical in shape, with size ranging from 10 to 50 nm. In addition, TEM analysis clearly indicated that the Au NPs were extracellularly surrounded by plant-mediated proteins and organic molecules in green synthesis approach (Paul et al. 2016). Similarly, Au NPs were

Table 2.1 Summary of green-synthesized nanoparticle from different plant extracts and their size and shape

Plant name	Nanoparticle	Size (nm)	Shape	References
Citrus lemon (fruit extract)	Au NPs	32	Spherical	Sujitha and Kannan (2013)
Citrus reticulata (fruit extract)	Au NPs	43	Spherical	Sujitha and Kannan (2013)
Citrus sinensis (fruit extract)	Au NPs	56	Spherical	Sujitha and Kannan (2013)
Pogostemon benghalensis (B) O.Ktz (leaf extract)	Au NPs	10–50	Spherical and triangular	Paul et al. (2015)
Momordica cochinchinensis (Lour.) Spreng (leaf extract)	Au NPs	10–80	Spherical, oval, and triangular	Paul et al. (2016)
Boerhavia diffusa (leaf extract)	Ag NPs	25	Spherical	Vijay Kumar et al. (2014)
Banana peel extract	Ag NPs	23	Spherical	Ibrahim (2015)
Saccharomyces cerevisiae	Ag NPs	10	Spherical	Roy et al. (2015)
Labeo rohita (fish scale extract)	Cu NPs	31	Spherical	Sinha and Ahmaruzzaman (2015)
Plantago asiatica (dried leaf powder)	Cu NPs	7–35	Spherical	Nasrollahzadeh et al. (2017)
Acalypha indica (dried leaf powder extract)	CONPs	31	Spherical	Sivaraj et al. (2014)
Aloe vera (leaf extract)	CONPs	2–30	Spherical	Kumar et al. (2015)
Bifurcaria bifurcate (brown algae extract)	CONPs	5–45	Spherical	Abboud et al. (2014)
Syzygium alternifolium (bark extract)	CONPs	5–13	Spherical	Yugandhar et al. (2017)
Punica granatum (leaf extract)	CONPs	40	Spherical	Ghidan et al. (2016)
Carica papaya (leaf extract)	CONPs	140	Rod	Sankar et al. (2014)
Gloriosa superba (leaf extract)	CONPs	5–10	Spherical	Naika et al. (2015)
Eucalyptus globulus (dried leaf extract)	ZnO NPs	10–20	Spherical	Siripireddy and Mandal (2017)
Coriandrum sativum (leaf extract)	ZnO NPs	9–18	–	Saad et al. (2015)
Corymbia citriodora (leaf extract)	ZnO NPs	20–120	Poly hedran	Zheng et al. (2015)
Gloriosa superba (leaf extract)	CeO$_2$ NPs	5	Spherical	Arumugam et al. (2015)
Azadirachta indica (leaf extract)	CeO$_2$ NPs	10–15	Spherical	Sharma et al. (2007)
Aloe vera (leaf extract)	CeO$_2$ NPs	2–3	Spherical	Dutta et al. (2016)

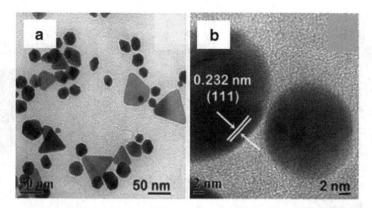

Fig. 2.2 HRTEM images of one-pot green synthesis of gold nanoparticles using *Pogostemon benghalensis (B) O.Ktz.* leaf extract. (Figures (**a**) and (**b**) were adopted from Paul et al. (2015) with permission by Elsevier Publishers)

synthesized using *Momordica cochinchinensis (Lour). Spreng* dry leaf powder. Surprisingly, this leaf (*Momordica cochinchinensis*) had produced various shaped nanoparticles like spherical, oval, and triangular with size ranging from 10 to 80 nm (Fig. 2.2a, b) (Paul et al. 2015). The abovementioned preparation techniques infer that each plant extract with varying concentrations contributes to the distinct size and morphological features of Au NPs depending on the phytochemicals present in the prepared extracts.

2.3.2   Green Synthesis of Silver Nanoparticles

The production of silver nanoparticles (Ag NPs) by green approach has drawn great attention as it is of single-step and economically beneficial synthesis method. In conventional chemical method, Ag NPs are produced by involving toxic chemicals (hydrazine hydrate, sodium borohydride, Dimethylformamide (DMF), and ethylene alcohols) which results in the absorption of toxic chemicals over the surface of nanoparticles (Iravani 2011). On the other hand, green synthesis of Ag NPs results in the absorption of plant-mediated phytochemicals such as proteins, amino acids, enzymes, polysaccharides, alkaloids, phenolics, tannins, saponins, and vitamins over the surface of the nanoparticles (Chithrani et al. 2006). It is reported that green synthesis-derived Ag NPs through *Boerhavia diffusa* plant extract exhibited spherical-shaped morphology with 25 nm. These Ag NPs had better bactericidal activity against fish pathogens (Vijay Kumar et al. 2014). Ag ions when combined with banana peel extracts displayed quick reduction process and revealed spherical morphology of 23.7 nm with good crystallinity. Further, it revealed strong bactericidal against both gram-positive and gram-negative bacteria (Fig. 2.3) (Ibrahim 2015). In another report, yeast (*Saccharomyces cerevisiae*) extract was used for

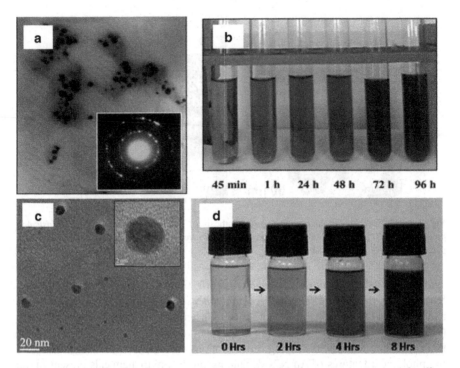

Fig. 2.3 (**a**) TEM image of silver nanoparticles synthesized from banana peel extract and (**b**) effect of reaction time on color intensity of synthesized silver nanoparticles (Figures **a**, **b** were adopted from (Ibrahim 2015) with permission by Elsevier Publishers); (**c**) TEM image and (**c**) of biogenic silver nanoparticles synthesized using yeast (*Saccharomyces cerevisiae*) extract (Figures **c**, **d** were adopted from Roy et al. (2015) with permission by Elsevier Publishers)

Ag NP production, and gradual dark brown color was observed after 4-h incubation time (Fig. 2.3b). Yeast extract-mediated silver nanoparticles were crystalline and were spherical in shape with an average diameter of 10 nm. The plausible mechanism which aided for silver nanoparticle formation on using yeast extract is that yeast is rich in functional organic molecules, and it has the ability to reduce the silver cations in the solvent to form well-crystalline nanoparticles (Ren et al. 2009; Roy et al. 2015).

2.3.3 Green Synthesis of Copper Nanoparticles

Copper nanoparticles (Cu NPs) are given much attention than gold, silver, and platinum nanoparticles as copper nanoparticles are less expensive to synthesize. Furthermore, Cu NPs are being utilized in numerous applications such as catalysis, sensors, field emission emitters, and inkjets (Ren et al. 2009; Zhang et al. 2008). Single-step synthesis process was employed for the synthesis of Cu NPs using fish

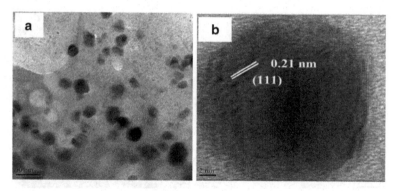

Fig. 2.4 TEM images of copper nanoparticles using household waste material (fish scales of *Labeo rohita*) as surfactant (**a**) 50 nm and (**b**) 2 nm scale. (Figures **a, b** were adopted from Sinha and Ahmaruzzaman (2015) with permission by Springer Publishers)

scale extract of *Labeo rohita* by Sinha and Ahmaruzzaman (2015). In this process, 0.5 g of $CuSO_4$ was dissolved with 10 mL of dis.H_2O, and 50 mL of fish scale extract was mixed under continuous stirring condition. The Cu NP formation was confirmed by color change after 1 h, and it displayed spherical nanoparticles with average size at 31 nm by High resolution transmission electron microscopy (HRTEM) analysis. According to the report, the formation of Cu NPs was due to the formation of instantaneous complex by the electrostatic interaction between the negatively charged group of gelatin and positively charged copper ions, when fish scale extract was combined with copper sulfate solution (Fig. 2.4). It is due to the collagen-rich *L. rohita* and some of the major components like glycine, amino acids, hydroxyproline, and hydroxyzine present in the fish scale extract (Ikoma et al. 2003). During the process, the fish scale extract was heated at 70 °C for 20 min. As a result, collagen got denatured and its physicochemical properties were changed due to the destruction of the triple helical structure, such denatured collagen called gelatin. Cu NPs were also synthesized using *Plantago asiatica* dried leaf powder which displayed uniform spherical NPs with size ranging from 7 to 35 nm. The stabilization and size reduction process in this case were within 5 min due to the presence of polyphenolics (biological compound) in the plant extract (Nasrollahzadeh et al. 2017). Hence, it is to be noted that the quick synthesis process, better stabilization, and size reduction differ based on the phytochemical nature and its environmental factors.

2.3.4 Green Synthesis of Copper Oxide Nanoparticles

Metal oxide nanoparticles are also been given greater attention due to their industrial importance and their wide applications such as solar devices, gas sensors, catalyst, and superconductors (Premkumar and Geckeler 2006; Ren et al. 2009; Zhang et al. 2012). Sivaraj et al. reported the production of copper oxide nanoparticles (CONPs)

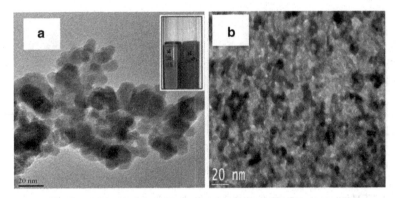

Fig. 2.5 HRTEM images of bioinspired green synthesis of copper oxide nanoparticles from (**a**) *Syzygium alternifolium* (Wt.) Walp and (**b**) *Gloriosa superba* L. extract. (Figure **a** is adopted from Yugandhar et al. (2017) with permission by Springer Publishers. Figure **b** is adopted from Naika et al. (2015) with permission by Elsevier Publishers)

using *Acalypha indica* which was found to be in the range of 26 nm with spherical shape (2014). Similarly, *Aloe vera* extract was used for the synthesis of spherical-shaped CONPs with size ranging between 20 and 30 nm (Kumar et al. 2015). In another report, *Syzygium alternifolium* (plant bark powder) was used for the synthesis of CONPs, producing 5–13 nm size of spherical-shaped NPs processed within 2 h (Yugandhar et al. 2017). Also, *Punica granatum* peel extract has produced CONPs within 10 min with average size of 40 nm (Ghidan et al. 2016). Sankar et al. used *Carica papaya* leaf extract as a stabilizing agent for CONP formation, and it exhibited rod-shaped particle after 24 h with average size range of 140 nm (2014). In addition, brown algae (*Bifurcaria bifurcata*) were used for CONPs formation, and it showed 5–45-nm-sized spherical particles in 24 h (Abboud et al. 2014). *Gloriosa superba*-mediated spherical-shaped CONPs synthesized using dried leaf powder were attained within 10 min (Naika et al. 2015). Furthermore, this report strongly indicated the production of CONPs in an eco-friendly manner. But some plants have efficiency for stabilization and size reduction of nanoparticles due to the presence of different phytochemicals. For instance, *Syzygium alternifolium*, *Punica granatum*, and *Gloriosa superba* plants reveal quick reduction process among which *Gloriosa superba* plant has high potential to attain minimal particle size (5–10 nm) (Fig. 2.5a, b). These reports infer the possibility of mass production of CONPs suitable for strong bactericidal activity, without the usage of hazardous chemicals.

2.3.5 Green Synthesis of Zinc Oxide Nanoparticles

Zinc oxide (ZnO) is a semiconducting metal oxide highly suitable in the field of electronics, optics, and biomedical systems, and it has been enlisted as a safe metal oxide by the US FDA (Anbuvannan et al. 2015; Jamdagni et al. 2018; Patil and

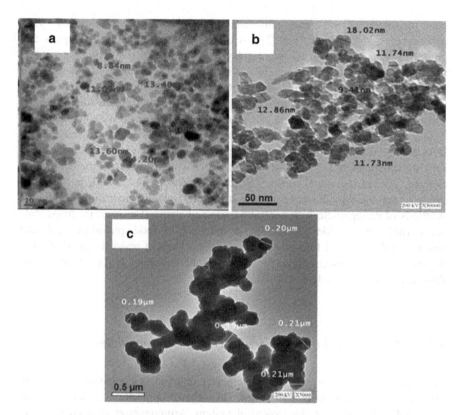

Fig. 2.6 TEM images of green-synthesized ZnO nanoparticles using (**a**) *Eucalyptus globulus*, (**b**) *Coriandrum sativum* leaf extract, and (**c**) conventional chemical. (Figure **a** adopted from Siripireddy and Mandal (2017) with permission by Elsevier publisher. Figures **b**, **c** were adopted from Saad et al. (2015) with permission by IOP publisher)

Taranath 2016; Vanathi et al. 2014). For the green synthesis of ZnO nanoparticles, *Eucalyptus globulus* plant leaf (dried powder) extract and zinc nitrate [Zn $(NO_3)_3.6H_2O$] were mixed with equal ratio (1:1 v/v) at 600 rpm for 3 h until the occurrence of brown color. It resulted in the crystalline spherical-shaped nanoparticles with size ranging between 10 and 20 nm (Fig. 2.6a). The plausible mechanism behind the reaction is that plant extracts are responsible for reduction of Zn^{2+} ions and stabilization of ZnO nanoparticles after 24 h due to the presence of polyphenols as confirmed by Fourier Transform Infrared Spectroscopy (FTIR) analysis (Balaji Reddy and Kumar Mandal 2017). In another report, zinc acetate dihydrate (0.2 g in 50 mL distilled H_2O) was used as a precursor and was stirred for 10 min, and 1.0 mL of *Coriandrum sativum* leaf extract was mixed with 2 .0 M NaOH until pH reaches 12. In contrast, for comparative analysis, chemically synthesized mixture was used (zinc acetate dihydrate and NaOH). Both aqueous solutions were processed for 2 h until precipitation is formed. The TEM analysis result displayed particle size ranging between 9 to 18 nm and 190 to 210 nm for plant-mediated (*Coriandrum sativum*) and chemically synthesized ZnO nanoparticles, respectively (Fig. 2.6b, c). Hence, this report clearly depicts that phytochemicals present in the plants play a vital role for size reduction process within a short duration in comparison with chemically synthesized nanoparticles (Saad et al. 2015). Yuhong et al. utilized *Corymbia citriodora* leaf extract for ZnO nanoparticle synthesis, and it was mixed with 0.5 M zinc nitrate solution under stirring process. After 48 h, white pale precipitate was observed which confirms the ZnO nanoparticle formation. The TEM micrograph analysis revealed size ranging from 20 to 120 nm with polyhedron-shaped particles. Also, *C. citriodora*-mediated ZnO nanoparticles have excellent dispersibility, which leads to sufficient surface charges between individual particles (Fig. 2.6d) (Zheng et al. 2015). In this green synthesis approach, plant sources act as a stabilizing as well as reducing agent for the synthesis of nanoparticle with controlled size and shape. The plant-mediated ZnO nanoparticles have wide applications in the field of food, pharmaceutical, and cosmetic industries and thus pave way to a major area of research. In addition to the bio-surfactant concentration, processing temperature also plays a vital role on ZnO nanoparticle size (Bala et al. 2015).

2.3.6   Green Synthesis of Cerium Oxide Nanoparticles

Cerium oxide (CeO_2) nanoparticles have been employed as a potential candidate in the field of nanotechnology with their wide range of applications as catalyst, fuel cells, and antioxidants in the biological system. Arumugam et al. used *Gloriosa superba* plant extract for CeO_2 nanoparticle synthesis. During the process, $CeCl_3$ was used as a precursor which was stirred continuously for 4–6 h along with the extract until white precipitate was observed. The formation of CeO_2 nanoparticles was confirmed by the color change from white to yellowish brown. The nanoparticles thus synthesized were in the size of 5 nm with spherical shape (Arumugam et al. 2015) (Fig. 2.6a). In another report, *Azadirachta indica* plant extract was used as a stabilizing agent, and cerous nitrate [$Ce(NO_3)_2.6H_2O$] surfactant was added under stirring condition, and the formation of CeO_2 NPs was confirmed by color change within 1–2 min. The *Azadirachta indica*-mediated CeO_2 nanoparticles were observed to be of polycrystalline nature having mixed fringes pattern with size ranging from 10 to 15 nm (Sharma et al. 2017). Dutta et al. reported yet another easy approach for green synthesis of spherical-shaped CeO_2 nanoparticles with size ranging from 2 to 3 nm, using *Aloe vera* extract at ambient temperature using cerium nitrate as a precursor. TEM image as in Fig. 2.6b clearly implies that these plant-based phytoconstituents act as a better stabilizing agent for CeO_2 nanoparticle formation within a short period (2016). The single-step synthesis process of CeO_2 nanoparticles is cost-effective and has a short-duration process with minimal energy requirement yet suitable for large-scale applications (Fig. 2.7).

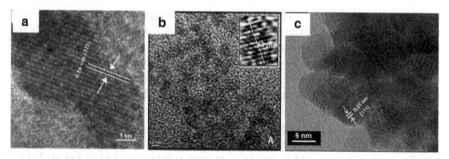

Fig. 2.7 TEM images of CeO$_2$ nanoparticles using (**a**) *Gloriosa superba* leaf extract, (**b**) *Aloe vera* leaf extract, and (**c**) *Azadirachta indica* plant extract. (Figure **a** adopted from Arumugam et al. (2015), Figure **b** adopted from Dutta et al. (2016), Figure **c** from Sharma et al. (2017). With permission by Elsevier Publishers)

2.4 Critics on Green Synthesis

2.4.1 Merits and Limitations of Green Synthesis Technique

The biologically synthesized nanoparticles are free from toxic contamination by-products. Also, the single-step process, cost-effectiveness, short-time preparation, and biocompatibility of the biogenic-mediated nanoparticles make it suitable candidate in the biomedical field. Furthermore, the method avoids the usage of external stabilizing agent as the biological components such as plants and microbes themselves act as capping and stabilizing agents (Singh et al. 2016). However, the disadvantage lies in the optimization of synthesis parameters such as concentration of salt, pH, temperature, contact time, and concentration of plant extract/microorganisms. These parameters will have adverse effect on the synthesis process and rate of production. For instance, at low pH, the nanoparticles get agglomerated which restricts the nucleation mechanism. This report is supported by Sujitha et al. that variation in the concentration of fruit extracts affects the morphology and size of the prepared nanoparticles (Sujitha and Kannan 2013).

2.4.2 How Green-Synthesized Nanoparticles Is Suitable for Photocatalytic Pollutant Degradation

Innovative "pollutant-degrading" technology for the water treatment is ineluctable for balancing the ecology in our planet since the organic toxic chemicals rushing out from the industries provide major contribution to the water pollution as per the US Environmental Protection Agency (US EPA). The entry of dye effluents from the textile industries exhibits severe blow on the environment and natural resources (Constantin et al. 2013; Jawad and Rashid 2016). These dark-colored chemicals are highly toxic and mutagenic and hence lead to murkiness in the industrial effluents, thus diminishing dissolved oxygen, thereby creating adverse effect on the aquatic ecosystem (Singh et al. 2009). To address this issue, various techniques as absorption, membrane-based separation (Purkait et al. 2006), electrocoagulation (Khansorthong and Hunsom 2009), biochemical degradation (Kagalkar et al. 2009), and oxidative degradation (Gözmen et al. 2009) have been employed among which absorption method has grabbed greater attention due to its economic benefits, easy operation, and insensitivity to toxic substances (Agarwal et al. 2016).

Various porous materials (metal organic frameworks (MOFs), carbon nanotubes, and nano-porous polymers) are being used as an absorbent which exhibits high affinity and rapid kinetics for dye removal. However, porous absorbents suffer from low-removal efficiency, and MOFs have low solubility in water, limiting their applications (Burtch et al. 2014). The upgrowth of the "Nano-world" offers numerous methodologies for the preparation of metal and metal oxide catalysts for the degradation of pollutants as mentioned earlier. However, involvement of high pressure, high temperature, high energy, and hazardous chemicals as capping agents which release venomous by-products through the methodology demands for the better efficient methodology. In recent days, green-synthesized metal and metal oxide nanoparticles using bio-products have been witnessed as an efficient alternative technique for the photocatalytic degradation due to its high adsorption and separation capacity (Asefa et al. 2009; Ghosh and Pal 2007). The schematic illustration of photocatalysis mechanism in organic dye pollutant treatment is presented in Scheme 2.1.

Under light illumination, the photocharge carriers (e^- and h^+) are generated at conduction band (CB) and valence band (VB), respectively. After photocharge carrier separation, the activated electrons react with oxidant to produce a reduced product and also a reaction between the generated holes with a reductant to produce an oxidized product. The photogenerated electrons at conduction band could reduce the dye pollutant or react with electron acceptors such as O_2 adsorbed on the photocatalyst surface or dissolved in water, reducing it to superoxide radical anion $O_2^{-\bullet}$. Conversely, the photogenerated holes will oxidize the organic molecule to form R^+ or react with OH^+ or H_2O oxidizing them into $OH^\bullet$ radicals. Combining

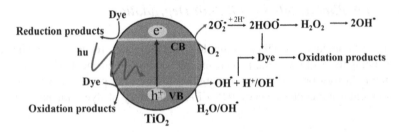

Scheme 2.1 Schematic illustration of photocatalytic dye pollutant degradation under light illumination

with other highly oxidant species, peroxide radicals will be responsible for the heterogeneous TiO_2 photodecomposition of organic substrates as dyes. Finally, $OH^{\bullet}$ radical, being a very strong oxidizing agent (standard redox potential +2.8 V), will oxidize organic pollutant (dyes) to the mineral end-products. Based on the above said mechanism, the chemical reactions at the semiconductor surface causing the degradation of dyes can be explained as follows (Ajmal et al. 2014; Schneider et al. 2014):

$$TiO_2 + h\upsilon\,(UV) \rightarrow TiO_2(e_{CB}^{\,-} + h_{VB}^{\,+}) \tag{2.1}$$

$$TiO_2(h_{VB}^{\,+}) + H_2O \rightarrow TiO_2 + H^+ + OH^{\bullet} \tag{2.2}$$

$$TiO_2(h_{VB}^{\,+}) + OH^- \rightarrow TiO_2 + OH^{\bullet} \tag{2.3}$$

$$TiO_2(e_{CB}^{\,-}) + O_2 \rightarrow TiO_2 + O_2^{\,-}\bullet \tag{2.4}$$

$$O_2^{\,-}\bullet + H^+ \rightarrow HO_2^{\bullet} \tag{2.5}$$

$$Dye + OH^{\bullet} \rightarrow degradation\ products \tag{2.6}$$

$$Dye + h_{VB}^{\,+} \rightarrow oxidation\ products \tag{2.7}$$

$$Dye + e_{CB}^{\,-} \rightarrow reduction\ products \tag{2.8}$$

Biosynthesis of metal and metal oxide nanoparticles using extracts from plants exhibits diversified applications such as biomedical, pharmaceutical, bactericidal, energy production, and electronics. Comparative study on the chemical vs green synthesis of gold nanoparticles (Au NPs) for the catalytic application by Ravishankar Rai et al. (Banerjee and Rai 2016) revealed improved efficiency for Au NPs synthesized through green approach. Francis et al. (2017) reported on the *Mussaenda glabrata*-mediated gold and silver nanoparticles and its efficient degradation on anthropogenic pollutants, rhodamine B and methyl orange, within a period of 4 and 7 min, respectively, through heterogeneous catalytic pathway. Also, Nilesh et al. (Sharma et al. 2007) reported on the catalytic reduction of aqueous 4-nitrophenol by Au NPs using *Sesbania* seedlings. Recent progress of green-synthesized photocatalyst in organic water pollutant treatment is summarized in Table 2.2.

Lebogang et al. (Katata-Seru et al. 2018) synthesized iron nanoparticles (Fe NPs) using leaf and seed extracts of *Moringa oleifera* for the removal of NO_3 ions from the groundwater and surface water. The increment in the percentage of removal of nitrate ions with increase in the pH of the *Moringa oleifera*-mediated Fe NPs revealed a possible alternative for degrading contaminants in the water bodies. Also, these NPs were reported to have dual properties such as coagulation and antibacterial activities, which are suitable for treating contaminated water. In a similar report by Shahwan et al. (2011), Fe NPs were synthesized using green tea leaves (GTL) extract. The prepared nanoparticles were tested as a Fenton-like catalyst, and the performance of the catalyst for the degradation of methylene blue and methyl orange dyes was studied. The degrading performance of green-synthesized Fe NPs was also compared with that of the iron nanoparticles reduced

Table 2.2 Summary of green-synthesized nanoparticles in water pollutant treatment

Bio-surfactant	Nanoparticles	Model pollutant	Retention period (minutes)	References
Aspergillus fischeri	Au	Methylene blue	9	Banerjee and Ravishankar Rai (2016)
Mussaenda glabrata	Ag and Au	Rhodamine B and methyl orange	7	Francis et al. (2017)
Sesbania drummondii seedlings	Au	Aqueous-4-nitrophenol	240	Sharma et al. (2007)
Moringa oleifera	Fe	Groundwater contamination	–	Saif et al. (2016)
Theobroma cacao L (seed extract)	Pd/CuO	4-Nitrophenol	1	Nasrollahzadeh et al. (2015)
Hibiscus rosa-sinensis	CuO	Potassium periodate	–	Vinay Kumar and Shantanu (2017)
Abutilon indicum	CuO	Acid Black 210	–	Ijaz et al. (2017)
Psoralea corylifolia	Fe_2O_3	Methylene blue	60	C. Nagajyothi et al. (2016)
Euphorbia heterophylla	Ag/TiO_2	Methyl orange, methylene blue, Congo red, and 4-nitrophenol	120	Atarod et al. (2016)
Carica papaya	CuO	Coomassie brilliant blue R-250	90	Sankar et al. (2014)
Saccharomyces cerevisiae	Ag	Methylene blue	6 (hours)	Roy et al. (2015)
Eucalyptus globulus	ZnO	Methyl orange	50	Siripireddy and Mandal (2017)
Coriandrum sativum	ZnO	Anthracene	240	Saad et al. (2015)
Tea	Ag	4-Nitrophenol	12	Wang et al. (2015)
Corymbia citriodora	ZnO	Methylene blue	90	Zheng et al. (2015)
Cocos nucifera	Ag	Malachite green	6 (hours)	Sumi et al. (2017)
Momordica cochinchinensis	Au	Congo red	4–8	Paul et al. (2016)
Pogostemon benghalensis	Au	Methylene blue	8	Paul et al. (2015)
Ulva lactuca	ZnO	Methylene blue	120	Ishwarya et al. (2018)
Ziziphus spina-christi (L)	Cu	Triphenylmethane	7.5	Khani et al. (2018)

using borohydride (chemical approach), and it was found that GTL-mediated Fe NPs possessed different morphological and structural characteristics than the NPs prepared by chemical method, and also, GTL-Fe NPs showed faster kinetics and better efficiency in the percentage of degradation.

Similar to metal nanoparticles, metal oxide nanoparticles also show significant contribution toward catalytic application. Mahmoud et al. (Nasrollahzadeh et al. 2015) used *Theobroma cacao L.* seed extracts for the synthesis of palladium/CuO nanoparticles (Pd/CuO NPs), and its catalytic behavior was tested toward the reduction of 4-nitrophenol, which is the most refractory pollutant generated by dye industries. The reduced 4-nitrophenol resulted in the production of 4-aminophenol which can be utilized as analgesic and antipyretic drugs, corrosion inhibitor, photographic developer, etc. The formation of high purity CuO nanorods of length 100 nm and average diameter 15–20 nm using the leaf extract of *Hibiscus rosa-sinensis* as capping agent as reported by Vinay et al. (Patel and Bhattacharya 2017) was recognized as a highly efficient catalyst for the thermal decomposition of potassium periodate microparticles. Also, Faheem et al. (Ijaz et al. 2017) contributed to the synthesis of CuO NPs using extract of *Abutilon indicum*. The prepared CuO NPs were evaluated for photocatalytic degradation of Acid Black 210 (AB) dye under sunlight. It also revealed good antimicrobial and antioxidant activities. Similarly, Nagajyothi et al. (2017) synthesized iron oxide nanoparticles (Fe_2O_3 NPs) using the extracts of *Psoralea corylifolia* seeds. The as-synthesized NPs were subjected to the catalytic study on the degradation of methylene blue dye and observed a reduction in dye within the period of 63 min. Another work deals with the preparation of mesoporous hematite-Fe_2O_3 NPs using *Camellia sinensis* as reported by Ahmmad et al. (2013). The photocatalytic activity of the prepared green-synthesized NPs was evaluated from the amount of hydroxyl radical formed when illuminated under visible light. It was observed and reported that the NPs prepared by green approach showed twice the efficiency when compared with the commercial NPs synthesized by chemical approach. Similarly, Atarod et al. (2016) synthesized Ag/TiO2 nanocomposite using leaf extract of *Euphorbia heterophylla* as a mediator. The same was tested for the degradation of various dyes such as methyl orange, methylene blue, Congo red, and 4-nitrophenol and was proved to be effective. Also, the prepared NPs exhibited 100% reduction of 4-nitrophenol even after getting recycled five times, thus proving the stability of the catalyst.

2.5 Conclusion

In summary, a wide range of plant leaf extract-based bio-surfactant have been discussed in nanomaterial synthesis. In particular, how these green-synthesized nanoparticles can be applied in photocatalysis water purification application is discussed in detail. The toxic-free, green bio-surfactant showed multifaceted advantages like particle size reduction and high internal surface area which result in high catalytic performance. However, interaction between the bio-surfactant and catalyst materials is not understood completely. Further research is required to analyze how organic moieties from bio-surfactant are reducing the particle size of the nanoparticle.

Acknowledgments S.P thanks to Sêr Cymru II-Rising Star Fellowship program for supporting this work through Welsh Government and European Regional Development Fund.

References

Abboud Y, Saffaj T, Chagraoui A, El Bouari A, Brouzi K, Tanane O, Ihssane B (2014) Biosynthesis, characterization and antimicrobial activity of copper oxide nanoparticles (CONPs) produced using brown alga extract (*Bifurcaria bifurcata*). Appl Nanosci 4:571–576. https://doi.org/10.1007/s13204-013-0233-x

Abdul Salam H, Sivaraj R, Venckatesh R (2014) Green synthesis and characterization of zinc oxide nanoparticles from *Ocimum basilicum* L. var. purpurascens Benth.-Lamiaceae leaf extract. Mater Lett 131:16–18. https://doi.org/10.1016/j.matlet.2014.05.033

Agarwal S, Tyagi I, Gupta VK, Mashhadi S, Ghasemi M (2016) Kinetics and thermodynamics of Malachite Green dye removal from aqueous phase using iron nanoparticles loaded on ash. J Mol Liq 223:1340–1347. https://doi.org/10.1016/j.molliq.2016.04.039

Ahmed M (2014) Plants: emerging as green source toward biosynthesis of metallic nanoparticles and its applications. J Bioprocess Chem Eng 2. https://doi.org/10.15297/JBCE.V2I1.02

Ahmed S, Ahmad M, Swami BL, Ikram S (2016) A review on plants extract mediated synthesis of silver nanoparticles for antimicrobial applications: a green expertise. J Adv Res 7:17–28. https://doi.org/10.1016/j.jare.2015.02.007

Ahmmad B, Leonard K, Islam MS, Kurawaki J, Muruganandham M, Ohkubo T, Kuroda Y (2013) Green synthesis of mesoporous hematite (α-Fe2O3) nanoparticles and their photocatalytic activity. Adv Powder Technol 24:160–167

Ajmal A, Majeed I, Malik RN, Idriss H, Nadeem MA (2014) Principles and mechanisms of photocatalytic dye degradation on TiO2 based photocatalysts: a comparative overview. RSC Adv 4:37003–37026. https://doi.org/10.1039/C4RA06658H

Alshehri SM, Naushad M, Ahamad T et al (2014) Synthesis, characterization of curcumin based ecofriendly antimicrobial bio-adsorbent for the removal of phenol from aqueous medium. Chem Eng J 254:181–189. https://doi.org/10.1016/j.cej.2014.05.100

Anbuvannan M, Ramesh M, Viruthagiri G, Shanmugam N, Kannadasan N (2015) Synthesis, characterization and photocatalytic activity of ZnO nanoparticles prepared by biological method. Spectrochim Acta A Mol Biomol Spectrosc 143:304–308. https://doi.org/10.1016/j.saa.2015.01.124

Arumugam A, Karthikeyan C, Haja Hameed AS, Gopinath K, Gowri S, Karthika V (2015) Synthesis of cerium oxide nanoparticles using Gloriosa superba L. leaf extract and their structural, optical and antibacterial properties. Mater Sci Eng C 49:408–415. https://doi.org/10.1016/j.msec.2015.01.042

Asefa T, Duncan CT, Sharma KK (2009) Recent advances in nanostructured chemosensors and biosensors. Analyst 134:1980–1990. https://doi.org/10.1039/B911965P

Atarod M, Nasrollahzadeh M, Mohammad Sajadi S (2016) *Euphorbia heterophylla* leaf extract mediated green synthesis of Ag/TiO2 nanocomposite and investigation of its excellent catalytic activity for reduction of variety of dyes in water. J Colloid Interface Sci 462:272–279. https://doi.org/10.1016/j.jcis.2015.09.073

Bala N, Saha S, Chakraborty M, Maiti M, Das S, Basu R, Nandy P (2015) Green synthesis of zinc oxide nanoparticles using Hibiscus sabdariffa leaf extract: effect of temperature on synthesis, anti-bacterial activity and anti-diabetic activity. RSC Adv 5:4993–5003. https://doi.org/10.1039/C4RA12784F

Balaji Reddy S, Kumar Mandal B (2017) Facile green synthesis of zinc oxide nanoparticles by Eucalyptus globulus and their photocatalytic and antioxidant activity. Adv Powder Technol 28 (3):785–797. https://doi.org/10.1016/j.apt.2016.11.026

Banerjee K, Rai VR (2016) Study on green synthesis of gold nanoparticles and their potential applications as catalysts. J Clust Sci 27:1307–1315

Banerjee K, Ravishankar Rai V (2016) Study on green synthesis of gold nanoparticles and their potential applications as catalysts. J Clust Sci 27:1307–1315. https://doi.org/10.1007/s10876-016-1001-3

Beydoun D, Amal R, Low G, McEvoy S (1999) Role of nanoparticles in photocatalysis. J Nanopart Res 1:439–458. https://doi.org/10.1023/a:1010044830871

Burtch NC, Jasuja H, Walton KS (2014) Water stability and adsorption in metal–organic frameworks. Chem Rev 114:10575–10612. https://doi.org/10.1021/cr5002589

Chakrabarti S, Dutta BK (2004) Photocatalytic degradation of model textile dyes in wastewater using ZnO as semiconductor catalyst. J Hazard Mater 112:269–278. https://doi.org/10.1016/j.jhazmat.2004.05.013

Chandrasekaran R, Gnanasekar S, Seetharaman P, Keppanan R, Arockiaswamy W, Sivaperumal S (2016) Formulation of Carica papaya latex-functionalized silver nanoparticles for its improved antibacterial and anticancer applications. J Mol Liq 219:232–238. https://doi.org/10.1016/j.molliq.2016.03.038

Chithrani BD, Ghazani AA, Chan WCW (2006) Determining the size and shape dependence of gold nanoparticle uptake into mammalian cells. Nano Lett 6:662–668. https://doi.org/10.1021/nl052396o

Constantin M, Asmarandei I, Harabagiu V, Ghimici L, Ascenzi P, Fundueanu G (2013) Removal of anionic dyes from aqueous solutions by an ion-exchanger based on pullulan microspheres. Carbohydr Polym 91:74–84. https://doi.org/10.1016/j.carbpol.2012.08.005

Dutta D, Mukherjee R, Patra M, Banik M, Dasgupta R, Mukherjee M, Basu T (2016) Green synthesized cerium oxide nanoparticle: a prospective drug against oxidative harm. Colloids Surf B: Biointerfaces 147:45–53. https://doi.org/10.1016/j.colsurfb.2016.07.045

Francis S, Joseph S, Koshy EP, Mathew B (2017) Green synthesis and characterization of gold and silver nanoparticles using Mussaenda glabrata leaf extract and their environmental applications to dye degradation. Environ Sci Pollut Res Int 24:17347–17357. https://doi.org/10.1007/s11356-017-9329-2

Ghidan AY, Al-Antary TM, Awwad AM (2016) Green synthesis of copper oxide nanoparticles using Punica granatum peels extract: effect on green peach. Aphid Environ Nanotechnol Monit Manag 6:95–98. https://doi.org/10.1016/j.enmm.2016.08.002

Ghosh SK, Pal T (2007) Interparticle coupling effect on the surface Plasmon resonance of gold nanoparticles: from theory to applications. Chem Rev 107:4797–4862. https://doi.org/10.1021/cr0680282

Gnanasekaran L, Hemamalini R, Naushad M (2018) Efficient photocatalytic degradation of toxic dyes using nanostructured TiO_2/polyaniline nanocomposite. Desalin Water Treat 108:322–328. https://doi.org/10.5004/dwt.2018.21967

Gözmen B, Kayan B, Gizir AM, Hesenov A (2009) Oxidative degradations of reactive blue 4 dye by different advanced oxidation methods. J Hazard Mater 168:129–136. https://doi.org/10.1016/j.jhazmat.2009.02.011

He Z, Liu D, Li R, Zhou Z, Wang P (2012) Magnetic solid-phase extraction of sulfonylurea herbicides in environmental water samples by Fe_3O_4@dioctadecyl dimethyl ammonium chloride@silica magnetic particles. Anal Chim Acta 747:29–35. https://doi.org/10.1016/j.aca.2012.08.015

Huy T, Quy N, Le A-T (2013) Silver nanoparticles: synthesis, properties, toxicology, applications and perspectives. Adv Nat Sci Nanosci Nanotechnol 4(3):033001. https://doi.org/10.1088/2043-6262/4/3/033001

Ibrahim HMM (2015) Green synthesis and characterization of silver nanoparticles using banana peel extract and their antimicrobial activity against representative microorganisms. J Radiat Res Appl Sci 8:265–275. https://doi.org/10.1016/j.jrras.2015.01.007

Ijaz F, Shahid S, Khan SA, Ahmad W, Zaman S (2017) Green synthesis of copper oxide nanoparticles using *Abutilon indicum* leaf extract: antimicrobial, antioxidant and photocatalytic dye degradation activities. Trop J Pharm Res 16:743–753

Ikoma T, Kobayashi H, Tanaka J, Walsh D, Mann S (2003) Microstructure, mechanical, and biomimetic properties of fish scales from *Pagrus major*. J Struct Biol 142:327–333

Iravani S (2011) Green synthesis of metal nanoparticles using plants. Green Chem 13:2638–2650. https://doi.org/10.1039/C1GC15386B

Ishwarya R et al (2018) Facile green synthesis of zinc oxide nanoparticles using *Ulva lactuca* seaweed extract and evaluation of their photocatalytic, antibiofilm and insecticidal activity. J Photochem Photobiol B 178:249–258. https://doi.org/10.1016/j.jphotobiol.2017.11.006

Jamdagni P, Khatri P, Rana JS (2018) Green synthesis of zinc oxide nanoparticles using flower extract of Nyctanthes arbor-tristis and their antifungal activity. J King Saud Univ Sci 30:168–175. https://doi.org/10.1016/j.jksus.2016.10.002

Jawad AH, Rashid R (2016) Adsorption of methylene blue onto activated carbon developed from biomass waste by H2SO4 activation: kinetic, equilibrium and thermodynamic studies. Desalin Water Treat 57(52):25194–25206. https://doi.org/10.1080/19443994.2016.1144534

Kagalkar AN, Jagtap UB, Jadhav JP, Bapat VA, Govindwar SP (2009) Biotechnological strategies for phytoremediation of the sulfonated azo dye Direct Red 5B using Blumea malcolmii Hook. Bioresour Technol 100:4104–4110. https://doi.org/10.1016/j.biortech.2009.03.049

Katata-Seru L, Moremedi T, Aremu OS, Bahadur I (2018) Green synthesis of iron nanoparticles using *Moringa oleifera* extracts and their applications: removal of nitrate from water and antibacterial activity against *Escherichia coli*. J Mol Liq 256:296–304

Khani R, Roostaei B, Bagherzade G, Moudi M (2018) Green synthesis of copper nanoparticles by fruit extract of Ziziphus spina-christi (L.) Willd.: application for adsorption of triphenylmethane dye and antibacterial assay. J Mol Liq 255:541–549. https://doi.org/10.1016/j.molliq.2018.02.010

Khansorthong S, Hunsom M (2009) Remediation of wastewater from pulp and paper mill industry by the electrochemical technique. Chem Eng J 151:228–234. https://doi.org/10.1016/j.cej.2009.02.038

Kumar PPNV, Shameem U, Kollu P, Kalyani RL, Pammi SVN (2015) Green synthesis of copper oxide nanoparticles using aloe vera leaf extract and its antibacterial activity against fish bacterial pathogens. BioNanoScience 5:135–139. https://doi.org/10.1007/s12668-015-0171-z

McCann J, Ames BN (1976) Detection of carcinogens as mutagens in the Salmonella/microsome test: assay of 300 chemicals: discussion. Proc Natl Acad Sci U S A 73:950–954

Nagajyothi PC, Pandurangan M, Kim DH, Sreekanth TV, Shim J (2016) Green synthesis of iron oxide nanoparticles and their catalytic and in vitro anticancer activities. J Clust Sci 28 (1):245–257. https://doi.org/10.1007/s10876-016-1082-z

Nagajyothi P, Pandurangan M, Kim DH, Sreekanth T, Shim J (2017) Green synthesis of iron oxide nanoparticles and their catalytic and in vitro anticancer activities. J Clust Sci 28:245–257

Naika HR, Lingaraju K, Manjunath K, Kumar D, Nagaraju G, Suresh D, Nagabhushana H (2015) Green synthesis of CuO nanoparticles using Gloriosa superba L. extract and their antibacterial activity. J Taibah Univ Sci 9:7–12. https://doi.org/10.1016/j.jtusci.2014.04.006

Nasrollahzadeh M, Sajadi SM, Rostami-Vartooni A, Bagherzadeh M (2015) Green synthesis of Pd/CuO nanoparticles by *Theobroma cacao* L. seeds extract and their catalytic performance for the reduction of 4-nitrophenol and phosphine-free Heck coupling reaction under aerobic conditions. J Colloid Interface Sci 448:106–113. https://doi.org/10.1016/j.jcis.2015.02.009

Nasrollahzadeh M, Momeni SS, Sajadi SM (2017) Green synthesis of copper nanoparticles using Plantago asiatica leaf extract and their application for the cyanation of aldehydes using K4Fe (CN)6. J Colloid Interface Sci 506:471–477. https://doi.org/10.1016/j.jcis.2017.07.072

Naushad M, Abdullah ALOthman Z, Rabiul Awual M et al (2016) Adsorption of rose Bengal dye from aqueous solution by amberlite Ira-938 resin: kinetics, isotherms, and thermodynamic studies. Desalin Water Treat 57:13527–13533. https://doi.org/10.1080/19443994.2015.1060169

Nishio J, Tokumura M, Znad HT, Kawase Y (2006) Photocatalytic decolorization of azo-dye with zinc oxide powder in an external UV light irradiation slurry photoreactor. J Hazard Mater 138:106–115. https://doi.org/10.1016/j.jhazmat.2006.05.039

Pandimurugan R, Thambidurai S (2016) Novel seaweed capped ZnO nanoparticles for effective dye photodegradation and antibacterial activity. Adv Powder Technol 27:1062–1072. https://doi.org/10.1016/j.apt.2016.03.014

Patel VK, Bhattacharya S (2017) Solid state green synthesis and catalytic activity of CuO nanorods in thermal decomposition of potassium periodate. Mater Res Express 4:095012

Patil B, Taranath T (2016) *Limonia acidissima* L. leaf mediated synthesis of zinc oxide nanoparticles: a potent tool against *Mycobacterium tuberculosis*. Int J Mycobacteriol 5:197–204. https://doi.org/10.1016/j.ijmyco.2016.03.004

Paul B, Bhuyan B, Dhar Purkayastha D, Dey M, Dhar SS (2015) Green synthesis of gold nanoparticles using Pogostemon benghalensis (B) O. Ktz. leaf extract and studies of their photocatalytic activity in degradation of methylene blue. Mater Lett 148:37–40. https://doi.org/10.1016/j.matlet.2015.02.054

Paul B, Bhuyan B, Purkayastha DD, Vadivel S, Dhar SS (2016) One-pot green synthesis of gold nanoparticles and studies of their anticoagulative and photocatalytic activities. Mater Lett 185:143–147. https://doi.org/10.1016/j.matlet.2016.08.121

Premkumar T, Geckeler KE (2006) Nanosized CuO particles via a supramolecular strategy. Small Weinheim Bergstr Ger 2:616–620. https://doi.org/10.1002/smll.200500454

Purkait MK, Dasgupta S, De S (2006) Micellar enhanced ultrafiltration of eosin dye using hexadecyl pyridinium chloride. J Hazard Mater 136:972–977. https://doi.org/10.1016/j.jhazmat.2006.01.040

Ren G, Hu D, Cheng EW, Vargas-Reus MA, Reip P, Allaker RP (2009) Characterisation of copper oxide nanoparticles for antimicrobial applications. Int J Antimicrob Agents 33:587–590. https://doi.org/10.1016/j.ijantimicag.2008.12.004

Robinson T, McMullan G, Marchant R, Nigam P (2001) Remediation of dyes in textile effluent: a critical review on current treatment technologies with a proposed alternative. Bioresour Technol 77:247–255

Roy K, Sarkar CK, Ghosh CK (2015) Photocatalytic activity of biogenic silver nanoparticles synthesized using yeast (*Saccharomyces cerevisiae*) extract. Appl Nanosci 5:953–959. https://doi.org/10.1007/s13204-014-0392-4

Saad SMH, Waleed IMEA, Hager RA, Mona SMM (2015) Green synthesis and characterization of ZnO nanoparticles for photocatalytic degradation of anthracene. Adv Nat Sci Nanosci Nanotechnol 6:045012

Saif S, Tahir A, Chen Y (2016) Green synthesis of iron nanoparticles and their environmental applications and implications. Nanomater Basel Switz 6(11):209. https://doi.org/10.3390/nano6110209

Sankar R, Manikandan P, Malarvizhi V, Fathima T, Shivashangari KS, Ravikumar V (2014) Green synthesis of colloidal copper oxide nanoparticles using Carica papaya and its application in photocatalytic dye degradation. Spectrochim Acta A Mol Biomol Spectrosc 121:746–750. https://doi.org/10.1016/j.saa.2013.12.020

Schneider J, Matsuoka M, Takeuchi M, Zhang J, Horiuchi Y, Anpo M, Bahnemann DW (2014) Understanding TiO2 photocatalysis: mechanisms and materials. Chem Rev 114:9919–9986. https://doi.org/10.1021/cr5001892

Shahwan T, Sirriah SA, Nairat M, Boyacı E, Eroğlu AE, Scott TB, Hallam KR (2011) Green synthesis of iron nanoparticles and their application as a Fenton-like catalyst for the degradation of aqueous cationic and anionic dyes. Chem Eng J 172:258–266

Sharma NC, Sahi SV, Nath S, Parsons JG, Gardea-Torresdey JL, Pal T (2007) Synthesis of plant-mediated gold nanoparticles and catalytic role of biomatrix-embedded nanomaterials. Environ Sci Technol 41:5137–5142

Sharma JK, Srivastava P, Ameen S, Akhtar MS, Sengupta SK, Singh G (2017) Phytoconstituents assisted green synthesis of cerium oxide nanoparticles for thermal decomposition and dye remediation. Mater Res Bull 91:98–107. https://doi.org/10.1016/j.materresbull.2017.03.034

Singh V, Sharma AK, Sanghi R (2009) Poly(acrylamide) functionalized chitosan: an efficient adsorbent for azo dyes from aqueous solutions. J Hazard Mater 166:327–335. https://doi.org/10.1016/j.jhazmat.2008.11.026

Singh P, Kim Y-J, Zhang D, Yang D-C (2016) Biological synthesis of nanoparticles from plants and microorganisms. Trends Biotechnol 34(7):588–599. https://doi.org/10.1016/j.tibtech.2016.02.006

Sinha T, Ahmaruzzaman M (2015) Green synthesis of copper nanoparticles for the efficient removal (degradation) of dye from aqueous phase. Environ Sci Pollut Res 22:20092–20100. https://doi.org/10.1007/s11356-015-5223-y

Sinha A, Jana NR (2012) Functional, mesoporous, superparamagnetic colloidal sorbents for efficient removal of toxic metals. Chem Commun 48:9272–9274. https://doi.org/10.1039/C2CC33893A

Siripireddy B, Mandal BK (2017) Facile green synthesis of zinc oxide nanoparticles by Eucalyptus globulus and their photocatalytic and antioxidant activity. Adv Powder Technol 28:785–797. https://doi.org/10.1016/j.apt.2016.11.026

Sivaraj R, Rahman PK, Rajiv P, Narendhran S, Venckatesh R (2014) Biosynthesis and characterization of Acalypha indica mediated copper oxide nanoparticles and evaluation of its antimicrobial and anticancer activity. Spectrochim Acta A Mol Biomol Spectrosc 129:255–258. https://doi.org/10.1016/j.saa.2014.03.027

Sujitha MV, Kannan S (2013) Green synthesis of gold nanoparticles using Citrus fruits (*Citrus limon*, Citrus reticulata and Citrus sinensis) aqueous extract and its characterization. Spectrochim Acta A Mol Biomol Spectrosc 102:15–23. https://doi.org/10.1016/j.saa.2012.09.042

Sumi MB, Devadiga A, Shetty KV, Saidutta MB (2017) Solar photocatalytically active, engineered silver nanoparticle synthesis using aqueous extract of mesocarp of *Cocos nucifera* (Red Spicata Dwarf). J Exp Nanosci 12:14–32. https://doi.org/10.1080/17458080.2016.1251622

Vanathi P, Rajiv P, Narendhran S, Rajeshwari S, Rahman PKSM, Venckatesh R (2014) Biosynthesis and characterization of phyto mediated zinc oxide nanoparticles: a green chemistry approach. Mater Lett 13:13. https://doi.org/10.1016/j.matlet.2014.07.029

Vijay Kumar PPN, Pammi SVN, Kollu P, Satyanarayana KVV, Shameem U (2014) Green synthesis and characterization of silver nanoparticles using Boerhavia diffusa plant extract and their antibacterial activity. Ind Crop Prod 52:562–566. https://doi.org/10.1016/j.indcrop.2013.10.050

Vinay Kumar P, Shantanu B (2017) Solid state green synthesis and catalytic activity of CuO nanorods in thermal decomposition of potassium periodate. Mat Res Express 4:095012

Wang Z, Xu C, Li X, Liu Z (2015) In situ green synthesis of Ag nanoparticles on tea polyphenols-modified graphene and their catalytic reduction activity of 4-nitrophenol. Colloids Surf A Physicochem Eng Asp 485:102–110. https://doi.org/10.1016/j.colsurfa.2015.09.015

Yugandhar P, Vasavi T, Uma Maheswari Devi P, Savithramma N (2017) Bioinspired green synthesis of copper oxide nanoparticles from Syzygium alternifolium (Wt.) Walp: characterization and evaluation of its synergistic antimicrobial and anticancer activity. Appl Nanosci 7:417–427. https://doi.org/10.1007/s13204-017-0584-9

Zhang X et al (2008) Different CuO nanostructures: synthesis, characterization, and applications for glucose sensors. J Phys Chem C 112:16845–16849. https://doi.org/10.1021/jp806985k

Zhang Z, Che H, Wang Y, Song L, Zhong Z, Su F (2012) Preparation of hierarchical dandelion-like CuO microspheres with enhanced catalytic performance for dimethyldichlorosilane synthesis. Cat Sci Technol 2:1953–1960. https://doi.org/10.1039/C2CY20199B

Zheng Y et al (2015) Green biosynthesis and characterization of zinc oxide nanoparticles using *Corymbia citriodora* leaf extract and their photocatalytic activity. Green Chem Lett Rev 8:59–63. https://doi.org/10.1080/17518253.2015.1075069

Chapter 3
Nanomaterials with Different Morphologies for Photocatalysis

P. Thangadurai, Rosalin Beura, and J. Santhosh Kumar

Contents

Abstract Photocatalysis is an important phenomenon for degrading the harmful products/by-products produced by various industrial and technological sectors. The wastes from this sector can be of any form, like solid, liquid, and gaseous. When it is in liquid form, in most of the cases, they are released untreated into the regular freshwater stream leading to water contamination, and therefore, they have to be properly treated before releasing them into water bodies. The treatment can be very well done with the advanced oxidation process, photocatalysis. This process depends on the type of photocatalyst, the type of irradiation, and the type of organic molecules to be degraded. The main player is the photocatalyst material, and its design and selection for best efficiency are challenging, because its performance depends on many parameters such as large surface area, high absorption of pollutant molecules, less charge recombination, high charge transfer rate, suitable band gap, and good light harvesting capability. Mostly, these photocatalysts are semiconductor-based material with a suitable band gap. Selecting a material with suitable parameters is highly challenging, and one such parameter is the morphology of the photocatalysts. This is because morphology of the material influences the

P. Thangadurai (✉) · R. Beura · J. S. Kumar
Centre for Nanoscience and Technology, Pondicherry University, Puducherry, India
e-mail: thangaduraip.nst@pondiuni.edu.in

© Springer Nature Switzerland AG 2020   47
M. Naushad et al. (eds.), *Green Photocatalysts*, Environmental Chemistry
for a Sustainable World 34, https://doi.org/10.1007/978-3-030-15608-4_3

above said parameters, to be suited as photocatalysts. In this regard, this chapter deals with the nanostructured materials with different morphologies prepared by different chemical/physical methods. Since the hydrothermal method can produce a variety of morphologies, this method has been largely used by the researchers. Different morphologies will yield different surface areas, for example, spherical particles with different particle sizes have different band gap influenced by the quantum confinement phenomenon, and similarly other parameters can also be influenced by morphology. There are reports in which particular planar arrangement of ZnO nanorods played a dominant role to have them the best photocatalytic activity. This is because the ZnO can have polar or nonpolar planes whose surfaces react differently with the reactant molecules. In some cases, the surface of the photocatalysts was decorated with metal nanoparticles such as Ag, Au, and Pd in order to increase the charge transfer rate so as to increase the lifetime of the charges and in turn enhanced photodegradation. Materials with porous morphology can have different activities. In line with these facts, the chapter presents the principle of photocatalysis, importance of photocatalysts and their design, and importance of nanomaterials with different morphologies followed by different types of photocatalysts based on quantum confinement, such as zero-dimensional, one-dimensional, two-dimensional, and three-dimensional materials. Various 0-, 1-, 2-, and 3-D nanomaterials were discussed along with their preparation, photocatalytic properties, and the reasons/mechanism for the improved photocatalytic activity with appropriate examples.

Keywords Nanomaterials · Morphology · Photocatalysis · Environment · Photocatalyst

3.1 Introduction

Photocatalysis is the process in which the oxidative and reductive reactions occur on the surface of the photocatalytic material in the presence of light. The photocatalytic material, usually a key player in initiating and promoting the photocatalysis, is known as photocatalyst. Various intrinsic factors such as superficial area, morphological structure, and crystalline phase of the photocatalyst play a rudimentary part in the reaction pathway of the photocatalytic reaction (Kumar et al. 2017; Sharma et al. 2015). In particular, morphology of the photocatalyst materials is very important in determining their various properties (such as optical, magnetic, electrical, catalytic, etc.), and therefore generating and controlling different morphologies of a particular material have received enormous attention (Xu et al. 2009). Each morphology of the same materials shows difference in their surface area, polar planes, or oxygen vacancies, and therefore, they show differential influence on their catalytic activity (Jin et al. 2018). In addition, other factors that can strongly affect the photocatalytic performance of the photocatalysts are adsorption ability of the reactants on their surface, active energy region of the irradiation light source according to the absorption ability of the photocatalyst, and efficient charge separation of the photogenerated charge carriers. The photocatalytic reactions usually occur at the interface between the catalyst's surface and the reactants, and the type of crystalline planes and surface structures significantly influences the photocatalytic activity (McLaren et al. 2009). Compared to its bulk counterparts, the nanoscale materials have proved to be a better photocatalyst because of the increased surface area and faster arrival of the photogenerated electrons or holes to the reaction sites of the photogenerated electrons or holes. But the problem with these nanomaterials is the difficulty to separate and recycle them from the reaction products after the photocatalytic experiments because of their smaller size. In addition, because of the high surface to volume ratio, these nanoscale materials have a tendency to aggregate with aging, which results in unwanted reduction of the active surface area. Hence, by taking into consideration the high energy conversion efficiency, large light harvesting capacity, and easy separation of solid/liquid, the micrometer-sized 3-D architectures with nanomaterials as building blocks can show superior photocatalytic performance (Li et al. 2014). The photocatalytic activity of the photocatalysts can be augmented by improving the light absorption and the pollutant adsorption by them (Dhiman et al. 2017; Elia et al. 2011). Already, in the literature, different morphologies have shown a difference in their photocatalytic activity, and various reasons have been explained in order to support the observations (Adán et al. 2016; Tang et al. 2012; Zhang et al. 2017a, b; Zheng et al. 2016). By taking these into account, this chapter explains the principle of photocatalysis and the role of photocatalysts in it. More specifically, the characteristics of the photocatalysts for enhanced photocatalytic activity are discussed based on the morphology of the photocatalysts. The methods of preparation of the photocatalysts are also discussed in the relevant places.

3.2 Photocatalysis

3.2.1 Principle of Photocatalysis

Photocatalysis is an advanced oxidation process where chemical reactions take place in the presence of light and catalyzed by appropriate catalyst. The photocatalysis process is unique in the sense that the chemical reactions occur only when both the photocatalyst and the light energy exist. Absence of any of these elements (i.e., light or photocatalyst) does not initiate the chemical reaction to occur. This process generally follows three different steps: (i) generation of electron-hole pairs upon illumination of light that has energy equal to or greater than the band gap energy of the photocatalyst; (ii) migration of these photogenerated electrons (in CB) and holes (in VB) to the surface of the photocatalyst, where they react with the surface oxygen and water molecules to generate radical species (like $^{\bullet}O_2^{-}$, $^{\bullet}OH$); and (iii) occurrence of secondary reactions where the created radical species react with the pollutant molecules and then degrade them into other harmless products (Beura and Thangadurai 2017). All these three processes are explained in Schemes 3.1 and

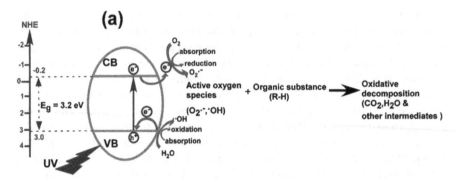

Scheme 3.1 Schematic representation portraying the mechanism of photocatalytic reaction in the presence of a homogeneous semiconductor photocatalyst (ZnO) with a band gap of 3.2 eV excited by UV light irradiation. (Reprinted with permission from Beura and Thangadurai 2017)

3.2. Scheme 3.1 portrays the mechanism of photocatalytic reaction occurring for a homogeneous photocatalyst (e.g., ZnO). Scheme 3.2 presents the schematic representation with the band diagram for the heterogeneous photocatalysts (a) Ag-loaded ZnO nanoflowers (Zhang et al. 2017a, b) and (b) Au-decorated V_2O_5-ZnO heteronanostructure (Yin et al. 2014). The reaction mechanisms occurring at each step in semiconductor photocatalysts (e.g., ZnO) are given in Eqs. 3.1, 3.2, 3.3, 3.4, 3.5, 3.6, and 3.7.

$$ZnO + h\nu(\text{light}) \rightarrow ZnO\left(e_{CB}^- + h_{VB}^+\right) \tag{3.1}$$

$$h_{VB}^+ + H_2O \rightarrow {}^\bullet OH + H^+ \tag{3.2}$$

$$e_{CB}^- + O_2 \rightarrow {}^\bullet O2^- \tag{3.3}$$

$${}^\bullet O2^- + H^+ \rightarrow {}^\bullet HO_2 \tag{3.4}$$

$$2{}^\bullet HO_2 \rightarrow H_2O_2 + O_2 \tag{3.5}$$

$$H_2O_2 + {}^\bullet O2^- \rightarrow {}^\bullet OH + OH^- + O_2 \tag{3.6}$$

$${}^\bullet HO + \text{Dye} \rightarrow \text{Oxidized Product} \tag{3.7}$$

3.2.2 Photocatalysts

Over the past decades, metal oxide materials have been used as photocatalysts due to their photochemical stability, lower charge recombination rate, high photocatalytic activity, non-toxicity, low cost, and environmental friendly characteristic (Bhande et al. 2015). These metal oxide photocatalysts are of two different types, namely, (1) homogeneous photocatalysts and (2) heterogeneous photocatalysts.

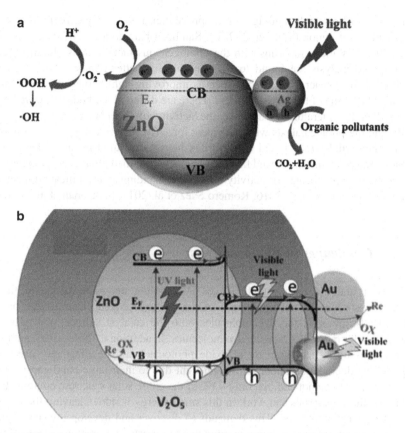

Scheme 3.2 Schematic representation portraying photocatalytic reaction in the heterogeneous photocatalysts (**a**) Ag-loaded ZnO nanoflowers and (**b**) Au-decorated V_2O_5-ZnO heteronanostructure. (Reprinted with permission from Zhang et al. 2017a, b and Yin et al. 2014)

Homogeneous types of photocatalysts are simple metal oxides such as V_2O_5, SnO_2, TiO_2, CuO, Bi_2O_3, ZnO, NiO, etc. that have been studied and utilized for the mineralization of organic pollutants present in water. These homogeneous photocatalysts have been widely used as the green photocatalysts because of their nontoxic nature that makes them environmentally benign. In addition to this, these green photocatalysts have been also developed by various green synthesis methods. For example, ZnO nanoparticles were synthesized by using *Coriandrum sativum* leaf extract and zinc nitrate that degraded 96% of anthracene in 240 min of UV irradiation (Hassan et al. 2015). Similarly, $Eu_2Ti_2O_7$ and CuO nanoparticles were prepared by green synthesis techniques that degraded 94% of MO and 96% of MB, respectively (Nasab and Behpour 2016; Ullah et al. 2017). However, the demerit in many of the homogeneous metal oxides is that the photoinduced charge carriers recombine at a faster rate which leads to weaken the photocatalytic performance and thus limit their applications in this process (Wang et al. 2011). The heterogeneous-type photocatalysts are combinations of metal and/or other metal oxides (Wang et al.

2014). A lot of research is going on to improve their photocatalytic performance in the visible light region (Tang et al. 2012; Santhosh Kumar et al. 2018; Rajeshwari et al. 2018). While comparing with the homogeneous metal oxides, the heterogeneous photocatalysts are found to be favorable for better photocatalytic activity because of their synergistic effects, such as improving light harvesting ability and prolonged lifetime of the charge carriers. There are different methods that have been reported in the literature to improve the activity of the photocatalysts such as (i) doping the photocatalysts with aliovalent ions, (ii) surface decorated with noble metal nanoparticles (NPs), and (iii) forming hetero-structured metal oxides (composites). Based on the above said facts, finding new/modified photocatalyst materials for visible light photocatalytic activity is highly demanding and significant (Qin et al. 2017; Gnanasekaran et al. 2016; Romero Sáez et al. 2017; Saravanan et al. 2018).

3.2.3   Challenges in Photocatalysts

The major and the foremost challenges in deciding the performance of these photocatalysts are fast recombination of the photogenerated electron-hole pairs, reduced adsorption of the dye molecules onto the catalyst surface, and reduced charge transfer rate to degrade the dye molecules. In addition to these issues, another important issue is that these metal oxide photocatalysts are effectively used for organic dye degradation under ultraviolet light due to their large band gap values. However, it is to be noticed that only ~4% of the entire solar spectrum contains UV light, and these materials can work in this small energy window, leaving the rest of the solar energy unutilized. This demands to discover and develop a visible light-responsive photocatalyst, keeping in mind that the substantial fraction (~45%) of the whole solar spectrum is visible light. Therefore, in order to overcome these existing problems, though many other methods have been reported by many researchers, this chapter will analyze the photocatalysts with different morphologies, where variation in the morphology can influence the surface and optical properties of them. As a result, the photocatalytic performance of them could be altered in a favorable way with enhanced activity.

3.3   Nanomaterials with Different Morphologies

When we talk about nanomaterials (as photocatalysts), two aspects can be considered. One is the increased surface area of the nanomaterials when compared to their respective bulk form. Increased surface area can increase the catalytic activity of the photocatalysts. The second aspect is altering their microstructure, namely, morphology of the nanomaterial photocatalysts so that their surface area, preferential planar orientation, and more active sites for catalysis can be modified/tuned. The latter part of obtaining different morphologies in nanomaterials by appropriate methods and tuning their properties to achieve the strong photocatalytic activity is discussed in this chapter. In general, based on the type of confinement, the nanomaterials can be classified into zero-, one-, two-, and three-dimensional nanomaterials. The 0-D materials are the ones in which electron movement is confined in all the three dimensions, whereas the same is confined in two and one dimensions in the 1- and 2-D nanomaterials. The 0-D materials are nanoparticles/quantum dots, and the 1- and 2-D materials are the wires/rods and nanosheets, respectively.

Among different methods used for the synthesis of photocatalysts, the hydrothermal method has been the most commonly and widely used. Hydrothermal method is a very popular method of chemical synthesis, where a chemical reaction occurs in water in a sealed container under controlled pressure and temperature. The term hydrothermal is of geological origin, where "hydro" means water and "thermal" means heat. In principle, any heterogeneous or homogeneous chemical reactions occurring in the presence of either aqueous or nonaqueous solvent, above room temperature, and at a pressure greater than 1 atm in a closed system can be termed as hydrothermal/solvothermal method (Byrappa and Yoshimura 2013). The major advantage of this method is that it is an environmentally benign method to develop a large variety of morphology (microstructure) in an easy and cost-effective way.

3.3.1 Zero-Dimensional Nanomaterials for Photocatalysis

The zero-dimensional materials are the ones confined in all the three dimensions. A variety of 0-D photocatalyst materials and various methods to synthesize them have been reported. Out of many semiconductor materials, ZnO is well known for making them in nanostructured form with many possible morphologies, and therefore, it is considered first. For example, nanostructures of ZnO in spherical shape and flowerlike morphologies were prepared by a microwave-assisted method (Fig. 3.1), where the heating rate of the solution was varied to get different morphologies (Kajbafvala et al. 2012). The photocatalytic activity of the nanostructured ZnO photocatalysts was investigated in methylene blue (MB) under UV light. Degradation performance of spherical and flowerlike ZnO nanostructures has shown 78% and 17% degradation of MB, respectively, after 4 h of UV irradiation. This enhanced activity of the spherical ZnO particles was attributed to the larger specific surface area (98 m^2/g) when compared to the same flowerlike ZnO morphology (22.9 m^2/g). Similarly, the large surface area leading to high adsorption rate and quantum confinement of photoinduced carriers causing slow recombination were stated as reasons for higher photocatalytic activity of ZnO dots among its three morphologies such as ZnO flakes, rods, and dots (Peter et al. 2017). Depending on the type of morphology, distinct dye adsorption behavior has been proposed and successfully observed in the photocatalytic activity of nanostructured ZnO dots (Peter et al. 2017). Their (ZnO flakes, rods, and dots) corresponding photodegradation of the floral dye solution and the linear kinetic plot of the photocatalytic activity (Peter et al. 2017) show that the ZnO dots show the highest activity compared to that of the flakes and rod morphologies, and this highest

Fig. 3.1 SEM images of ZnO nanostructures. (**a**) Low-magnification and (**b**) high-magnification images of ZnO with flowerlike morphology and (**c**) low-magnification and (**d**) high-magnification images of ZnO with spherical particle morphology. (Reprinted with permission from Kajbafvala et al. 2012)

activity of the dots was reported to be because of the highest absorbance of the dye molecules onto the surface by the dots. The concept of microstructure with high surface area showing higher photocatalytic activity was also proved by Jin et al., where the ZnO with varied morphology was achieved by varying the zinc precursor (Jin et al. 2018). In this case, the highest degradation of Rhodamine B (RhB) was 99.43% shown by ZnO with hollow sphere morphology that was prepared by using 0.171 mol/L zinc acetate solution, and the increasing photodegradation was accounted to the maximum specific surface area as well as the increased transformation efficiency by multiple reflection of UV light within the cavity.

In another case, the ZnO nanostructures of varied morphologies have been made through three wet chemical methods such as co-precipitation, hydrothermal, and sonochemical methods (Bordbar et al. 2016). The co-precipitation method has yielded spherical-shaped ZnO particles, whereas the hydrothermal and sonochemical methods have yielded irregular-shaped agglomerated nanoparticles (NPs) of ZnO. When looking at their photocatalytic performance against methyl orange dye, the spherical-shaped ZnO nanoparticles synthesized by sonochemical method have shown the best activity (78%) because of their uniform size distribution, whereas the irregular-shaped agglomerated ZnO nanoparticles prepared through the co-precipitation and hydrothermal methods have shown lower efficiency (58% for co-precipitation and 60% for hydrothermal). Another research group (Akir et al. 2016) had synthesized spherical NPs, nanosheets, and hexagonal prismatic nanoparticles of ZnO, by the eco-friendly co-precipitation process by changing the

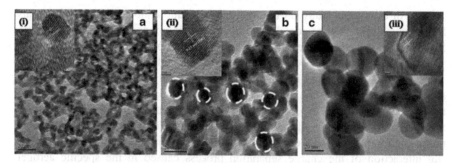

Fig. 3.2 TEM microphotographs of SnO$_2$ NPs and HRTEM image (inset of each samples): (**a**) (S1), (**b**) S2, and (**c**) S3. The respective inserts (i, ii, and iii) are the high-magnification images of the corresponding SnO$_2$ nanoparticles. (Reprinted with permission from Bhattacharjee et al. 2016)

precursor mixing. How the optical and photocatalytic characteristics got influenced by the ZnO particle morphology was investigated. Photocatalytic degradation of RhB (Rhodamine B) expressed that among the three different morphologies, the spherical particles showed the best photocatalytic performance with a rate constant ($k = 0.022$ min^{-1}) comparable to that for nanosheets ($k = 0.02$ min^{-1}). This enhancement by spherical particles is two times larger than that shown by the hexagonal prismatic ZnO nanoparticles. This enhanced catalytic performance by spherical particles was attributed to increased oxygen vacancies, which can adsorb more quantity of reactants onto the ZnO surface and in turn allow more electrons to be injected into ZnO from the excited dye molecules. This resulted in the generation of a lot of ˙OH radicals that are responsible for the degradation and mineralization of the RhB dye.

Another interesting material for photocatalysis is tin (II) oxide, SnO$_2$. The SnO$_2$ NPs were prepared by a green synthesis method (microwave heating method) in which samples with three different water to glycerol volumetric ratios such as 1:1 (S1), 1:2 (S2), and 1:3 (S3) were synthesized. Figure 3.2 illustrates the TEM (Transmission electron microscopy) micrographs of those three different SnO$_2$ NPs with average particle size ranges of 8–10 nm (S1), 13–17 nm (S2), and 8–30 nm (S3). Generally for semiconductors, the band gap varies with the varying particle size, i.e., the band gap energy decreases with the increasing particle size of the SnO$_2$ NPs. The band gap energy of the three photocatalysts was calculated to be 4.17, 4.00, and 3.78 eV for S1, S2, and S3 NPs, respectively. Photocatalytic degradation of methyl violet 6B (MV6B) and MB dyes was carried out under direct sunlight irradiation. The photocatalyst with lower band gap (S3) was used for the photodegradation of these dyes as degradation is inversely proportional to the band gap of the material. The SnO$_2$ NPs (S3) showed the maximum photocatalytic efficiency of 96.2% and 96% for MV6B and MB dyes for the irradiation times of 270 min and 240 min, respectively (Bhattacharjee et al. 2016).

The other popular photocatalyst with a band gap lying in the UV range is TiO$_2$. Different morphologies of TiO$_2$ such as NPs, nanotubes, and aerogels forms were studied for hydrogen evolution through photocatalytic water splitting (Elia et al.

2011). An elaborated study on the hydrogen evolution showed that the materials have shown different behaviors depending upon their composition and morphology. Since the nanoparticles made of pure anatase structural phase have a smaller particle size, they have a higher specific surface area compared to the commercial anatase TiO_2, and this has resulted in a higher catalytic activity. It was observed that the specific surface area was the major factor for photocatalysis, but when compared to the enhancement of charge separation, the area is not that important. For example, in this case of TiO_2, among all the microstructures, the TiO_2 aerogels were the most promising morphology that exhibited excellent performance because of the beneficial interaction of the charge separation process caused in the specific aerogel morphology. The NPs of TiO_2 have higher surface area, but the TiO_2 aerogels have better charge separation capability due to their morphology, and therefore, the latter has been a better photocatalyst. In another case, the TiO_2 quantum dots (QDs) synthesized by sol-gel method were used for the photodegradation of MO and MB dye molecules upon UV light irradiation (Gnanasekaran et al. 2015). Figure 3.3 shows that the MO and MB degradation efficiencies were 97.1% and 97.9%, respectively, after 80 min of irradiation by using TiO_2 QDs. The TiO_2 QDs exhibited higher photodegradation than the same shown by commercially available Degussa P-25 TiO_2 powder, and their performance is compared in Fig. 3.3a, b.

3.3.2 One-Dimensional Nanomaterials for Photocatalysis

One-dimensional materials are confined in two dimensions, and the charge transport is restricted in only one direction. This particular morphology can give wider active sites along their long surface for the fast rate of photocatalysis. For instance, 1-D ZnO nanorods were prepared by solvothermal method without any surface capping agent, and its photocatalytic activity to degrade MO dye under UV radiation was studied (Liu et al. 2013). Its photocatalytic performance was compared with the ZnO with two other microstructures, namely, ZnO multilayer disks (MDs) and truncated hexagonal cones (THCs) that were also prepared by the same method, and their photocatalysis experimental results are presented in Fig. 3.4. The photodegradation results show that the photocatalytic reaction rate was different for different morphologies of ZnO nanostructures, and it followed the order of NRs > THCs > MDs (Fig. 3.4). This highest catalytic activity for ZnO NRs was accounted to the highest surface area of the NRs (~ 9.5 m^2 g^{-1}) when compared to its counterparts THCs (~ 4.2 m^2 g^{-1}) and MDs (~ 7.3 m^2 g^{-1}). This result seemed contradictory when considering the photodegradation behavior of semiconductor nanomaterials. This is because ZnO MDs have high surface oxygen vacancies compared to the other two nanostructures and therefore are expected to show higher catalytic activity. But, since ZnO suffers from photocorrosion, the crystal planes with more oxygen vacancies and other interparticle defects, are more prone to photocorrosion. Depending upon the type of exposed crystal planes of the ZnO to the reactant solution, the photocorrosion behavior also varies. Hence, depending on the dominating factors,

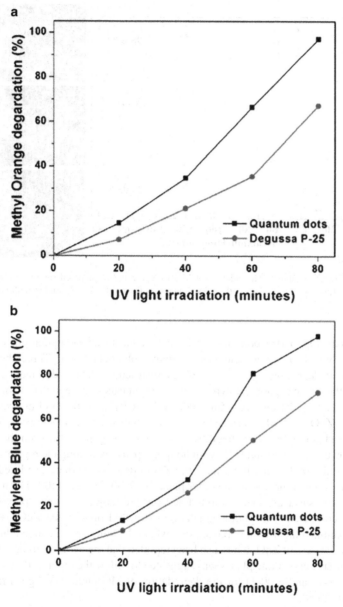

Fig. 3.3 The time varying photodegradation of (**a**) MO and (**b**) MB dyes by the TiO_2 QDs photocatalyst and the commercially available Degussa P 25 TiO_2 powder. (Reprinted with permission from Gnanasekaran et al. 2015)

whether photocorrosion or photocatalysis process, the dye is accordingly photodegraded. Since the photocorrosion is predominant in MDs, it has resulted in a reduced photodegradation of MO. In case of NRs, the rods are surrounded by polar

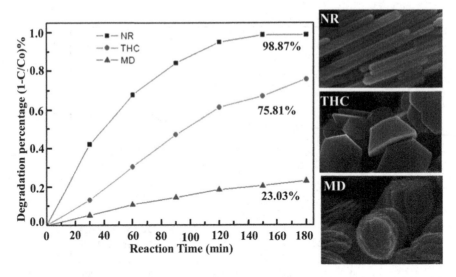

Fig. 3.4 The photocatalysis degradation of MO by using the as-prepared samples of ZnO with different morphologies as shown in the SEM images on the right side. (Reprinted with permission from Liu et al. 2013)

planes on the top (0001), bottom (000-1), and the most stable nonpolar planes on the six sides, which are inert and stable against photocorrosion. Thus, the planar arrangement depending on the type of microstructure plays a dominant role and presents the highest photocatalytic activity. Photocatalytic degradation of methyl orange for two polar surfaces ZnO (0001)-Zn and ZnO (000-1)-O as well as the nonpolar ZnO (10-10) surface was also studied, which showed that the photodegradation rate was differed by one order of magnitude between the highest and the lowest active crystal faces with nonpolar plane exhibiting the highest activity (Kislov et al. 2009). This is because the ZnO (100-1)-O surface exhibited higher photolysis when compared to the other planes of ZnO, i.e., ZnO (0001)-Zn and the ZnO (10-10) surfaces, that exhibited localized dissolution. Since photolysis is inversely correlated with photodegradation of methyl orange, the ZnO (000-1)-O surface exhibited the lowest photoactivity, where the holes are consumed during the photolysis process of ZnO. It has also been reported that the photocatalytic efficiency of various ZnO crystalline planes was suggested to follow the order as (0001) > {10-10} > {101-1} and (000-1) for the degradation of MO under UV light irradiation (Han et al. 2009).

In addition to these three morphologies synthesized by solvothermal method, other different morphologies of ZnO such as hexagonal disks, dumbbell-like bipods, rice-like, and rods have also been fabricated by managing the pH of the reactants (such as 7.5, 8, 9, and 10, respectively) in the hydrothermal method, which showed that the reactant solution's pH influenced both morphology and defect content of the nanostructured product (Flores et al. 2014). The UV light-induced photocatalytic degradation of MB showed that the catalytic performance of ZnO was strongly influenced by microstructure, specific surface area, and surface defect content. The dye MB was best degraded by the rice-like ZnO because this particular morphology had evolved with the highest surface area of 18.88 m^2 g^{-1}. In addition, the rice-like ZnO have posed more defects that are a type of electron trap compared to other ZnO microstructures. Larger content area is available because of the higher surface area of these materials that give more space or more dye absorption on their surface. Similarly, a variety of hierarchical microstructures such as rods, peanuts, dumbbells, and notched spherical shapes of ZnO were synthesized by hydrothermal method (Kim and Huh 2011). Their photocatalytic activity showed that the morphology had strongly influenced the performance of ZnO where the rodlike structure showed the best performance and the corresponding rate constant was 1.1×10^{-2} min^{-1}.

The growth of ZnO nanorod-like structures on different rectangular seeds such as ZnO and SnO$_2$ was done by using the hydrothermal method (Silva et al. 2016). The photocatalytic activity of three samples namely SnO$_2$, ZnO-ZnO, and ZnO-SnO$_2$ was tested for degrading the MB dye under UV light irradiation. It was concluded that all samples are photocatalytically active for MB photodegradation. However, the extent of photocatalytic activity followed the order as ZnO-SnO$_2$ > ZnO-ZnO > SnO$_2$. The ZnO-SnO$_2$ had shown the highest (30%) MB degradation in 4 h when irradiated with UV light compared to that of the ZnO-ZnO and SnO$_2$ samples (Silva et al. 2016). This enhanced photocatalytic performance was imputed to the presence of oxygen defects and good charge separation. The other interesting materials of interest are the magnetic nanostructures. The nanorods of Fe$_2$O$_3$ with porous structure have been prepared by simple chemical method and used as photocatalyst (Liu et al. 2015). The photocatalytic activity of the porous Fe$_2$O$_3$ nanorods was compared with the commercially available Fe$_2$O$_3$ to degrade five different types of organic dye molecules such as Rhodamine B, MB, MO, p-nitrophenol, and eosin B. About 15.2% MB, 7.9% MO, 62.4% eosin B, and 11.3% pNP were degraded by the Fe$_2$O$_3$ commercial photocatalyst under solar irradiation for180 min. The degradation efficiencies for eosin B, pNP, MB, RhB, and MO and in the presence Fe$_2$O$_3$ nanorods with porous structure are 81%, 16.7%, 22.5%, 82.6%, and 12.8%, under solar irradiation for 180 min, respectively. The porous Fe$_2$O$_3$ nanorods showed higher photocatalytic activity that is attributed to their large surface area due to the porous morphology.

Many times, graphene nanosheets (GNS) have been used as a good material to make a composite with homogeneous or heterogeneous materials in order to get a well-performing photocatalyst. The V$_2$O$_5$/TiO$_2$ core-shell nanorods were prepared by a facile hydrothermal process, and these nanorods were made composite with GNS to get the GNS-V$_2$O$_5$/TiO$_2$ nanocomposites by a sol-gel technique (Rakkesh et al. 2015). The microstructure studied by HRTEM of V$_2$O$_5$, TiO$_2$, and V$_2$O$_5$-TiO$_2$ core-shell nanorods and GNS-V$_2$O$_5$/TiO$_2$ can be seen in the relevant paper published (Rakkesh et al. 2015). The core V$_2$O$_5$ and the shell TiO$_2$ are clearly seen in the core-shell architecture, and the graphene sheets are also highly visible. The photocatalytic activity of these GNS-V$_2$O$_5$/TiO$_2$ photocatalysts had been studied on acridine orange (AO) dye under sunlight irradiation. The AO dye was more or less 40% degraded by pure V$_2$O$_5$ nanorods in 60 min, 95% by V$_2$O$_5$-TiO$_2$ core-shell nanorods within 60 min, and very successfully 100% by GNS-V$_2$O$_5$/TiO$_2$ photocatalyst within just 20 min under direct sunlight irradiation. The GNS-V_2O_5/TiO_2 has exhibited the highest photocatalytic activity than that of pure V_2O_5 and V_2O_5/TiO_2 core-shell nanomaterials because of the increased adsorption of the dye onto its surface, increased interfacial charge transfer (assisted by GNS), high electron mobility, and enhanced charge separation. A similar explanation is also valid for the ternary photocatalyst consisting of Au-decorated V_2O_5@ZnO heteronanorods (Fig. 3.5) prepared by a four-step process involving thermal evaporation of ZnO powders, CVD of intermediate on ZnO, solution deposition of Au NPs, and finally thermal oxidization. The microstructure of Au nanoparticles, pure ZnO nanorods, V_2O_5@ZnO heteronanorods, and the latter decorated with Au nanoparticles are given in the electron micrographs shown in Fig. 3.5 (Fig. 3.5a is a TEM micrograph and 3.5b–f are the SEM micrographs). The photocatalytic behavior of these ternary materials was checked on MB dye under UV-vis irradiation (Yin et al. 2014). In Fig. 3.5g, h, the Au-decorated V_2O_5@ZnO nanorods show the highest photocatalytic activity within 150 min of irradiation. The photocatalytic perfor-mance followed the order, such that Au-decorated V_2O_5@ZnO nanorods > V_2O_5@ZnO nanorods > ZnO nanorods. That is, the degradation efficiency for MB degradation was 22.6%, 46.5%, and 82.6% respectively (Yin et al. 2014), by ZnO nanorods, V_2O_5@ZnO nanorods, and Au-decorated V_2O_5@ZnO nanorods.

A detailed report on TiO_2 nanotubes electrode correlating the morphology with the photocatalytic activity (PC) and photoelectrocatalytic activity (PEC) can be seen in the literature (Adán et al. 2016). It was observed that the photodegradation efficiency was influenced by the length of the nanotubes. An enhancement in both PC and PEC processes was obtained for longer nanotubes because of the higher surface area that was in contact with the electrolyte and also increased light absorp-tion. But further increase in the rod length showed a reduction in the activity. This is because beyond a certain length, the higher resistance for the electrons reaches the back contact, and diffusional restrictions to the mass transport of the reactants along the tubes reduce the activity. This optimal length of the nanotubes was because of the compromise between the reactivity and transport properties.

Nanorods and platelets of pure V_2O_5 and Sn^{4+}-doped V_2O_5 (Sn:V_2O_5) were prepared by hydrothermal method, and their visible light photocatalytic activity was evaluated on MB (Rajeshwari et al. 2018). Here, two different morphologies, i.e., rods and platelets, were formed due to a change in the Sn ion concentrations. The change of V_2O_5 nanorods to the platelets can be due to the SnO_2 secondary phase formed above 10% Sn doping in V_2O_5. The estimated degradation efficiencies to degrade the MB dye were 92%, 95%, 96%, 93%, 91%, and 88% for pure V_2O_5, 1Sn-V_2O_5, 3Sn-V_2O_5, 5Sn-V_2O_5, 15Sn-V_2O_5, and 20Sn-V_2O_5, respectively. The 3Sn-V_2O_5 photocatalyst which showed the best degradation with 96% of MB was degraded in 180 min compared to pure V_2O_5 and other photocatalysts (Rajeshwari et al. 2018).

Santhosh et al. have reported the photocatalytic activity of the hydrothermally prepared V_2O_5 nanorods tested for MO, Rhodamine 6G (Rh-6G), and MB under visible light irradiation. The photodegradation efficiencies were 85%, 48%, and 24% for Rh-6G, MO, and MB, respectively. Figure 3.6a shows the irradiation time

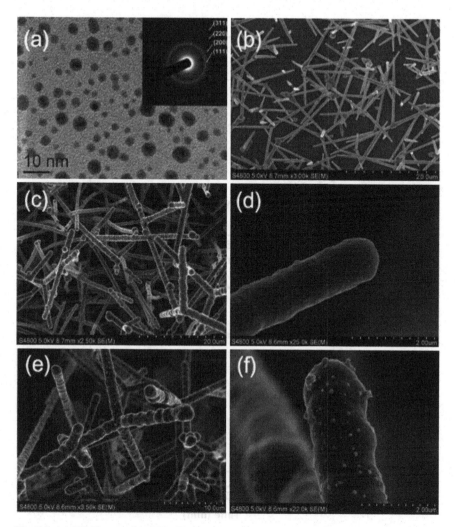

Fig. 3.5 (**a**) TEM image of Au NPs. The inset is the corresponding SAED pattern. (**b**) SEM image of pure ZnO nanorods. (**c**) Low- and (**d**) high-magnification SEM images of V_2O_5@ZnO nanorods. (**e**) Low- and (**f**) high-magnification SEM images of Au-decorated V_2O_5@ZnO nanorods. (**g**) Decrease (C/C_0) and (**h**) natural logarithm [$\ln(C_0/C)$] of the normalized concentration vs. irradiation time for MB solutions containing different photocatalysts: pure ZnO nanorods, V_2O_5@ZnO nanorods, Au-decorated V_2O_5@ZnO nanorods. The dashed lines in panel (**h**) represent the linear fitting results corresponding to the three different photocatalysts. (Reprinted with permission from Yin et al. 2014)

evolution of the UV absorption spectra of the degradation Rh-6G in the presence of V_2O_5 nanorod photocatalyst under visible light irradiation. Respectively the concentration-time profile and the reaction kinetics of the photocatalysis by V_2O_5 nanorods on three dyes (MB, MO, and Rh-6G) are presented in Fig. 3.6b, c.

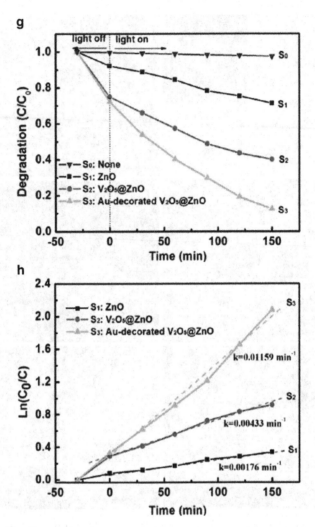

Fig. 3.5 (continued)

The V_2O_5 nanorods showed higher degradation efficiency of Rh-6G (85%) in 300 min compared to other dye molecules (Santhosh Kumar et al. 2018). The corresponding reaction rate constant was obtained to be 0.603 h^{-1}. The same group has prepared different concentrations of Pd NPs-decorated V_2O_5 nanorods by hydrothermal method and evaluated the visible light degradation performance on Rh-6G dye. The photocatalytic activity of 5Pd-V_2O_5 photocatalyst was obtained to be the highest with 98% degradation, and the reason was attributed to its high reaction rate constant (1.8544 h^{-1}) and high surface area (19 m^2/g) that have enhanced the visible light harvesting and reduced the electron-hole recombination lifetime (35.64 ns) (Kumar and Thangadurai 2018).

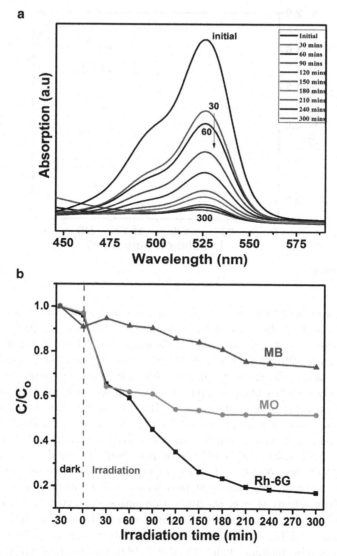

Fig. 3.6 (**a**) Absorption spectra demonstrating the photodegradation of the dye Rh-6G under the irradiation of visible light in the presence of V_2O_5 nanorods. (**b**) The normalized degradation intensity C/C0 versus light irradiation time (in minutes) for Rh-6G, MO, and MB, (**c**) scaled plot to find the rate constant for Rh-6G, MO, and MB. (Reprinted with permission from Santhosh Kumar et al. 2018)

Nanowires of pure ZnO, In_2O_3/ZnO binary, and In_2O_3/ZnO@Ag ternary composites have been developed, and their application in the visible light degradation of methyl orange (MO) and 4-nitriphenol (4NP) had been studied (Liu et al. 2017). The SEM images (Liu et al. 2017) of all the photocatalyst samples showed that the surface of ZnO is decorated with the In_2O_3 nanoparticles. The degradation

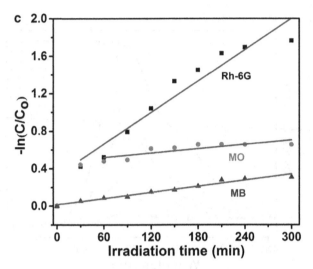

Fig. 3.6 (continued)

performance of these photocatalysts (ZnO, Ag/ZnO, In_2O_3/ZnO) to degrade MO was obtained to be 10%, 43%, and 67% in 90 min of visible light irradiation, respectively. In addition to this, the degradation efficiencies for 4NP showed an increase in the efficiency to 40%, 44%, 68%, and 92% in 240 min for Ag/ZnO, In_2O_3/ZnO/Ag, ZnO/In_2O_3, and In_2O_3/ZnO@Ag under visible light irradiation, respectively. The ternary In_2O_3/ZnO@Ag composite photocatalyst showed the highest photocatalytic activity for the degradation of MO and 4NP. This increase is because of the increased electron-hole separation and therefore reduced charge recombination. Its band alignment and the mechanism of photocatalysis were also explained schematically by Liu et al. This result and the mechanism have proposed that the addition of Ag into ZnO/In_2O_3 composites can lead to sustain the recombination of electron-hole pairs, and therefore, this prolonged recombination had favorably improved the photocatalytic activity (Liu et al. 2017). Combination of two different methods, i.e., hydrothermal and wet chemical routes at room temperature, was used to prepare one-dimensional SnO_2/V_2O_5 core-shell nanowires (their microstructures are shown in Fig. 3.7). The toluidine blue "O" dye (TBO) was used to investigate the photocatalytic performance of the SnO_2/V_2O_5 core-shell nanowires under UV exposure (Shahid et al. 2010), and their performance was compared with the bulk V_2O_5. The SnO_2/V_2O_5 core-shell structure showed the highest photocatalytic performance, and it was much higher than that of the bulk V_2O_5 and pure nanowire counterparts.

Ultra-long nanowires (NW) of V_2O_5 were prepared by the structural intercalative properties using a pyridine-nitric acid (Py-HNO_3) system. In Fig. 3.8, the FE-SEM images show uniform shape and size distribution of NW morphology of different V_2O_5 systems. The photocatalytic activity of these NWs was tested by the photodegradation of MB dye solution. The V-1 catalyst (HNO_3-pyridine-mediated

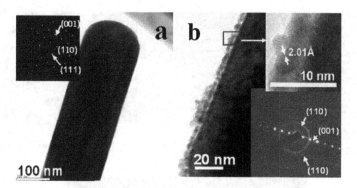

Fig. 3.7 TEM image of V_2O_5 nanowires (**a**) inset shows the SAED pattern. (**b**) TEM image of SnO_2/V_2O_5 nanowire with insets showing the HRTEM image on the top right and SAED pattern on the lower right. (Reprinted with permission from Shahid et al. 2010)

synthesis) has degraded 89% of the MB dye, and the V-2 catalyst (only pyridine-mediated synthesis) had degraded 63% (Roy et al. 2014).

A well-dispersed Pt QDs were used to decorate the TiO_2 nanotube arrays (TNTAs) by a modified photoirradiation-reduction method (Wang et al. 2015) with different loading concentrations of Pt QDs. The Rh-B was degraded to an efficiency of 31.61%, 23.76%, 73.47%, and 38.71% by pure TNTAs, Pt(0.3)-TNTAs, Pt(1.0)-TNTAs, and Pt(2.0)-TNTAs, respectively, under UV light irradiation for 180 min. For the Pt(0.3)-TNTAs sample, the observed photocurrent was lower than the pure TNTA catalyst, which is due to the availability of more number of trapped holes, formed at the interface formed by highly dense Pt QDs, to recombine with the electrons. The Pt(1.0)-TNTAs sample showed higher photocurrent value because of the lower photogenerated electron-hole recombination. Thus, higher photocurrent value could also enhance the photocatalytic activity (Fig. 3.9a). Figure 3.9b presents the electron paramagnetic resonance (EPR) spectral signal showing a well-resolved four peaks of DMPO-$^{\bullet}$OH with an intensity ratio of 1:2:2:1 which clearly clarifies that the $^{\bullet}$OH radical was generated on the catalyst (Pt(1.0)-TNTAs) when irradiated with UV light. In addition to this, six-peak signals were obtained because the radical of $O_2^{-\bullet}/HO_2^{\bullet}$ adducts with an intensity ratio of 1:1:1:1:1:1 was also confirmed (marked with red dot). It was found that the intensity of all the peaks was enhanced with increasing of irradiation time. These EPR results indicate the generation of both $^{\bullet}$OH and $O_2^{-\bullet}$ radicals that are involved in the degradation process (Wang et al. 2015).

The Co_3O_4 nanomaterials in nanorods, nanosheets, nanotubes, and nanoparticles morphology were synthesized by hydrothermal method for photocatalytic applications (Pan et al. 2014). Their photocatalytic degradation properties under visible light were investigated on methyl orange with the assistance of hydrogen peroxide. Each microstructure of Co_3O_4 has shown different catalytic activities, where particularly the nanosheets (97.5%) and nanotubes (99.1%) exhibited the highest activity

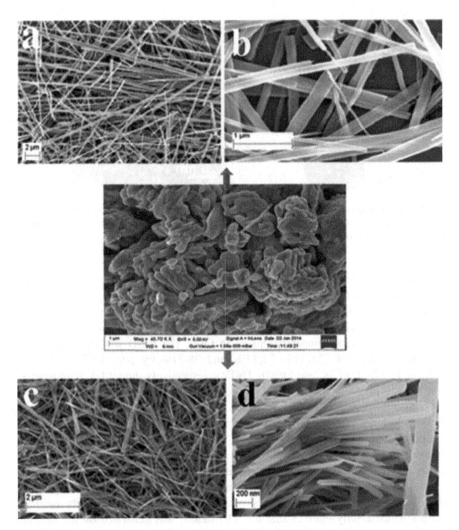

Fig. 3.8 High- and low-magnification FESEM images of V_2O_5 NWs obtained from their bulk shown in the middle. (**a**), (**b**) For V-2 (only pyridine-mediated synthesis) (up arrow) and (**c**), (**d**) for V-1 (HNO_3-pyridine-mediated synthesis) system (down arrow). (Reprinted with permission from Roy et al. 2014)

with excellent repeatability for catalytic degradation of MO. The highest activity for the nanotubes was accounted because of the novel tubelike structure and the highest surface area of them (67.2 m^2/g), and the second largest degradation was shown by the nanosheets (46.7 m^2/g), which is because of the highest surface area possessing more active sites.

Fig. 3.9 (a) Photocurrent responses under UV light irradiation on pure TNTAs, Pt(0.3)-TNTAs, Pt(1.0)-TNTAs, and Pt(2.0)-TNTAs. (b) The EPR spectra of DMPO-·OH adducts and DMPO-O$_2^{-\cdot}$/ HO$_2^{\cdot}$ adducts. (Reprinted with permission from Wang et al. 2015)

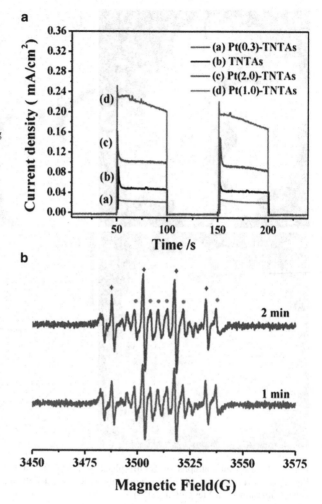

3.3.3   Two-Dimensional Nanomaterials for Photocatalysis

The two-dimensional nanostructures are the materials with confinement in one direction, and the charge carriers can move in any of the other two directions. The best known examples are graphene and molybdenum disulfide (MoS$_2$) sheets. The 2-D structured platelets of BiOBr nanostructures were prepared with their surface decorated with Pd nanoparticles by two-step synthesis method: first the BiOBr was prepared by hydrothermal method and then the Pd NPs were deposited on it by photodeposition. Figure 3.10 shows the SEM micrographs of pure BiOBr and Pd NPs-decorated BiOBr, where a platelet-like structure of BiOBr is well noticed (Fig. 3.10). The Pd particles that decorate the surface of the BiOBr are clearly seen in Fig. 3.10d. Photocatalytic activity of pure BiOBr and Pd NPs-decorated

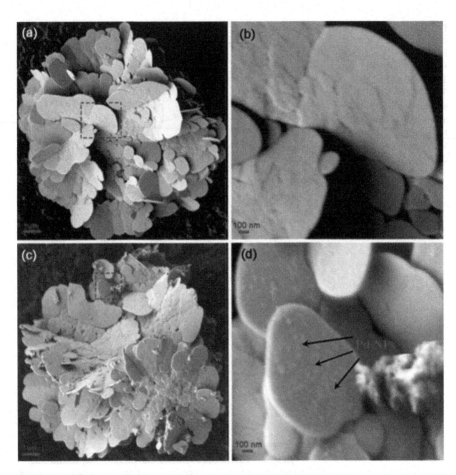

Fig. 3.10 SEM micrographs of BiOBr platelets at (**a**) low and (**b**) high magnifications, 0.5 Pd-BiOBr at (**c**) low and (**d**) high magnifications for pure BiOBr and 0.5 Pd-BiOBr. (Reprinted with permission from Meng et al. 2018)

BiOBr was studied for phenol degradation by visible light irradiation. The efficiency of pure BiOBr was 67% in 300 min, whereas the Pd-BiOBr showed better activity due to the improved visible light harvesting capacity via the SPR of Pd NPs. The photocurrent measurement of Pd-BiOBr and BiOBr was also done, and it was observed that the photocurrent response of Pd-BiOBr is five times better than the pure BiOBr, because of the enhanced electron-hole pair recombination rate, which has further improved the visible light absorption (SPR effect). This activity was found to be Pd content dependent; the photocatalytic efficiency for phenol degradation has increased from 77% to 100% under visible light (Meng et al. 2018) when the Pd content was increased from 0.1% to 0.5%.

In another work, the nanosheets-like morphology of Sr-doped Bi_2O_3 with 0–10 wt% of Sr contents was synthesized. Figure 3.11a, f shows the FE-SEM

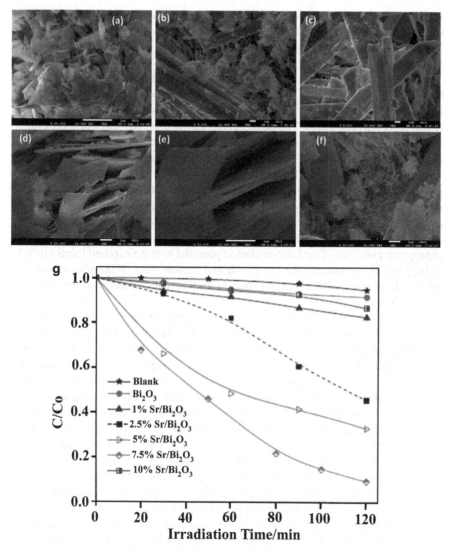

Fig. 3.11 Typical FE-SEM images of undoped Bi_2O_3 (**a**), 1% Sr-doped Bi_2O_3 (**b**), 2.5% Sr-doped Bi_2O_3 (**c**), 5% Sr-doped Bi_2O_3 (**d**), 7.5% Sr-doped Bi_2O_3 (**e**), and 10% Sr-doped Bi_2O_3 (**f**). Plot for comparison of change in concentration vs. irradiation time in the presence of undoped Bi_2O_3 and various % of Sr-doped Bi_2O_3 (MB concentration [0.022 mmol/l], volume of MB = 100 mL, catalyst loading 1 g/L) (**g**). (Reprinted with permission from Faisal et al. 2014)

micrographs of the pure Bi_2O_3 and Sr-doped Bi_2O_3 where pure Bi_2O_3 is possessing smooth plate-like structure (Fig. 3.11a). Then, with increasing the concentrations of Sr, the Bi_2O_3 morphologies were changed to sheets from flakes and a rod-shaped structures that can be seen clearly from Fig. 3.11d, f (Faisal et al. 2014). As shown in Fig. 3.11g, the photocatalytic activity to degrade MB dye under visible light irradiation has significantly been improved with increasing the Sr doping levels from 0% to 7.5% Sr, and then gradually the activity has been decreased at 10% Sr. The MB is almost completely degraded by 7.5% Sr-doped Bi_2O_3 after 120 min of visible light irradiation (Faisal et al. 2014).

A similar reason of high surface area of the photocatalyst is also given for the higher degradation activity shown by hydrothermally synthesized ferroelectric $Bi_3Fe_{0.5}Nb_{1.5}O_9$ (BFNO-H) as compared to that synthesized by solid-state method (BFNO-S) (Yin et al. 2018). The BFNO-H has shown a hierarchical morphology that is stacked by an intersecting single crystal nanosheet with {001} and {110} exposed facets, whereas the BFNO-S has shown disorganized micron scale morphology. It is well known that the hydrothermal method can yield well-ordered nanostructured materials, whereas the solid-state reaction method yields only micron-sized materials. The higher photocatalytic performance of BFNO-H was because of the hierarchical structure and the large surface area that can shorten the photogenerated carrier migration distance and increase the reactive sites. It is also observed that the appropriate matching of the built-in electric potential with the exposed facets ({001}, {110}) has also improved the separation and transmission of the photogenerated electron-hole pairs, which has resulted in further enhancement of the catalytic activity (Yin et al. 2018).

A composite of Sn-doped ZnO with graphene nanosheets was prepared by hydrothermal method with varied Sn ion content from 0 to 10 wt%, and they were tested for their photocatalytic performance of anionic dyes (MO and methyl red) and cationic dyes (MB and Rh-B) under UV and sunlight irradiation (Beura and Thangadurai 2018). The highest degradation for MO was 89.3% and 99.1% under UV and sunlight irradiation, respectively. The same group has also developed graphene-ZnO composite for the photodegradation of MO under UV and direct sunlight irradiation, and the MO degradation efficiency was 97% and 98%, respectively, by the 5 wt% ZnO-graphene nanocomposite. The 5 wt% was optimum to attain the enhanced photocatalytic activity (98%). The enhanced photocatalytic activity was ascribed to the large surface area and enhanced PL lifetime (Beura and Thangadurai 2017). Another research group has developed the MoS_2-based photocatalysts, in which ultrathin-layered MoS_2 nanosheets were developed by hydrothermal method, and the photocatalytic performance of them was studied on MB and RhB under visible light irradiation. Photocatalytic degradation efficiencies of 95.3% and 41.1% were obtained for MB and RhB, respectively (Abinaya et al. 2018).

Similarly, the MoS_2/carbon nanofibers (CNFs) hybrid structure has been fabricated by electrospinning and hydrothermal method. Pure MoS_2 and MoS_2/CNFs hybrid structures were tested for photocatalytic activity on the degradation of Rhodamine B (Rh-B) when irradiated with visible light. The Rh-B was degraded 50% and 67% by pure MoS_2 and MoS_2/CNFs photocatalysts, respectively, after 5 h of visible light irradiation. The enhancement in the photocatalytic activity in MoS_2/CNFs has been attributed to the synergistic effects. In this case, the photoelectrons are moved to the CNFs and thus enhanced the separation of electron-hole to reduce the rate of recombination (Wang et al. 2017a, b). Hierarchical structure of MoS_2 on TiO_2-coated Si nanowires was synthesized by hydrothermal method, and their visible light photodegradation of Rh-B was studied (Hamdi et al. 2016). The photocatalytic activities of Si NW, TiO_2/Si NW, and MoS_2 on TiO_2/Si NW, for the Rh-B degradation, were 70%, 72%, and 90% in 180 min, respectively. The MoS_2 on TiO_2/SiNW exhibited the highest photocatalytic activity because the MoS_2 was trapping the electrons and acted as an electron sink. Thus, the suppression of electron-hole recombination has enhanced the photocatalytic activity.

3.3.4   Three-Dimensional Photocatalyst Materials

The 3-D materials are the materials not confined in any directions, and the charge transfer is possible in all the three dimensions. For example, microstructures for this type of materials are flower, microspheres, cone-shaped heterostructure, etc. Although a lot of works have been reported on the development and effect of nanostructures on photocatalysis, an interesting work has been reported on the development of ZnO hierarchical micro-/nanostructure (Lu et al. 2008), which composes of microsized conic-like particles that are built on many alternating nanosheets of 10 nm thickness standing on hexagonal, pyramid-like core microcrystals. The model for the growth sequence to form this particular 3-D structure was also dealt in their paper. Comparison of the photocatalytic performance of this special structure with Degussa P25 titania and other ZnO structures like nanoneedles and nanosheets was carried out to study the structure-induced enhancement in the photocatalytic performance in the MO dye under UV irradiation. The photocatalytic experimental results showed that the ZnO micro-/nanoarchitectures have shown significantly enhanced photocatalytic performance because of the special structure that exhibits a high surface to volume ratio, creating a high specific surface area (>180 m^2 g^{-1}) and stability against aggregation.

The flowerlike Ag/ZnO photocatalysts were prepared for various Ag/ZnO ratios by hydrothermal method without surfactants (Zhang et al. 2017a, b). The Ag/ZnO nanocomposites of different Ag/Zn ratios such as 1:50, 1:40, 1:30, 1:20, and 1:10 were obtained, which were labeled as Ag/ZnO-x (x = 1, 2, 3, 4, and 5). The photocatalytic activity of pure ZnO and Ag/ZnO samples was tested for the degradation of MB solution under visible light irradiation. The SEM micrographs of pure ZnO and Ag/ZnO samples are shown in Fig. 3.12. Absence of Ag particles on the surfaces of ZnO implies that the Ag particles may be wrapped by ZnO nanosheets. Increasing the Ag content has maintained the flowerlike structure in all the Ag/ZnO samples, implying that deposition of Ag does not affect the morphology. Figure 3.13a shows the degradation efficiency of MB by using pure ZnO and Ag/ZnO photocatalysts. Compared to pure ZnO, the Ag/ZnO showed higher photocatalytic activity. The degradation efficiencies of ZnO, Ag/ZnO-1, Ag/ZnO-2, Ag/ZnO-3, Ag/ZnO-4, and Ag/ZnO-5 are, respectively, 40%, 51%, 64%, 77%, 93%, and 71% to degrade MO solution under visible light in 180 min. The kinetics of the Ag/ZnO photocatalysts for degradation of MB is presented in Fig. 3.13b. Compared to pure ZnO, the rate constant k of all Ag/ZnO is higher. On increasing of Ag content, the activity of Ag/ZnO samples initially increased and then

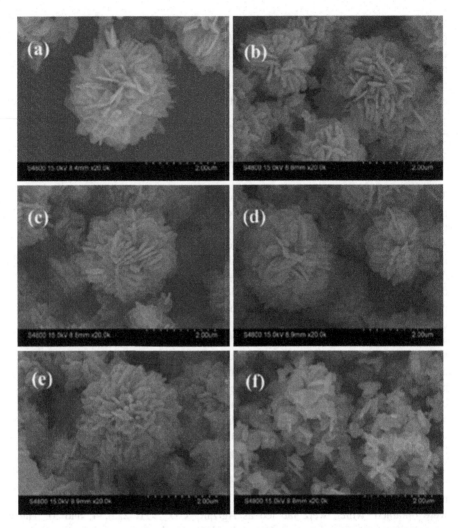

Fig. 3.12 SEM images of ZnO (**a**), Ag/ZnO-1 (**b**), Ag/ZnO-2 (**c**), Ag/ZnO-3 (**d**), Ag/ZnO-4 (**e**), and Ag/ZnO-5 (**f**). (Reprinted with permission from Zhang et al. 2017a, b)

decreased. The rate constant $k = 0.01363$ min^{-1} of Ag/ZnO-4 showed the highest photocatalytic activity. Thus, it is clear that the presence of Ag nanoparticles in ZnO has significantly increased the performance of ZnO, and an optimum Ag loading was necessary to improve the photocatalytic activity. However, further increasing of Ag content not only damages the flowerlike morphology but also leads to an aggregation of Ag particles and decrease in specific surface that together spoil the recombination of the photoexcited electron-hole pairs and result in the decreased photocatalytic activity (Zhang et al. 2017a, b).

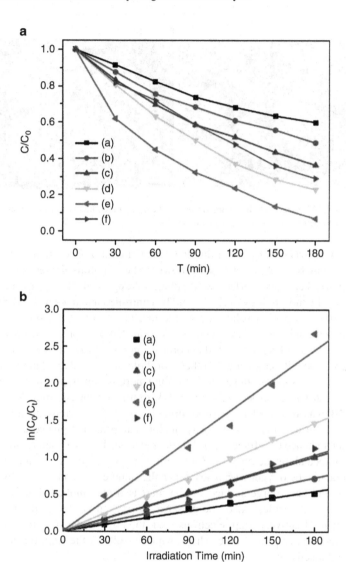

Fig. 3.13 Photocatalytic activity (**a**) and chemical kinetics study (**b**) of ZnO (a), Ag/ZnO-1 (b), Ag/ZnO-2 (c), Ag/ZnO-3 (d), Ag/ZnO-4 (e), and Ag/ZnO-5 (f). (Reprinted with permission from Zhang et al. 2017a, b)

Different microstructures of ZnO (porous flower, solid sphere, branchy, and porous microsphere morphologies) were prepared through a facile one-step hydro-thermal method by varying the concentration of citric acid (CA) (Wang et al. 2017a, b). The photocatalytic performance of these different ZnO microstructures was investigated by photodegradation of MB under UV light irradiation (Wang et al. 2017a, b). Among all the microstructures, the branchy structure was obtained in the

a   b

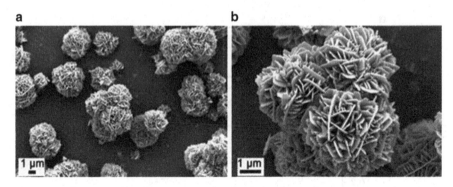

Fig. 3.14 (a) SEM image and (b) enlarged view of the as-prepared hierarchical ZnO nanoflowers. (Reprinted with permission from Xu et al. 2015)

absence of CA, whereas by increasing the concentration of CA, the morphology of the samples has been changed to solid flower sphere, porous flower, solid flower, porous microsphere, and perfect solid sphere (Wang et al. 2017a, b). In a more generic way, the morphology of ZnO could be controlled and that can be varied from porous nature to a perfect solid. A possible mechanism of formation of this morphology in ZnO microstructures with and without CA was proposed (Wang et al. 2017a, b). It was observed that the porous ZnO structure showed the highest photocatalytic activity compared to other morphologies. The degradation efficiency of MB is 90.15%, 86.11%, and 58.57% for ZnO, porous flower, porous microsphere, and branchy structure, respectively, under UV irradiation for 180 min. No degradation of MB was observed for a solid structure of ZnO. Various parameters like morphology, surface area, and porosity of ZnO catalysts had played an important role that can enhance the efficiency of MB degradation. The two main reasons for the highest photocatalytic activity of ZnO with porous structures are the increased surface area and the oxygen vacancies. Since the sample was constructed by several nanosheets and has a large surface area, it increased the number of active sites to absorb the dye molecules. Also, the presence of oxygen vacancies in their crystal structure can effectively trap electrons and thereby hinder the recombination between electrons (e_{cb}^{-}) and holes (h_{vb}^{+}), which results in the improvement in the photocatalytic activity.

There is another morphology, named as ZnO nanoflower, that was prepared by one-step hydrothermal method (Xu et al. 2015). Their photocatalytic performance was evaluated by degrading metamitron under 4 h of UV-vis irradiation, where the nanoflower has exhibited the highest photocatalytic activity of about 97%. Figure 3.14a, b shows the SEM micrographs of the ZnO nanoflower morphology.

The porous-structured heterostructural Ag/ZnO composites with 3-D flowerlike morphology had been made with different molar concentrations of Ag (0–20%) by hydrothermal and photochemical deposition methods. No reagents for pore directing have been used, and no surfactants were used for the preparation (Liang et al. 2015). Figure 3.15 presents the morphologies of these pure ZnO and Ag/ZnO composites.

Fig. 3.15 SEM images of (**a, b**) the porous flowerlike pure ZnO and (**c, d**) 15% Ag/ZnO heterostructure composites. (Reprinted with permission from Liang et al. 2015)

Figure 3.15a shows the ZnO 3-D flowerlike morphology, and complete observation of a single ZnO flower is shown in Fig. 3.15b. Many numbers of self-assembled porous nanoplates have formed these flowerlike morphologies. In addition, presence of pores can be easily visualized (inset of Fig. 3.15b) in the nanosheets. These porous arrangements in these structures probably increased the surface area for improved reactant diffusivity that is very much favorable for photocatalytic applications. Figure 3.15c, d shows the Ag/ZnO (Ag particle-decorated ZnO) flowerlike structure. It can be noted that the morphology of ZnO has not been changed with the addition of Ag. As shown in Fig. 3.16, the photodegradation efficiency of these Ag/ZnO on Rh-B under UV light for 120 min is about 27%, 21%, 56%, 63%, 79%, and 41% for P25, pure ZnO, 5% Ag/ZnO, 10% Ag/ZnO, 15% Ag/ZnO, and 20% Ag/ZnO, respectively. The 15% Ag/ZnO exhibited the highest photodegradation efficiency.

A wide range of morphologies of ZnO like cauliflower, hourglass-like, truncated tubular, hexagonal conical, nanorods, and spherical shape were synthesized via solvothermal method with different solvents like decane, THF, toluene, ethanol, water, and acetone, respectively (Xu et al. 2009), and their microstructures are shown in Fig. 3.17. Various morphologies of ZnO were then utilized to evaluate the influence of morphology on phenol photocatalytic degradation. The ZnO cauliflower morphology obtained with THF solvent has shown the best catalytic activity having the rate constant ($k = 0.1496$ min^{-1}), followed by ZnO obtained with other

Fig. 3.16 Photodegradation curves of Rh-B by different Ag/ZnO catalysts under UV light irradiation for different loadings of Ag. (Reprinted with permission from Liang et al. 2015)

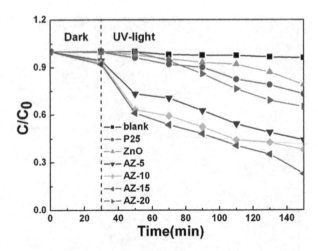

morphologies like truncated hexagonal conical ($k = 0.0867$ min^{-1}), tubular ($k = 0.0651$ min^{-1}), hourglass-like ($k = 0.0518$ min^{-1}), nanorods ($k = 0.0468$ min^{-1}), and spherical ($k = 0.0385$ min^{-1}). Results showed no correlation between the catalytic activity and the surface areas in these microstructures, implying that there are other important factors that influence the catalytic activity. Generally, better performance is shown by the catalysts with higher surface energy. Since higher surface energy is possessed by the nonfaceted particles, the higher catalytic activity is shown by the cauliflower-shaped ZnO because they are formed with nonfaceted nanoparticles. Again, both conical and hourglass-like ZnO having similar crystal facets showed differences in the catalytic performance that is because of the large surface area and the interior cavities of conical ZnO compared to the hourglass-like structure. The least catalytic activity of spherical-shaped ZnO was due to the formation of particles with smooth facets having lower surface energy.

Microspheres of bismuth vanadate (BiVO$_4$) with different surface morphologies in monoclinic structural phase have been prepared by varying pH values. Photocatalytic degradation of MB was obtained to be 69%, 80%, 92%, 49%, 57%, and 53% in the presence of microspheres of BiVO$_4$ prepared at pH $= 1, 3, 5, 7, 9$, and 11, respectively. The high catalytic activity shown by the microsphere synthesized at pH 5 was backed by several reasons. First, these hierarchical mesoporous structures have smaller and larger mesopores, which could provide more chances for diffusion and mass transport of the reactant molecules to hit the active sites. Second, the charge transfer in the materials is promoted by the surface to volume ratio of size-quantized nanosheets, thereby increasing the charge transfer rate and thus decreasing the recombination of the photogenerated electron-hole pairs, which promotes the degradation of MB (Li et al. 2014).

Among the various semiconductor materials used as photocatalysts, bismuth-subcarbonates (Bi$_2$O$_2$CO$_3$) have gained much interest toward photocatalytic appli-

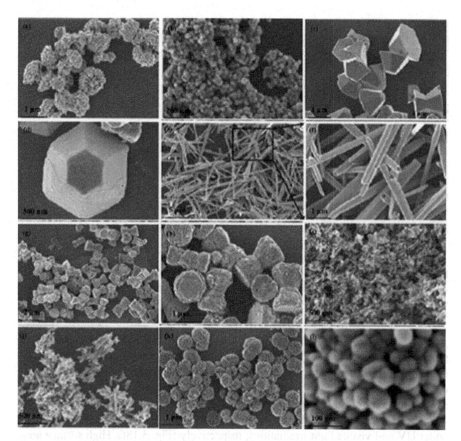

Fig. 3.17 FE-SEM micrographs of ZnO samples with different microstructures prepared in different solvents by hydrothermal method: (**a**, **b**) THF, (**c**, **d**) decane, (**e**, **f**) H$_2$O, (**g**, **h**) toluene, (**i**, **j**) ethanol, (**k**, **l**) acetone. (Reprinted with permission from Xu et al. 2009)

cation for dye degradation with their tunable response to morphology variation. The Bi$_2$O$_2$CO$_3$ with various morphologies (Tang et al. 2012) like irregular nanoplates, uniform nanoplates, and nanocubes have been synthesized, and their photocatalytic properties were studied to degrade various organic dyes. Degradation performance is strongly dependent on the morphology. Although a number of works focused on the morphology-dependent photocatalysis, only a few works have focused on how the crystallinity of different morphologies affects the photocatalytic performance. Cen et al. reported that the carbonate ions play an important role that controls the morphology and crystallinity, thereby affecting the degradation performance (Cen et al. 2014). Lower concentration of CO$_3^{2-}$ created microsphere structure, whereas increasing the CO$_3^{2-}$ concentration increases the nucleation and

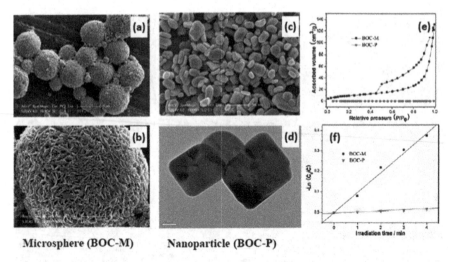

Microsphere (BOC-M)   Nanoparticle (BOC-P)

Fig. 3.18 (a) and (b) SEM micrographs of the as-prepared $Bi_2O_2CO_3$-M (microsphere), (c) SEM, (d) TEM of as-prepared $Bi_2O_2CO_3$-P, (e) nitrogen adsorption-desorption isotherms, and (f) reaction rate constants k of the $Bi_2O_2CO_3$-M and $Bi_2O_2CO_3$-P samples under visible light irradiation. (Reprinted with permission from Cen et al. 2014)

growth turning into a kinetic control, which induces preferred growth orientation, resulting in thick nanoparticles and thus generating sheet morphology (Fig. 3.18a, d). Investigating the performance of these materials to remove NO (nitrous oxide) showed that microsphere structure exhibited enhanced photocatalytic activity of 42.6% and 36.1% compared with that of the sheet morphology (24.8% and 3.3%) under UV and visible light irradiations, respectively (Fig. 3.18f). High surface area and presence of pores in the microsphere structure have favored diffusion of the reaction intermediates and the products, and improve photocarrier separation efficiency, thus accelerating the reaction rate. In addition to this, the low crystallinity generates defects which increase the light absorption, and the high surface area (34.5 m^2/g) behaves as an important adsorption and active site for the photocatalytic reaction (see Fig. 3.18 for surface area).

The microcrystals of $BiVO_4$ with butterfly-like morphology have been prepared in monoclinic structure by alcohol-hydrothermal method (Wang et al. 2016). Photocatalytic studies on MO revealed that the butterfly-like $BiVO_4$ showed higher degradation efficiency compared to that by the irregular-shaped $BiVO_4$. These $BiVO_4$ showed a great photocatalytic activity to be 96%, and this performance was attributed to their special butterfly-like morphology, enhanced surface area, and decreased band gap energy (compared to their counterpart with irregular morphology). Nanostructured SiC with morphologies such as whiskery, worm-like, and particulate have been used for the production of hydrogen by the process of photocatalytic water splitting by irradiating them with visible light in the presence of these photocatalysts (Hao et al. 2013). The carbothermal reduction method (using carbon and silica as precursor) was used to prepare these SiC photocatalysts. Pure water was used to evaluate their photocatalytic hydrogen evolution. Different morphologies of SiC present different photocatalytic hydrogen evaluation performance, i.e., worm-like SiC and particulates showed much higher performance (83.9 μL/g h and 82.8 μL/g h, respectively) than that of the same shown by SiC whiskers for which it was 45.7 μL/g h. This large improvement in the photocatalytic performance was attributed to two factors, namely, high specific surface area and crystallinity induced differently by different morphologies. Higher is the surface area, more is the active sites and surface roughness, that provides more steps and sharp edges with more dangling bonds, thus resulting in the increased water splitting and enhanced photocatalytic activity. At the same time, high crystallinity provides lower defect concentrations that are generally acting as a recombination centers for the photoexcited electrons, thus avoiding higher recombination. Table 3.1 summarizes various photocatalyst materials with different morphologies that have been used in the literature for degrading a given dye molecule under light irradiation. The efficiency obtained by each of them is also presented in it.

3.4 Conclusions

In this chapter, the effect of morphology on the photocatalytic performance of the photocatalysts has been discussed. The basics of photocatalysis and types of photocatalysis were explained in the introduction. Importance of the advance oxidation processes for the photodegradation of many organic dyes/molecules that are harmful to the environment has been presented. Necessity and challenges of designing and developing the photocatalyst materials with required properties have been discussed. On the basis of the photocatalysts requirement, the role and significance of nanomaterials were explained. Possible photocatalyst materials used so far and the nanomaterials with different confinements were incorporated. Majorly, photocatalysts classified based on 0-, 1-, 2-, and 3-D confinements were explained with suitable examples. Later, separate and dedicated sections were included for quantum dots (0-D materials), nanowires/nanorods (1-D materials), nanosheets (2-D materials), and bulk materials (3-D materials). However, in each case, morphology of the nanostructured materials that have influenced the act of photocatalysis on the model dyes has been thoroughly elucidated with example microstructures and their photocatalytic activity. In particular, results of the catalytic activity for 0-D (QDs, nanoparticles), 1-D (nanowires, nanorods, nanotubes), 2-D (nanosheets), and 3-D

Table 3.1 Detailed list of the recent photocatalyst materials, their morphologies, irradiation source, dyes, and the efficiency achieved in photodegrading the selected dyes

S. no	Materials	Morphology	Dye	Irradiation source	Efficiency (%)	References
1	ZnO	Spherical	Methylene blue	UV light	78	Kajbafvala et al. (2012)
		Flower			17	
2	ZnO	Hollow sphere	Rhodamine B	UV light	99	Jin et al. (2018)
3	ZnO	Spherical	Methyl orange	UV light	78	Bordbar et al. (2016)
		Irregular shaped			58	
		Irregular shaped			60	
4	SnO_2	Spherical	Methyl violet 6B (MV6B)	Sunlight	96	Bhattacharjee et al. (2016)
			Methylene blue		96	
5	TiO_2	Quantum dots	Methyl orange	UV light	97	Gnanasekaran et al. (2015)
			Methylene blue		98	
6	ZnO	Nanorods	Methyl orange	UV light	98	Liu et al. (2013)
		Multilayer disks			23	
		Truncated hexagonal cones			76	
7	Fe_2O_3	Nanorods	Rhodamine B	Sunlight	83	Liu et al. (2015)
			Methylene blue		22	
			Methyl orange		13	
			P-nitrophenol		17	
			Eosin B		81	
8	V_2O_5	Nanorods	Acridine orange	Sunlight	40	Rakkesh et al. (2015)
9	V_2O_5-TiO_2				95	
10	GNS-V_2O_5-TiO_2				100	
11	Au-V_2O_5-ZnO	Nanorods	Methylene blue	Sunlight	82	Yin et al. (2014)

	Material	Morphology	Dye/Pollutant	Light source	Efficiency (%)	Reference
12	$Sn-V_2O_5$	Nanorods	Methylene blue	Visible light	96	Rajeshwari et al. (2018)
13	V_2O_5	Nanorods	Rhodamine 6G	Visible light	85	Santhosh Kumar et al. (2018)
14	$In_2O_3-ZnO-Ag$	Nanowires	4-Nitrophenol	Visible light	92	Liu et al. (2017)
			Methyl orange		67	
15	V_2O_5	Nanowires	Methylene blue	Visible light	89	Roy et al. (2014)
16	$Pt-TiO_2$	Nanotubes	Rhodamine B	UV light	73	Yanfei Wang et al. (2015)
17	$Pd-BiOBr$	Platelets	Phenol	Visible light	100	Meng et al. (2018)
18	$Sn-ZnO-graphene$	Nanosheets	Methyl orange	UV light	89	Beura and Thangadurai (2018)
				Sunlight	99	
19	$ZnO-graphene$		Methyl orange	UV light	97	Beura and Thangadurai (2017)
				Sunlight	98	
20	MoS_2	Nanosheets	Methylene blue	Visible light	95	Abinaya et al. (2018)
			Rhodamine B		41	
21	MoS_2-GNS	Nanowires	Rhodamine B	Visible light	67	Wang et al. (2017a, b)
22	MoS_2-TiO_2-Si	Nanosheets	Rhodamine B	Visible light	90	Hamdi et al. (2016)
23	$Ag-ZnO$	Nanoflower	Methylene blue	Visible light	93	Zhang et al. (2017a, b)
24	ZnO	Porous flower	Methylene lue	UV light	90	Wang et al. (2017a, b)
		Porous Microsphere	Methylene lue	UV light	86	
25	ZnO	Nanoflower	Metamitron	UV light	97	Xu et al. (2015)
26	$Ag-ZnO$	Flower	Rhodamine B	UV light	79	Liang et al. (2015)
27	$BiVO_4$	Microspheres	Methylene blue	Visible light	92	Li et al. (2014)
28	$Bi_2O_2CO_3$	Microspheres	Nitrous oxide	UV light	42	Cen et al. (2014)
				Visible light	36	
29	$BiVO_4$	Butterfly-shaped	Methyl orange	Visible light	96	Wang et al. (2016)
30	$Cds-CuS$	Microflower	Methyl orange	Visible light	93	Deng et al. (2017)
31	$ZnTiO_3$	Nanoparticles	Methyl orange	UV light	70	Selvamani et al. (2013)
32	Ag_2O/TiO_2	Nanobelts	Methyl orange	Visible light	89	Zhou et al. (2010)

(continued)

Table 3.1 (continued)

S. no	Materials	Morphology	Dye	Irradiation source	Efficiency (%)	References
33	$ZnO/CuInSe_2/CuInS_2$	Microspheres	Rhodamine B	Mercury lamp	81	Shen et al. (2012)
34	$NaNbO_3/CdS$	Core-shell	Methylene blue	Visible light	98	Sandeep Kumar et al. (2014)
35	$Se-TiO_2$	Nanoparticles	Rhodamine B	Visible light	91	Xie et al. (2018)
36	Fe-Co-ZnO	Nanoparticles	Rhodamine B	Visible light	76	Neena et al. (2018)
			Methylene blue		82	
37	$Bi_2MoO_6/$ $Bi_{3.64}Mo_{0.36}O_{6.55}$	Platelets	Rhodamine B	Visible light	99	Ren et al. (2011)
38	CdSe/Au	Quantum dots	Methylene blue	UV light	92	Soni et al. (2014)
39	$V_2O_5/BiVO_4/TiO_2$	Nanorods	Toluene	Visible light	91	Sun et al. (2014)
40	$g-C_3N_4$- MoS_2/graphene	Nanosheets	Rhodamine B	Visible light	95	Tian et al. (2017)
41	$Pd-V_2O_5$	Nanorods	Rhodamine 6G	Visible light	98	Kumar and Thangadurai (2018)

(flower, microspheres, cone-shaped heterostructure) have been discussed along with the possible reasons and mechanism to support the observations. It has been observed that particle size, phase composition, specific surface area, and geometric structures have been important factors that varied for different morphologies and have strongly influenced the photocatalytic performance. In conclusion, the morphology of the materials that influences various factors decides the performance of the materials to be used as photocatalysts in photocatalysis. By tuning the required morphology, the photocatalytic activity of a given material can be altered/improved.

Acknowledgments The DST-SERB (EMR/2016/005795), India, and UGC-DAE-CSR, India (CSR-KN/CRS-89/2016-17/1130), are acknowledged for the research grants.

References

Abinaya R, Archana J, Harish S, Navaneethan M, Muthamizhchelvan C, Shimomura M, Hayakawa Y (2018) Ultrathin layered MoS_2 nanosheets with rich active sites for enhanced visible light photocatalytic activity. RSC Adv 8:26664–26675. https://doi.org/10.1039/c8ra02560f

Adán C, Marugán J, Sánchez E, Pablos C, Van Grieken R (2016) Understanding the effect of morphology on the photocatalytic activity of TiO_2 nanotube array electrodes. Electrochim Acta 191:521–529. https://doi.org/10.1016/j.electacta.2016.01.088

Akir S, Alexandre B, Yannick C, Mohamed B, Rabah B, Amel DO (2016) Eco-friendly synthesis of ZnO nanoparticles with different morphologies and their visible light photocatalytic performance for the degradation of Rhodamine B. Ceram Int 42:10259–10265. https://doi.org/10.1016/j.ceramint.2016.03.153

Beura R, Thangadurai P (2017) Structural, optical and photocatalytic properties of graphene-ZnO nanocomposites for varied compositions. J Phys Chem Solids 102:168–177. https://doi.org/10.1016/j.jpcs.2016.11.024

Beura R, Thangadurai P (2018) Effect of Sn doping in ZnO on the photocatalytic activity of ZnO-graphene nanocomposite with improved activity. J Environ Chem Eng 6:5087–5100. https://doi.org/10.1016/j.jece.2018.07.049

Bhande SS, Ambade RB, Shinde DV et al (2015) Improved photoelectrochemical cell performance of tin oxide with functionalized multiwalled carbon nanotubes-cadmium selenide sensitizer. ACS Appl Mater Interfaces 7:25094–25104. https://doi.org/10.1021/acsami.5b05385

Bhattacharjee A, Ahmaruzzaman M, Th BD, Nath J (2016) Photodegradation of methyl violet 6B and methylene blue using tin-oxide nanoparticles (synthesized via a green route). J Photochem Photobiol A 325:116–124. https://doi.org/10.1016/j.jphotochem.2016.03.032

Bordbar M, Solmaz FP, Ali YF, Bahar K (2016) Effect of morphology on the photocatalytic behavior of ZnO nanostructures: low temperature sonochemical synthesis of Ni doped ZnO nanoparticles. J Nanostruct 6:190–198. https://doi.org/10.7508/JNS.2016.03.003

Byrappa K, Yoshimura M (2013) Handbook of hydrothermal technology, 2nd edn. William Andrew Publishing, Oxford, pp 615–762

Cen W, Ting X, Chiyao T, Shandong Y, Fan D (2014) Effects of morphology and crystallinity on the photocatalytic activity of $(BiO)_2$ CO_3 nano/microstructures. Ind Eng Chem Res 53:15002–15011. https://doi.org/10.1021/ie502670n

Deng X, Chenggang W, Hongcen Y, Minghui S, Shouwei Z, Xiao W, Meng D, Jinzhao H, Xijin X (2017) One-pot hydrothermal synthesis of CdS decorated CuS microflower-like structures for enhanced photocatalytic properties. Sci Rep 7:32–36. https://doi.org/10.1038/s41598-017-04270-y

Dhiman P, Naushad M, Batoo KM et al (2017) Nano FeXZn1-XO as a tuneable and efficient photocatalyst for solar powered degradation of bisphenol A from aqueous environment. J Clean Prod 165:1542–1556. https://doi.org/10.1016/j.jclepro.2017.07.245

Elia DD, Christian B, Jean FH, Arnaud R, Marie HB, Nicolas K, Valérie KS, Yoshikazu S, Mourad V, Jean CB, Patrick A (2011) Impact of three different TiO_2 morphologies on hydrogen evolution by methanol assisted water splitting: nanoparticles, nanotubes and aerogels. Int J Hydrog Energy 36:14360–14373. https://doi.org/10.1016/j.ijhydene.2011.08.007

Faisal M, Ahmed AI, Houcine B, Al-sayari SA, Al-assiri MS, Ismail AA (2014) Hydrothermal synthesis of Sr-doped alpha-Bi_2O_3 nanosheets as highly efficient photocatalysts under visible light. J Mol Cat A: Chem 387:69–75. https://doi.org/10.1016/j.molcata.2014.02.018

Flores NM, Umapada P, Reina G, Alberto S (2014) Effects of morphology, surface area, and defect content on the photocatalytic dye degradation performance of ZnO nanostructures. RSC Adv 4:41099–41110. https://doi.org/10.1039/c4ra04522j

Gnanasekaran L, Hemamalini R, Ravichandran K (2015) Synthesis and characterization of TiO_2 quantum dots for photocatalytic application. J Saudi Chem Soc 19:589–594. https://doi.org/10.1016/j.jscs.2015.05.002

Gnanasekaran L, Hemamalini R, Saravanan R, Ravichandran K, Gracia F, Gupta VK (2016) Intermediate state created by dopant ions (Mn, Co and Zr) into TiO_2 nanoparticles for degradation of dyes under visible light. J Mol Liq 223:652–659. https://doi.org/10.1016/j.molliq.2016.08.105

Hamid A, Luc B, Pascal R, Ahmed A, Hatem E, Rabah B, Yannick C (2016) Hydrothermal preparation of $MoS_2/TiO_2/Si$ nanowires composite with enhanced photocatalytic performance under visible light. J Mater Design 109:634–643. https://doi.org/10.1016/j.matdes.2016.07.098

Han X, Hui-zhong H, Qin K, Xi Z, Xian-hua Z, Tao X, Zhao-Xiong X, Lan-Sun Z (2009) Controlling morphologies and tuning the related properties of nano/microstructured ZnO crystallites. J Phys Chem C 113:584–589. https://doi.org/10.1021/jp808233e

Hao JY, Ying YW, Xi LT, Guo QJ, Xiang YG (2013) SiC nanomaterials with different morphologies for photocatalytic hydrogen production under visible light irradiation. Catal Today 212:220–224. https://doi.org/10.1016/j.cattod.2012.09.023

Hassan SSM, Azab WIME, Ali HR, Mansour MSM (2015) Green synthesis and characterization of ZnO nanoparticles for photocatalytic degradation of anthracene. Adv Nat Sci:Nanosci Nanotech 6:045012. https://doi.org/10.1088/2043-6262/6/4/045012

Jin C, Kexin Z, George P, Zengyun J, Gang X, Yongxing W, Chenghai G, Jinhua L (2018) Morphology dependent photocatalytic properties of ZnO nanostructures prepared by a carbon-sphere template method. J Nanosci Nanotechnol 18:5234–5241. https://doi.org/10.1166/jnn.2018.15471

Kajbafvala A, Hamed G, Asieh P, Joshua PS, Ehsan K, Sadrnezhaad SK (2012) Effects of morphology on photocatalytic performance of Zinc oxide nanostructures synthesized by rapid microwave irradiation methods. Superlattice Microst 51:512–522. https://doi.org/10.1016/j.spmi.2012.01.015

Kim D, Young-duk H (2011) Morphology-dependent photocatalytic activities of hierarchical microstructures of ZnO. Mater Lett 65:2100–2103. https://doi.org/10.1016/j.matlet.2011.04.074

Kislov N, Jayeeta L, Himanshu V, Yogi GD, Elias S, Matthias B (2009) Photocatalytic degradation of methyl orange over single crystalline ZnO: orientation dependence of photoactivity and photostability of ZnO. Langmuir 25:3310–3315. https://doi.org/10.1021/la803845fCCC

Kumar JS, Thangadurai P (2018) Enhanced visible-light-driven photodegradation of Rh-6G by surface engineered Pd-V_2O_5 heterostructure nanorods. J Environ Chem Eng 6:5320–5331. https://doi.org/10.1016/j.jece.2018.08.028

Kumar S, Sunita K, Meganathan T, Ashok KG (2014) Achieving enhanced visible-light-driven photocatalysis using type- II $NaNbO_3/CdS$ core/shell heterostructures. Appl Mater Interfaces 6:13221–13233. https://doi.org/10.1021/am503055n

Kumar A, Naushad M, Rana A et al (2017) ZnSe-WO$_3$ nano-hetero-assembly stacked on Gum ghatti for photo-degradative removal of Bisphenol: symbiose of adsorption and photocatalysis. Int J Biol Macromol 104:1172–1184. https://doi.org/10.1016/j.ijbiomac.2017.06.116

Kumar S, Vishwanathan S, Thirumurugan A, Thangadurai P (2018) Enhanced photocatalytic activity of V$_2$O$_5$ nanorods for the photodegradation of organic dyes: a detailed understanding of the mechanism and their antibacterial activity. Mater Sci Semicond Process 85:122–133. https://doi.org/10.1016/j.mssp.2018.06.006

Li J, Zhan-yun G, De-fang W, Hui L, Juan D, Zhen-feng Z (2014) Effects of pH value on the surface morphology of BiVO$_4$ microspheres and removal of methylene blue under visible light. J Exp Nanosci 9:616–624. https://doi.org/10.1080/17458080.2012.680931

Liang Y, Guo N, Linlin L, Ruiqing L, Guijuan J, Shucai G (2015) Fabrication of porous 3D flower-like Ag/ZnO heterostructure composites with enhanced photocatalytic performance. Appl Surf Sci 332:32–39. https://doi.org/10.1016/j.apsusc.2015.01.116

Liu T, Qi W, Peng J (2013) Morphology-dependent photo-catalysis of bare zinc oxide nanocrystals. RSC Adv 3:12662–12670. https://doi.org/10.1039/c3ra41399c

Liu X, Kaikai JS, Jiarui H (2015) Facile synthesis of porous Fe$_2$O$_3$ nanorods and their photocatalytic properties. J Saudi Chem Soc 19:479–484. https://doi.org/10.1016/j.jscs.2015.06.009

Liu H, Chunjie H, Haifa Z, Jien Y, Xuguang L (2017) Fabrication of In$_2$O$_3$/ZnO@Ag nanowire ternary composites with enhanced visible light photocatalytic activity. RSC Adv 7:37220–37229. https://doi.org/10.1039/C7RA04929C

Lu BF, Weiping C, Yugang Z (2008) ZnO hierarchical micro/nanoarchitectures: solvothermal synthesis and structurally enhanced photocatalytic performance. Adv Funct Mater 18:1047–1056. https://doi.org/10.1002/adfm.200700973

McLaren A, Valdes-Solis T, Li G, Shik CT (2009) Shape and size effects of ZnO nanocrystals on photocatalytic activity. J Am Chem Soc 131:12540–12541. https://doi.org/10.1021/ja9052703

Meng X, Zizhen L, Jie C, Hongwei X, Zisheng Z (2018) Enhanced visible light-induced photocatalytic activity of surface-modified BiOBr with Pd nanoparticles. Appl Surf Sci 433:76–87. https://doi.org/10.1016/j.apsusc.2017.09.103

Nasab AS, Behpour M (2016) Synthesis, characterization, and morphological control of Eu$_2$Ti$_2$O$_7$ nanoparticles through green method and its photocatalyst application. J Mater Sci Mater Electron 27:11946–11951. https://doi.org/10.1007/s10854-016-5341-4

Neena D, Kiran KK, Han B, Dingze L, Pravin K, Dwivedi RK, Vasiliy OP, Xing ZZ, Wei G, Dejun F (2018) Enhanced visible light photodegradation activity of RhB/MB from aqueous solution using nanosized novel Fe-Cd co-modified ZnO. Sci Rep 8:10691. https://doi.org/10.1038/s41598-018-29025-1

Pan LU, Li L, Dong T, Chen L, Juan W (2014) Synthesis of Co$_3$O$_4$ nanomaterials with different morphologies and their photocatalytic performances. J Miner Met Mater Soc 66:1035–1042. https://doi.org/10.1007/s11837-014-0983-2

Peter JI, Praveen E, Vignesh G, Nithiananthi P (2017) ZnO nanostructures with different morphology for enhanced photocatalytic activity. Mater Res Express 4:124003. https://doi.org/10.1088/2053-1591/aa9d5d

Qin J, Yang C, Cao M, Zhang X, Saravanan R, Limpanart S, Mab M, Liu R (2017) Two-dimensional porous sheet-like carbon-doped ZnO/g-C$_3$N$_4$ nanocomposite with high visible-light photocatalytic performance. Mater Lett 189:156–159. https://doi.org/10.1016/j.matlet.2016.12.007

Rajeshwari S, Santhosh KJ, Rajendrakumar RT, Ponpandian N, Thangadurai P (2018) Influence of Sn ion doping on the photocatalytic performance of V$_2$O$_5$ nanorods prepared by hydrothermal method. Mater Res Express 5:025507. https://doi.org/10.1088/2053-1591

Rakkesh AR, Durgalakshmi D, Balakumar S (2015) Nanostructuring of a GNS-V$_2$O$_5$–TiO$_2$ core–shell photocatalyst for water remediation applications under sun-light irradiation. RSC Adv 5:18633–18641. https://doi.org/10.1039/c5ra00180c

Ren J, Wenzhong W, Meng S, Songmei S, Erping G (2011) Heterostructured bismuth molybdate composite: preparation and improved photocatalytic activity under visible-light irradiation. Appl Mater Interfaces 3:2529–2533. https://doi.org/10.1021/am200393h

Romero-Sáez M, Jaramillo LY, Saravanan R, Benito N, Pabón E, Mosquera E, Gracia F (2017) Notable photocatalytic activity of TiO_2-polyethylene nanocomposites for visible light degradation of organic pollutants. Express Polym Lett 11(11):899–909. https://doi.org/10.3144/expresspolymlett.2017.86

Roy A, Mukul P, Chaiti R, Ramkrishna S, Soumen D, Tarasankar P (2014) Facile synthesis of pyridine intercalated ultra-long applications in selective dye degradation. Cryst Eng Comm 16:7738–7744. https://doi.org/10.1039/c4ce00708e

Saravanan R, Manoj D, Qin J, Naushad M, Gracia F, Lee AF, MansoobKhan MM, Gracia-Pinilla MA (2018) Mechanothermal synthesis of Ag/TiO_2 for photocatalytic methyl orange degradation and hydrogen production. Process Saf Environ Prot 120:339–347. https://doi.org/10.1016/j.psep.2018.09.015

Selvamani T, Abdullah MA, Abdulrahman OA, Sambandam A (2013) Emergent synthesis of bismuth subcarbonate nanomaterials with various morphologies towards photocatalytic activities – an overview. Mater Sci Forum 764:169–193. https://doi.org/10.4028/www.scientific.net/MSF.764.169

Shahid M, Imran S, Seok-jo Y, Dae J (2010) Facile synthesis of core–shell SnO_2/V_2O_5 nanowires and their efficient photocatalytic property. Mater Chem Phys 124:619–622. https://doi.org/10.1016/j.matchemphys.2010.07.023

Sharma G, Naushad M, Kumar A et al (2015) Lanthanum/Cadmium/Polyaniline bimetallic nanocomposite for the photodegradation of organic pollutant. Iran Polym J (English Ed) 24:1003–1013. https://doi.org/10.1007/s13726-015-0388-2

Shen F, Wenxiu Q, Yucheng H, Yuan Y, Xingtian Y, Gangfeng W (2012) Enhanced photocatalytic activity of ZnO microspheres via hybridization with $CuInSe_2$ and $CuInS_2$ nanocrystals. Appl Mater Interfaces 4:4087–4092. https://doi.org/10.1021/am3008533

Silva DF, Osmando FL, Ariadne CC, Avansi J, Bernardi IB, Siu LM, Caue R, Elson L (2016) Hierarchical growth of ZnO nanorods over SnO_2 seed layer: insights into electronic properties from photocatalytic activity. RSC Adv 6:2112–2118. https://doi.org/10.1039/C5RA23824B

Soni U, Puspanjali T, Sameer S (2014) Photocatalysis from fluorescence-quenched CdSe/Au nanoheterostructures: a size-dependent study. J Phys Chem Lett 5:1909–1916. https://doi.org/10.1021/jz5006863

Sun J, Xinyong L, Qidong Z, Jun K, Dongke Z (2014) Novel $V_2O_5/BiVO_4/TiO_2$ nanocomposites with high visible-light-induced photocatalytic activity for the degradation of toluene. J Phys Chem C 118:10113–10121. https://doi.org/10.1021/jp5013076

Tang J, Gang C, Huamin Z, Hao Y, Zhong L, Rong C (2012) Shape-dependent photocatalytic activities of bismuth subcarbonate nanostructures. J Nanosci Nanotechnol 12:4028–4034. https://doi.org/10.1166/jnn.2012.6168

Tian H, Ming L, Weitao Z (2017) Constructing 2D graphitic carbon nitride nanosheets/layered MoS_2/graphene ternary nanojunction with enhanced photocatalytic activity. Appl Catal B 225:468–476. https://doi.org/10.1016/j.apcatb.2017.12.019

Ullah H, Ullah Z, Fazal A, Irfan M (2017) Use of vegetable waste extracts for controlling microstructure of CuO nanoparticles: green synthesis, characterization, and photocatalytic applications. J Chem 2017:1–5. https://doi.org/10.1155/2017/2721798

Wang Y, Su YR, Qiao L, Liu LX, Su Q, Zhu CQ, Liu XQ (2011) Synthesis of one-dimensional TiO_2/V_2O_5 branched heterostructures and their visible light photocatalytic activity towards Rhodamine B. Nanotechnology 22:225702. https://doi.org/10.1088/0957-4484/22/22/225702

Wang H, Lisha Z, Zhigang C, Junqing H, Shijie L (2014) Semiconductor heterojunction photocatalysts: design, construction, and photocatalytic performance. Chem Soc Rev 43:5234–5244. https://doi.org/10.1039/c4cs00126e

Wang Y, Yuxiang Z, Xiaoyu Z, Xu Y, Xianchao L, Zhi C, Libin Y, Liang Z (2015) Surface & coatings technology improving photocatalytic Rhodamine B degrading activity with Pt quantum dots on TiO_2 nanotube arrays. Surf Coat Technol 281:89–97. https://doi.org/10.1016/j.surfcoat.2015.09.048

Wang Q, Haiyan J, Shuting D, Hyeon MN, Byung KM (2016) Butterfly-like $BiVO_4$: synthesis and visible light photocatalytic activity. Synth React Inorg Met Org Nano Met Chem 46:483–488. https://doi.org/10.1080/15533174.2014.988801

Wang Y, Jaka S, Fengxiang W, Bo Z, Xuewen L, Guihua C (2017a) Electrospinning and hydrothermal synthesis of recyclable MoS_2/CNFs hybrid with enhanced visible-light photocatalytic performance. Ceram Int 43:11028–11033. https://doi.org/10.1016/j.ceramint.2017.05.145

Wang C, Yindi G, Lei W, Ping L (2017b) Morphology regulation, structural and photocatalytic properties of ZnO hierarchical microstructures synthesized by a simple hydrothermal method. Phys Status Solidi A 1600876:1–8. https://doi.org/10.1002/pssa.201600876

Xie W, Rui L, Qingyu X (2018) Enhanced photocatalytic activity of Se-doped TiO_2 under visible light irradiation. Sci Rep 8:8752. https://doi.org/10.1038/s41598-018-27135-4

Xu L, Yan LH, Candice P, Chun HC, Lei J, Hui H, Shanthakumar S, Mark A, Raymond J, Steven LS (2009) ZnO with different morphologies synthesized by solvothermal methods for enhanced photocatalytic activity. Chem Mater 21:2875–2885. https://doi.org/10.1021/cm900608d

Xu Y, Jingjie J, Xianliang L, Yide H, Hao M, Tianyu W, Xia Z (2015) Simple synthesis of ZnO nanoflowers and its photocatalytic performances toward the photodegradation of metamitron. Mater Res Bull 76:235–239. https://doi.org/10.1016/j.materresbull.2015.11.062

Yin H, Ke Y, Changqing S, Rong H, Ziqiang Z (2014) Synthesis of Au-decorated V_2O_5@ZnO heteronanostructures and enhanced Plasmonic photocatalytic activity. Appl Mater Interfaces 6:14851–14860. https://doi.org/10.1021/am501549n

Yin X, Xiaoning L, Wen G, Wei Z, Huan L, Liuyang Z, Zhengping F, Yalin L (2018) Morphology effect on photocatalytic activity in $Bi_3Fe_{0.5}Nb_{1.5}O_9$. Nanotechnology 29:265706. https://doi.org/10.1088/1361-6528/aabdba

Zhang H, Hui Z, Pengfei Z, Fei H (2017a) Morphological effect in photocatalytic degradation of direct blue over mesoporous TiO_2 catalysts. Chem Selec 2:3282–3288. https://doi.org/10.1002/slct.201601346

Zhang X, Yuxin W, Fulin H, Hongxin L, Yang Y, Xinxin Z, Yiqiong Y, Yin W (2017b) Effects of Ag loading on structural and photocatalytic properties of flower-like ZnO microspheres. Appl Surf Sci 391:476–483. https://doi.org/10.1016/j.apsusc.2016.06.109

Zheng Y, Cui L, Lei L, Zhixian Y, Suimin R, Yaxuan C, Juan X (2016) Synthesis of hierarchical TiO_2/SnO_2 photocatalysts with different morphologies and their application for photocatalytic reduction of Cr(VI). Mater Lett 181:169–172. https://doi.org/10.1016/j.matlet.2016.06.031

Zhou W, Hong L, Jiyang W, Duo L, Guojun D, Jingjie C (2010) Ag_2O/TiO_2 nanobelts heterostructure with enhanced photocatalytic activity. Appl Mater Interfaces 2:2385–2392. https://doi.org/10.1021/am100394x

Chapter 4
Metal and Non-metal Doped Metal Oxides and Sulfides

Poonam Yadav, Pravin K. Dwivedi, Surendar Tonda, Rabah Boukherroub, and Manjusha V. Shelke

Contents

P. Yadav · P. K. Dwivedi · M. V. Shelke (✉)
Physical & Material's Chemistry Division, CSIR-National Chemical Laboratory, Pune, India
e-mail: mv.shelke@ncl.res.in

S. Tonda
Department of Environmental Engineering, Kyungpook National University, Daegu,
South Korea

R. Boukherroub
University Lille, CNRS, Centrale Lille, ISEN, University Valenciennes, UMR 8520–IEMN,
Lille, France

© Springer Nature Switzerland AG 2020
M. Naushad et al. (eds.), *Green Photocatalysts*, Environmental Chemistry
for a Sustainable World 34, https://doi.org/10.1007/978-3-030-15608-4_4

Abstract Increasing demand of clean energy, growth in global population, and tremendous global warming necessitate sustainable energy production. Photochemistry or photocatalysis using semiconductor is one direct way of converting solar light into fuel without CO_2 emission. Use of cheap and renewable energy sources like sunlight and water is an efficient way toward green energy generation. Semiconductors like metal oxides and metal sulfides are materials of interest for photocatalysis because of their good optical and electronic properties. However, most of the semiconductors do not have suitable band alignment to absorb sunlight efficiently. In this line, use of metal oxides and sulfides can be an advantageous strategy for desirable band tuning for visible light photocatalysis. In this chapter, we present highlights on metal- and nonmetal-doped metal oxides and sulfides for photocatalysis.

Photocatalysis is an important field for hydrogen production, environment remediation, and electronic manufacturing. Since the introduction of TiO_2 as photocatalyst in the late 1960s by Honda and Fujishima, many other semiconductors gained attention due to their good optical, electronic, and photocatalytic properties. To absorb whole solar spectrum, proper VB and CB alignment with respect to water oxidation and reduction potential, respectively, is required for photocatalysis. All semiconductors do not have suitable band gap and band alignment to exhibit efficient or visible light photocatalysis. Metal and nonmetal doping of semiconductors like metal oxides and sulfides can be an effective strategy to push the band gap in the optimum region to be suitable for overall photocatalysis. The present chapter is an attempt to provide a detailed description of various aspects for the development of suitable photocatalyst, i.e., preferred band gap materials, metal doping, nonmetal doping, and challenges associated with them like broad absorption capability and low recombination probability.

Keywords Visible light · Photocatalysis · Metal · Nonmetal · Oxides · Sulfides · Doping · Green photocatalysis · Nanostructure

4.1 Introduction

Solar spectrum contains 45% visible light, 50% infrared (IR) light, and 5% ultraviolet (UV) light. To use most of the solar spectrum for photocatalysis, visible and IR regions are of interest (Zou et al. 2001). But in case of IR region, due to small band gap, recombination of charges leads to inefficient photocatalysis. So, visible light photocatalysis is of more importance and interest. Photocatalysis includes mainly three research areas, i.e., water splitting, dye degradation, and CO_2 reduction (Fig. 4.1). Semiconductor materials are the preferred choice as photocatalyst. Semiconductors are characterized by two separated energy bands: a filled low-energy valence band (VB) and an empty high-energy conduction band (CB). A forbidden region of energy, called the band gap (E_g), is located between the conduction and valence bands. Incident light on the semiconductor promotes excitation of electrons from the

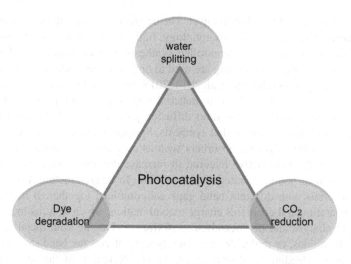

Fig. 4.1 Applications of photocatalysis

VB to the CB. The empty level in the valence band is called a hole. Electrons and holes are called charge carriers. A photocatalyst is capable of light absorption, producing electron–hole pairs that enables chemical transformations of species involved after every cycle of such interactions. Photocatalysis is categorized into two types: homogenous and heterogeneous photocatalysis (Kim et al. 2004). For heterogeneous photocatalysis, mostly semiconductor materials are used due to their merits like electronic features, light absorption properties, good charge transport, large charge lifetime, and reusability. Upon light incidence, electrons are excited from the VB of a semiconductor to CB provided that the energy of the irradiated photon exceeds the energy of the band gap (E_g) of the semiconductor (Hoffmann et al. 1995). After photoexcitation, the charge carriers are separated and travel to the particles surface, and then charges may undergo redox catalytic reactions. The photoexcited negative charge carriers are strong reducing agents and possibly employed to generate, e.g., hydrogen from protons, and the positive charge carriers in the VB are strong oxidizing agents, which can be employed to oxidize surface-adsorbed molecules, e.g., water to generate oxygen.

The photocatalytic activity of metal oxides originates from two sources: (i) generation of OH radicals by oxidation of OH^- and (ii) generation of $O_2^{\cdot-}$ by reduction of O_2. For an energetically favorable water splitting reaction, the CB of the semiconductor should be more negative than the reduction potential of H^+/H_2, and the VB maximum should be more positive than the oxidation potential of O_2/H_2O. This means that the E_g of the photocatalyst should be more than 1.23 eV with suitable band positions.

The constituting steps for a photocatalytic reaction mechanism are light absorption, charge carrier separation and migration to the surface of the photocatalyst material, and the further surface redox reactions with adsorbed species at the active sites of the photocatalyst material. These steps are regulated by bulk, surface, and electronic structure of the photocatalyst. The desirable features of photocatalysts are the suitable band gap, suitable morphology, high surface area, stability, and reusability. Various strategies like controlled synthesis, surface modification/sensitization, heterostructuring, and doping are used to optimize semiconductor materials for visible light absorption (Kumar et al. 2014, 2018; Saravanan et al. 2013, 2014, 2016, 2018). Controlled synthesis leads to material with desired size, phase, and crystallinity. Crystalline material with short diffusion path length and abundant reaction sites can be obtained by controlled synthesis. Surface modification includes attachment of molecular/polymeric absorbers such as co-catalyst or carbon materials on the surface of the photoactive material to increase the absorption. This limits the recombination of photogenerated charge carriers. In heterostructure, two semiconductor materials with different band gaps are combined together. It increases the charge separation and decreases charge recombination, which in turns increases the lifetime of charges and photocatalytic performance. Doping with metallic cations, nonmetallic anions, or molecules also significantly affects the band structure of the semiconductor, leading to absorption in the visible region of the solar spectrum (Gnanasekaran et al. 2016). In this chapter, we have focused on the metal and nonmetal doping of semiconductors like metal oxides and metal sulfides for band tuning into the visible region of the spectrum and better photocatalytic performance.

4.1.1 Types of Semiconductor Heterojunction

Depending upon the relative band alignment of VB and CB, semiconductor–semiconductor interface is classified into three types: type I (straddling), type II (staggered), and type III (broken), which is shown in Fig. 4.2. In type I semiconductor heterojunction, CB of A is higher and VB of B is lower than semiconductor B (Fig. 4.2a). Under light irradiation, electrons and holes get accumulated at semiconductor B, which leads to charge recombination. In type I heterojunction, there is no effective charge transfer due to unavoidable charge recombination. In type II heterojunction, CB and VB of A are higher than CB and VB of semiconductor B. Upon light irradiation, electrons travel from CB of A to CB of B and holes travel from VB of B to VB of A. Photogenerated charges travel in different direction, so there is less chances of recombination. Type II heterojunction is preferred for photocatalysis. In type III heterojunction, band gaps are separated largely, so bands do not overlap and hence no charge separation and transfer take place.

4.2 Principle of Photocatalysis

Photocatalysis defines the relationship between photochemistry and catalysis. The light and the photocatalyst are essential to a chemical reaction. It includes irradiation of a semiconductor for inducing a reaction at the solid–solution interface in

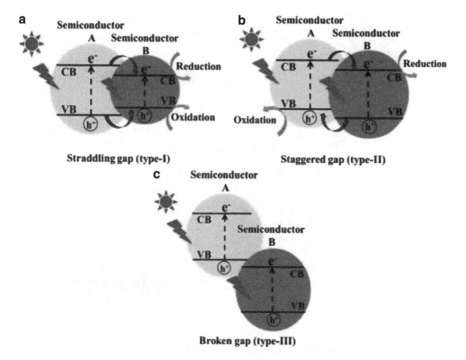

Fig. 4.2 Types of semiconductor heterojunction. (This figure was obtained from https://doi.org/10.1002/adma.201601694)

heterogenous catalysis. When a semiconductor comes into contact with a liquid electrolyte, charge transfer takes place through the interface to stabilize the potentials/energy of the two phases involved (Fig. 4.3). Photocatalysis takes place in the following four steps:

   I. Light absorption by photocatalyst to generate electron–hole pairs
  II. Separation of generated charge carriers (excited charges)
 III. Transfer of electrons and holes to the surface of the photocatalyst
  IV. Consumption of charges on the surface for redox reactions

  Lifetime of photogenerated charge carriers should be large enough to reach the catalyst surface before recombination (Linsebigler et al. 1995).

4.3 Effect of Doping

Semiconductor materials are attractive candidates for various applications due to size quantization effect (Gao 2011), tunable band gap, and capture of photogenerated charge carriers. Doping is one of the promising ways to modulate optical and electronic properties of the semiconductor, like desirable band gap and current

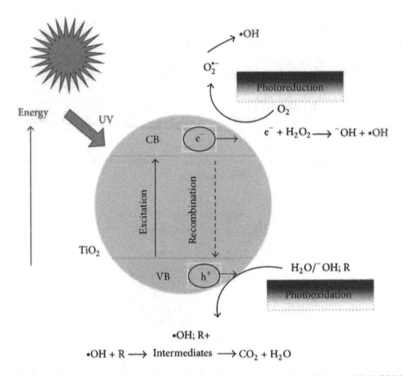

Fig. 4.3 Principle of photocatalysis. (This figure was obtained from https://doi.org/10.1155/2013/940345)

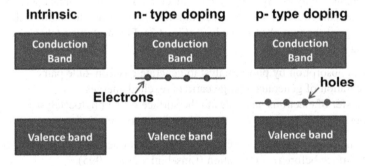

Fig. 4.4 Effect of doping

densities. Doping affects the charge carrier concentration, band gap, crystal structure, etc. Doping allows energy states inside the band gap of the semiconductor near energy band of the dopant type. Electron donor dopant creates states near the CB, while electron acceptor dopant creates states near the VB (Fig. 4.4). Energy difference between introduced states is the new band gap, which is smaller than the original band gap.

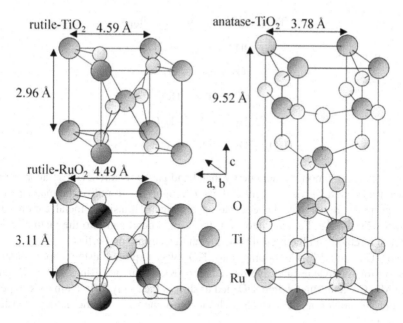

Fig. 4.5 Crystal phases of TiO$_2$. (This figure was obtained from https://doi.org/10.1149/2. 040201jes)

4.3.1 Metal-Doped Metal Oxides

Metal oxides play a substantial role in different kinds of applications due to their compositional simplicity and stoichiometric diversity. Photoinduced water splitting was first studied using TiO$_2$ (Fujishima and Honda 1972), and it was commercialized due to low cost, no adverse effect, less chemical reactivity, and high photocatalytic performance. Many other metal oxides like ZnO (Ong et al. 2018), SnO$_2$ (Al-Hamdi et al. 2017), MoO$_3$ (Liu et al. 2017), WO$_3$ (Dong et al. 2017), ZrO$_2$ (Polisetti et al. 2011), Fe$_2$O$_3$ (Mishra and Doo-Man 2015), SrTiO$_3$ (Zou et al. 2012), ZnTiO$_3$ (Li et al. 2011), FeVO$_4$ (Feng et al. 2018), CeO$_2$ (Montini et al. 2016), etc., emerged as photocatalysts due to their suitability for photocatalysis. Different metal-doped metal oxides are discussed in the following section in detail.

4.3.2 Transition Metal-Doped TiO$_2$

TiO$_2$ is the most widely explored metal oxide for photocatalysis. It exists in three crystalline phases: anatase, rutile, and brookite (Fig. 4.5). Anatase phase is favored in the photocatalysis field. The TiO$_2$ exhibits band gap of 3.2 eV, which corresponds to 380 nm. The mechanism of photocatalysis in TiO$_2$ is represented by the following equations:

$$TiO_2 + h\nu \rightarrow TiO_2\left(e_{CB}^- + h_{VB}^+\right)$$

$$OH^- + h_{VB}^+ \rightarrow \dot{O}H$$

$$O_2 + e_{CB}^- \rightarrow \dot{O}_2^-$$

$$\dot{O}_2^- + H^+ \rightarrow \dot{H}O_2$$

$$2\,\dot{H}O_2 \rightarrow O_2 + H_2O_2$$

$$H_2O_2 + \dot{O}_2^- \rightarrow OH^- + \dot{O}H + O_2$$

For photoexcitation, higher energy than band gap is required. But solar spectrum contains almost 45% visible light and 5% UV light. So, to utilize the major part of the spectrum, researchers have explored doping of transition metal cations like Mn, V, Cu, Ni, Ag, Fe, etc., into TiO_2 to push its band gap into the visible region because of change in the crystal structure and electronic properties.

Mn, Co, and Mn–Co doping into TiO_2 was explored using coprecipitation method for methylene blue dye degradation (Kiriakidis and Binas 2014). Doping with Mn, Co, and Mn–Co into TiO_2 led to shift in the X-ray diffraction (XRD) peak position to lower angle, which indicates the increase in the lattice constant (Fig. 4.6a).

Lattice constant increased from 35.5 nm in undoped TiO_2 to 40.1 nm in doped TiO_2. Broad absorption was observed from 400 to 800 nm in UV–Vis spectra (Fig. 4.6b). The band gap value was also decreased from 3.1 eV to 2.8 eV with sub-bands around 1.6 eV after doping (Fig. 4.6c). At pH 10, photocatalytic efficiency increased to 70% in doped TiO_2 cases (Fig. 4.6d).

It has been proposed that the dopants play the role of electron hunter to increase the electron–hole separation or lifetime of charges. This makes charge carriers available for redox reactions. Ag doping into TiO_2 has been extensively explored. Yu et al. (2009) reported TiO_2 doped with Ag showing enhanced photocatalytic activity (Fig. 4.6).

Possibility of electron–hole pair recombination is found to be reduced and charge transfer to oxidant was increased. They synthesized visible light active photocatalyst $Ag/AgCl/TiO_2$ by deposition of $AgCl$ particles on TiO_2 surface and further reduction of some of the Ag^+ ions to Ag species. A new plasmonic photocatalytic mechanism is explained on the basis that Ag nanoparticles are photoexcited due to plasmon resonance and charges are separated by the migration of photoexcited electrons from the Ag nanoparticles to the TiO_2 CB and the migration of electrons from a chloride donor to the Ag nanoparticles.

Fe^{3+} is a suitable dopant into TiO_2 as its radius is close to Ti^{4+}. It is reported that Fe^{3+} acts as a shallow charge trapper in the TiO_2 lattice and increases photocatalytic performance by promoting interfacial charge transfer process (Wang et al. 2012). Fe^{3+} doping leads to induction of more oxygen deficiencies in the crystal lattice and on the TiO_2 surface, which leads to increase in the absorption of water (Neren Ökte and Akalın 2010). Fan et al. in 2014 reported $Ag–Fe–TiO_2$ and $Fe–TiO_2$ for photocatalytic water splitting. $Ag–Fe–TiO_2$ composite is less crystalline than pure TiO₂. In UV–Vis spectrum, absorption edge shifted to red region in doped cases, and photo-absorption is enhanced significantly (Fig. 4.7).

Solar energy conversion efficiency can be improved by extending the band gap to much longer wavelength because of SPR (surface plasmon resonance) effect of the Ag nanoparticles, increasing light scattering and photo-generating carriers in the semiconductor by transferring the plasmonic energy from the Ag^0 to the TiO₂ semiconductor. Photoluminescence (PL) emission spectroscopy is used to analyze the lifetime of electron–hole pairs in photocatalyst, and it provides details about

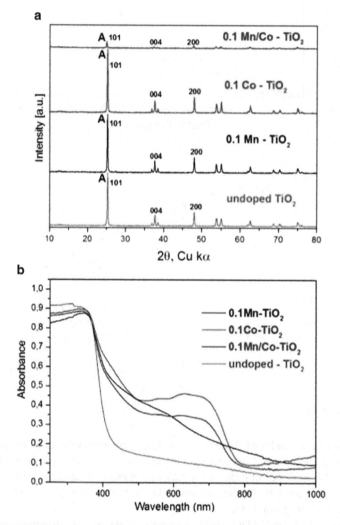

Fig. 4.6 (**a**) XRD of Mn-, Co-, and Mn–Co-doped TiO₂ (**b**) UV–Vis spectra (**c**) Kubelka–Munk plot (**d**) dye degradation at pH 10 under visible light. (This figure was obtained from https://doi.org/10.3938/jkps.65.297)

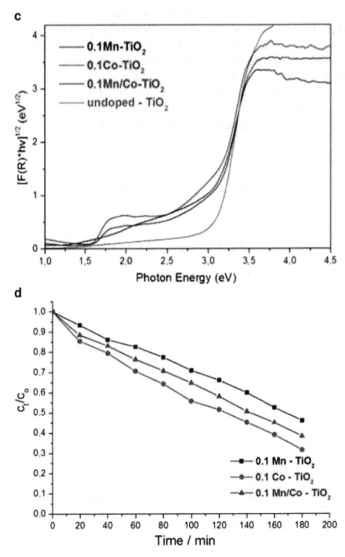

Fig. 4.6 (continued)

separation of photoinduced charge carriers. Recombination of excited electron–hole pairs gives PL emission. Less emission intensity indicates a reduced recombination rate. With the incidence of light on the catalyst, an electron can be excited and travel from the VB to the CB. Because of instability at higher energy state, the electrons will come back to ground state.

PL emission of Ag–Fe/TiO$_2$ shows a decrease in the photoemission intensity after doping, which results in increase in the lifetime of photogenerated charge carriers (Fig. 4.8a). Photocatalytic H$_2$ production result is shown in Fig. 4.8b, indicating the better H_2 production in the case of 0.2 mM Ag –0.3 mM Fe doping into TiO_2. Mechanism of water splitting on Ag–Fe/TiO_2 is shown in Fig. 4.9. Doping with Fe^{3+} leads to the introduction of impurity band beneath the CB of TiO_2. With the incidence of light on the catalyst, an electron is excited from the VB to the Fe impurity band, which generates an electron vacancy (hole) in the VB. The holes in Fe^{4+} can diffuse to the surface and produce hydroxyl radical. Further, Fe^{2+} is

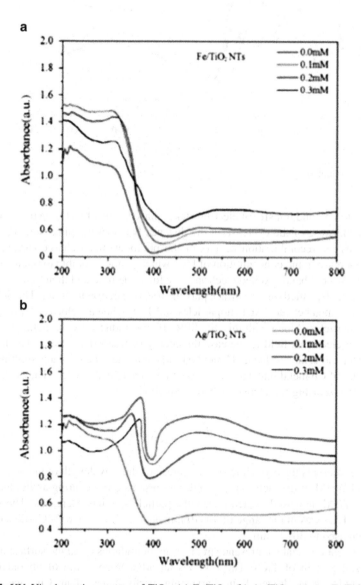

Fig. 4.7 UV–Vis absorption spectra of TiO_2: (a) Fe/TiO_2, (b) Ag/TiO_2, (c) Ag–Fe/TiO_2. (This figure was obtained from https://doi.org/10.1016/j.ceramint.2014.07.119)

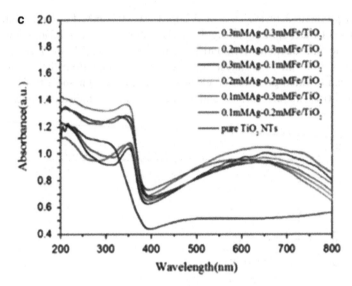

Fig. 4.7 (continued)

oxidized to Fe^{3+} and transfer of electrons takes place to absorbed O_2 on the surface of catalyst. Because of the decrease in the distance between trapping sites, Fe^{3+} may also act as the recombination centers of the photogenerated charge carriers upon further increase in the concentration of Fe^{3+} which is not favorable to photocatalytic reaction. At Fe^{3+} band, excited electrons stay for some time and then transfer. Metal Ag particles deposited on the surface play the role of electron trapping. The Schottky barrier is formed between Ag nanoparticles and TiO_2 when the electrons travel to the Ag nanoparticles. Along with this, the SPR effect is induced by the light, producing high local electronic field to increase the energy of trapped electrons. The h^+ react with water molecules to form H^+ and OH radical at the time of water splitting.

Doping of Cu metal into TiO_2 was studied by Yang et al. (2015). They calculated crystallite size using the Debye–Scherrer equation:

$$d = \frac{k\lambda}{\beta \cos \theta}$$

where d represents the crystallite size, λ represents the wavelength of incident X-ray, β is the FWHM of diffraction peak, and θ represents the scattering angle. In doped cases, FWHM increased, which means the particle size has decreased. The size of undoped TiO_2 was in the range of 40–70 nm, and the size of 1 wt% Cu-doped TiO_2 was decreased to 30–45 nm.

It was reported that Cu presence in the grain boundaries or on the surface inhibits the crystal growth of TiO_2. The UV–Vis absorbance spectrum of undoped TiO_2 nanoparticles have lower light harvesting performance than Cu-doped TiO_2 nanoparticles with the narrow UV absorbance peak (Fig. 4.10a) and evident

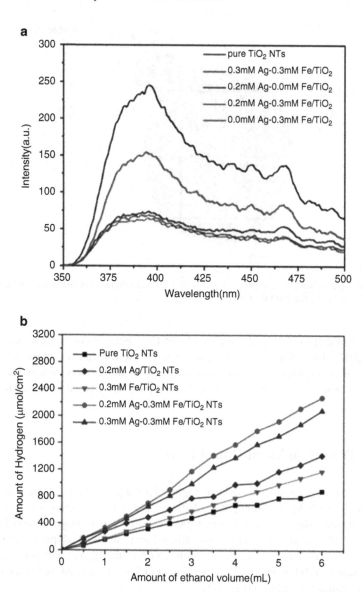

Fig. 4.8 (a) Photoluminescence of Ag–Fe/TiO$_2$. (b) H$_2$ production on catalyst surface. (This figure was obtained from https://doi.org/10.1016/j.ceramint.2014.07.119)

decrement in the light absorbance. The band gap red shifted from 3.28 eV for the undoped TiO$_2$ to 3.02–3.17 eV for the doped TiO$_2$ (Fig. 4.10b). Photodegradation of methyl orange is depicted in Fig. 4.10c. The doped TiO$_2$ show increased photocatalytic performance because of the formation of Schottky barrier in the metal–semiconductor contact region, the charge separation is increased, and photocatalytic performance of TiO$_2$ is increased. The formal charge of Cu^{2+} ion is

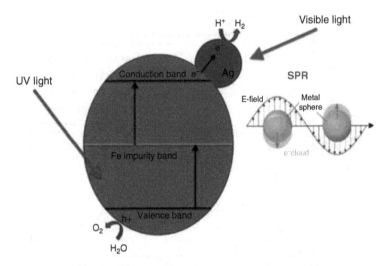

Fig. 4.9 Mechanism of water splitting on Ag–Fe/TiO$_2$. (This figure was obtained from https://doi. org/10.1016/j.ceramint.2014.07.119)

lower than that of Ti^{4+}, and the presence of Cu induces oxygen vacancies (Wang et al. 2012). Dissociation of water takes place at these vacancies on the surface of TiO$_2$ and also captures the holes to restrict the recombination of photogenerated hole–electron pairs, which results in enhanced photocatalytic performance of TiO$_2$.

4.4 Transition Metal-Doped ZnO

ZnO-based photocatalysts have been applied for the degradation of organic pollutants through UV light or sunlight irradiation. ZnO is a very famous semiconductor photocatalyst exhibiting a wide band gap of 3.36 eV with large exciton binding energy of 60 meV at room temperature (Klingshirn et al. 2007). Doping with various transition metals has been explored by researchers to enhance the photocatalytic performance of ZnO semiconductor. Fe-doped ZnO has been used for photocatalysis by Bousslama et al. (2017) (Fig. 4.11).

With the increase of Fe doping into ZnO, the full width at half maximum (FWHM) of the XRD peaks increased. It hints about the decrement in the crystallinity of the composite as well as a decrease in crystallite size. Introduction of lattice disorder and strain induced by interstitial Fe atoms leads to size reduction. As the Fe content increases, lattice strain increases and inhibits the grain growth. Distribution of lattice constants resulting from crystal imperfections leads to lattice strain. It can be calculated by Williamson–Hall method (Urbach 1953):

$$\beta \cos \theta = 0.9 \frac{\lambda}{D} + 4\varepsilon \sin \theta$$

where ε is the strain associated with the nanocrystals. The strain ε can be estimated from the slope of the graph ($\beta \cos \theta$) versus ($4 \sin \theta$) and the average crystallite size from the intersection of the linear fit with the vertical axis. Kubelka–Munk equation was used to calculate the band gap energy:

$$F(R) = \frac{(1 - R)^2}{2R} = \frac{\alpha}{s}$$

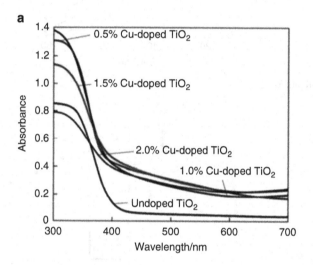

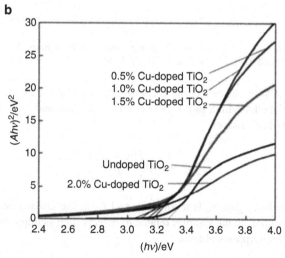

Fig. 4.10 (a) UV–Vis absorption spectra, (b) Kubelka–Munk plot, (c) photodegradation of methyl orange. (This figure was obtained from https://doi.org/10.1016/S1003-6326(15)63631-7)

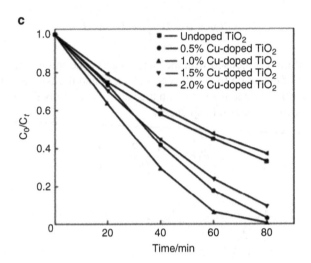

Fig. 4.10 (continued)

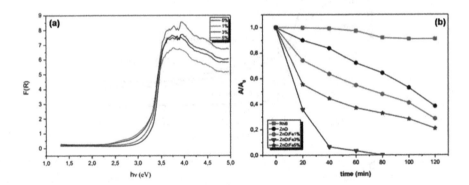

Fig. 4.11 (a) Kubelka–Munk plot. (b) Photodegradation of Rhodamine B. (This figure was obtained from https://doi.org/10.1016/j.ijleo.2017.01.025)

where R, α, and S are the reflection, the absorption, and the scattering coefficients, respectively. Kubelka–Munk plot is between $F(R)$ and hv, and the band gap (E_g) is calculated by Urbach model:

$$\alpha = \alpha_0 \exp\left(\frac{\mathrm{hv} - E_g}{E_u}\right)$$

where Eg is the band gap, α_0 is a constant, and E_U is the Urbach energy, which is related to the disorder in the samples related to defects such as impurities/doping and the presence of amorphous phases.

4.4.1   Transition Metal-Doped WO₃

Due to lack of visible light-responsive photocatalysis in TiO_2, strategies like hybrid-ization of wide band gap and narrow band gap semiconductors and exploration of new visible light responsive photocatalysts were researched. WO_3 is one such example of visible light responsive photocatalyst having an absorbance up to 480 nm with a band gap ranging between 2.5 and 3 .0eV. It was first reported in 1976. It is a low-cost and stable photocatalyst in acidic and oxidative environments. Among various crystal phases, gamma phase is stable at room temperature and photocatalytically active. Even though the band gap of WO_3 is suitable for photocatalysis, the CB edge of WO_3 is lower than the standard H^+/H_2 redox couple, so there is no spontaneous H_2 formation. Single electron reaction is not possible, which leads to the collection of electrons on the surface of WO_3 increasing the chances of recombination. Lower CB and high charge recombination rate decrease the photocatalytic efficiency of WO_3. Various strategies like surface modification, tuning morphology, and structure and doping have been explored to improve its performance. Fe-doped WO_3 was used for photocatalysis by Song et al. (2015).

The photocatalytic activity of WO_3 nanostructures for Rhodamine B (RhB) degradation is enhanced by doping with Fe impurity (Fig. 4.12).The red shift of absorption edge and the trapping effect of the Fe-doped WO_3 nanostructures lead to increase in performance. Fe-doped WO_3 with 10 nm grain size, high crystallinity, and high surface area of 225 m^2/g were obtained. It was observed that the band gap of WO_3 can be tuned by Fe content, and photocatalytic effect was studied by the decomposition of RhB dye.

Figure 4.12a summarizes the synthesis of hollow Fe-doped WO_3.The UV–Vis diffuse reflectance spectra are displayed in Fig. 4.12b.These doped cases showed absorption in UV as well as visible light regions. A redshift and increased in the absorption are observed in all Fe-doped WO_3 nanostructures. The increase of Fe contents leads to a decrease in the band gap. Fe doping possibly creates a donor level over the original VB of WO_3 which decreases the WO_3 band gap. The RhB concentration for samples is changed with different contents of Fe as shown in Fig. 4.12c. Because of the large band gap of 3.10 eV, the pure hollow WO_3 exhibited an inferior photocatalytic performance, compared to Fe-doped WO_3. The photocatalytic performances of the products increased, along with the increasing Fe content, even though further increment in Fe content led to a decrease in the photocatalytic activity. The proposed mechanism of photocatalytic degradation of RhB revealed that the dye degradation is due to OH• radicals which are produced by Fe^{3+} and holes (Fig. 4.12d). When a hole scavenger was added, it decreases the concentration of OH• radicals, and hence dye degradation decreased from 93% to 25% (Fig. 4.13). Further, Mn-, Nb-, Mo-doped WO_3 were used for photocatalysis in various other studies.

4.4.2 Other Transition Metal-Doped Metal Oxides

Other than TiO_2, ZnO, and WO_3, various other metal oxides have been investigated for photocatalysis including SnO_2, Fe_2O_3, $FeVO_4$, $SrTiO_3$, and $ZnTiO_3$.

Doped FeVO4

$FeVO_4$ is a visible light active photocatalyst exhibiting a band gap ranging between 2.5 and 2.0 eV. The low absorption capability and high photogenerated charge recombination rate in $FeVO_4$ decrease photocatalytic activity. Dutta et al. (2017) reported Ti^{4+}-, Mn^{2+}-, and Zn^{2+}-doped $FeVO_4$ nanoparticles. The undoped $FeVO_4$ nanocrystals exhibit a band gap of 2.14 eV. Doping of $FeVO_4$ with Ti, Mn, and Zn ions decreases the band gap to 2.08 eV for Ti^{4+}, 2.03 eV for Zn^{2+}, and 1.98 eV for Mn^{2+} (Fig. 4.14).

The CB edge of $FeVO_4$ is composed of Fe 3d and V 3d orbitals. The VB edge is consisting of O 2p with partial sharing of Fe 3d and V 3d orbitals (Li et al. 2015).

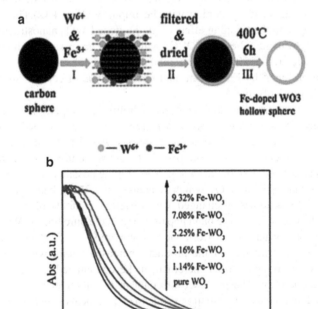

Fig. 4.12 (**a**) Synthesis of hollow Fe-doped WO_3, (**b**) UV–Vis absorption spectra, (**c**) photodegradation of RhB, (**d**) mechanism of dye degradation. (This figure was obtained from https://doi.org/10.1016/j.apcatb.2014.11.020)

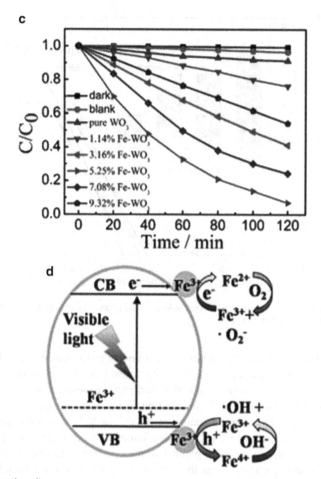

Fig. 4.12 (continued)

Fig. 4.13 The degradation rates of RhB after 2 h using S4 (5.25% Fe-doped WO_3) in the absence and presence of BuOH and EDTA. (This figure was obtained from https://doi.org/10.1016/j.apcatb.2014.11.020)

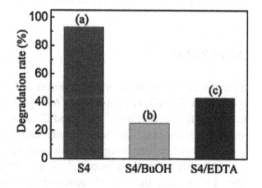

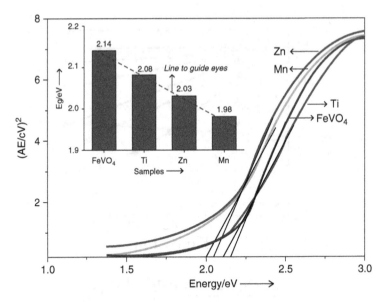

Fig. 4.14 Band gap of undoped and doped FeVO$_4$. (This figure was obtained from https://doi.org/ 10.1016/j.jphotochem.2016.11.022)

Doping of Mn^{2+} cation at the site of Fe^{3+} cation in FeVO$_4$ lowers the CB, which in turn lowers the band gap of the material. Zn^{2+} also has a lower formal charge than Fe^{3+}, like Mn^{2+}; hence, diminution in the band gap was observed in the case of Zn^{2+} ion doping also. In case of Ti^{4+} substitution in Fe^{3+}site, formal charge increases, which is followed by a partial reduction of V^{5+} to V^{3+}, which compensates the change in formal charge, which led to slight diminution in band gap observed in the Ti^{4+} case. The difference in the band gap between the materials can be attributed to the different d orbitals participating in the hybridization of the energy levels. Doped and undoped FeVO$_4$ were used to study photocatalytic degradation of malachite green dye under light irradiation (Fig. 4.15). An enhancement in the photocatalytic activity of FeVO4, Zn-doped FeVO4, and Mn-doped FeVO4 was found under visible light irradiation compared to UV light irradiation. This is because of their smaller band gap.

Doped ZnTiO$_3$

Surendar et al. (2014) reported La-doped and non-doped ZnTiO$_3$ for photocatalysis application. ZnTiO$_3$ is a perovskite (ABO$_3$) structure, which provides the possibility to change the proportion of the A and B cations to form a substituted perovskite. In an ABO$_3$ perovskite, varying the composition or doping with a cation of a different charge or valence can change the basic structure, band gap, and electronic properties, which mainly affect the photocatalytic activity of the material. It was reported that after La doping, the crystal structure remains cubic even after increasing the

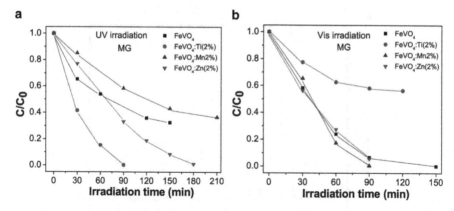

Fig. 4.15 Photocatalytic degradation of malachite green (MG) using undoped and doped FeVO$_4$. (This figure was obtained from https://doi.org/10.1016/j.jphotochem.2016.11.022)

annealing temperature. But in non-doped case, the crystal phase changed from cubic to hexagonal, which indicates that doping blocks the crystal growth and phase transformation (Fig. 4.16).

The photocatalytic degradation of Rhodamine B by non-doped ZnTiO$_3$ and La-doped ZnTiO$_3$ is depicted in Fig. 4.16. Optimum photocatalytic activity was achieved in the case of 2% La-doped ZnTiO$_3$. As doping concentration further increases, photocatalytic activity diminishes due to rise in the recombination of charge carriers and agglomeration of the nanoparticles, which blocks the active sites for the photocatalytic activity. PL spectrum gives information about charge migration and recombination rate of photogenerated charge carriers.

It is observed that the non-doped ZnTiO$_3$ sample has a strong PL peak. Interestingly, decrease in recombination rate of photogenerated charges led to decrease in the PL intensity in the La-doped ZnTiO3 nanoparticles.

Doped SrTiO$_3$

The perovskite SrTiO$_3$ is an efficient photocatalyst due to its high chemical stability, abundance, and good biological compatibility, exhibiting a band gap of 3.2 eV.

Owing to its large band gap, it absorbs in the UV region. Many attempts have been made to dope SrTiO$_3$ to shift its absorption in the visible region of the spectrum. It was found that doping with Cr^{3+} enhances the photocatalytic activity, but due to multiple valence states of Cr, the presence of Cr^{+6} reduces its performance. A way to address this issue is to codope the SrTiO$_3$. Photocatalytic degradation of Rhodamine B by undoped SrTiO$_3$ and La–Cr-codoped SrTiO$_3$ was investigated by Tonda et al. (2014); see Fig. 4.17. It was found that RhB degradation using 1% of Cr-, La-codoped SrTiO3 nanoparticles showed better performance than Cr-doped SrTiO3 nanoparticles (six times) and SrTiO3 nanoparticles (three times).

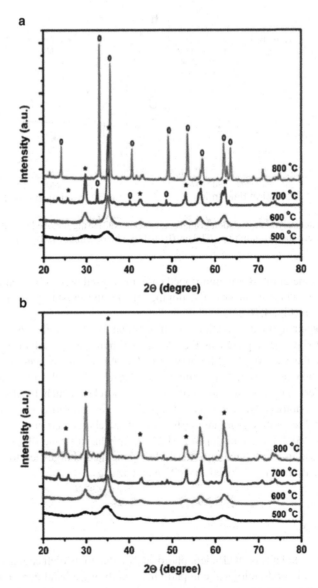

Fig. 4.16 (a) Non-doped ZnTiO₃, (b) La-doped ZnTiO₃, (c) photocatalytic activity. (This figure was obtained from https://doi.org/10.1039/C3CP53855A)

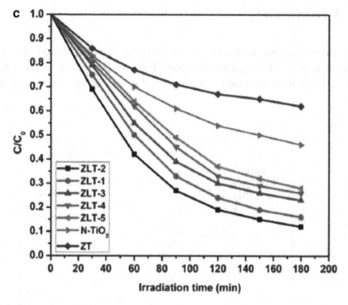

Fig. 4.16 (continued)

4.5   Nonmetal-Doped Metal Oxides

Asahi et al. (2001) estimated the outcome of the compositional doping of C, N, F, P, and S for O in anatase TiO_2 using a full-potential linearized augmented plane wave (FLAPW) formalism. To ensure better charge transfer of photoexcited charge carriers to reactive surface sites within lifetime, after doping, the produced states should effectively overlap with the TiO_2 band states. It was observed that integration of 2p states of N with 2p states of O produces new VB with shift in VB edge upward, narrowing the band gap of TiO_2. Doping of S ensued decrease in band gap, but the ionic radius of S was not small enough to be internalized into the lattice of TiO_2. Dopants C and P seemed to be not that effective due to deep internalization of photogenerated charge carriers, which were not easy to be transferred to the surface of the catalyst. After this study, N doping was intensively explored for visible light photocatalysis. The formation energy of oxygen voids decreases from 4.2 to 0.6 eV upon nitrogen doping, which indicates that oxygen vacancies are the result of nitrogen doping (Valentinet al. 2005). These results of the role of oxygen voids upon nitrogen doping in photocatalysis were confirmed by Rumaiz et al. (2009) using combined experimental and DFT.

4.5.1 N-Doped TiO₂

Many researchers explored N doping into TiO_2 for dye degradation and H_2 generation. Parida et al. (2013) reported N-doped TiO_2 for photocatalytic H_2 generation. Different annealing temperatures were used for the sample. UV–Vis spectrum

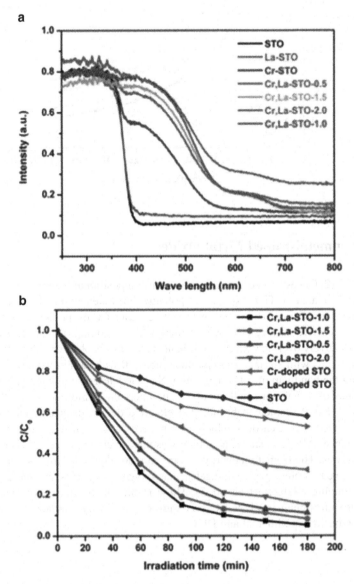

Fig. 4.17 (a) UV spectrum, (b) photocatalytic degradation of Rhodamine B, (c) reaction mechanism. (This figure was obtained from https://doi.org/10.1039/c4cp02963a)

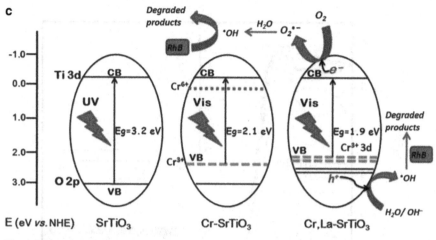

Fig. 4.17 (continued)

reveals an increase in absorbance in the case of N-doped-400 (Fig. 4.18). The meso-macroporous N–TiO$_2$ exhibit highest photocatalytic activity (H$_2$ evolution, 364.3 mmol/h) with annealing temperature of 400 °C.

4.5.2 N- and S-Codoped TiO$_2$

Wei et al. (2008) reported N- and S-codoped TiO$_2$ (N-S-TO) with enhanced photocatalytic activity after doping. The diffuse reflectance spectra (DRS) of the as-prepared samples are displayed in Fig. 4.19. It is observed that the absorption of N-S-TO in the UV–Vis region is higher than that of S-doped TiO$_2$ (S-TO) and TiO$_2$ and red shifted. The increase of absorbance in the UV–Vis region raises the number of photogenerated charge carriers to undergo phase change in the photocatalytic reaction to increase the photocatalytic activity of doped TiO$_2$. The N-S-TO exhibited the best photocatalytic activity in comparison to undoped TiO$_2$ with the rate constant of about 0.267 h^{-1} and 1.3854 h^{-1} under sunlight and UV light irradiation, respectively.

4.5.3 N-Doped SrTiO$_3$

Zou et al. (2012) reported N-doped SrTiO$_3$ for the photodegradation of various dyes under visible light irradiation. After N doping, the band gap diminished from 3.2 eV to 2.9 eV due to incorporation of sub-states in the band gap. N-doped SrTiO$_3$ was observed to be efficient for methylene blue dye degradation (Fig. 4.20b).

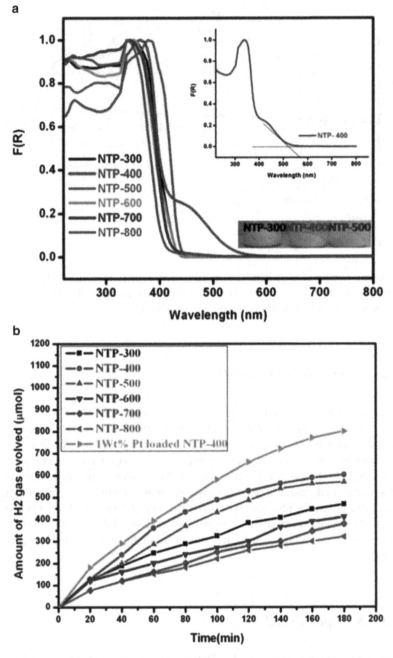

Fig. 4.18 (**a**) UV spectrum, (**b**) photocatalytic H$_2$ generation. (This figure was obtained from https://doi.org/10.1016/j.ijhydene.2012.12.118)

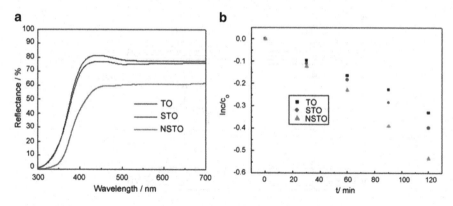

Fig. 4.19 (a) UV spectrum, (b) photocatalytic methyl orange degradation. (This figure was obtained from https://doi.org/10.1016/j.jhazmat.2007.12.018)

4.5.4   F-Doped Bi_2WO_6

Fu et al. (2008) reported F-doped Bi_2WO_6 for dye degradation. In earlier reports it was indicated that F^- doping increases the crystallinity of TiO_2 and the photocatalytic reactivity. F-doped $TiO_{2-x}F_x$ exhibits lesser anion deficiencies due to doping and is more photo-resistant against corrosion. Yu et al. projected that the doping fluoride ions convert Ti^{4+} to Ti^{3+} by charge equilibration and that a certain amount of Ti^{3+} presence decreases the photogenerated charge recombination rate and increases the photocatalytic activity.

UV spectrum and photocatalytic degradation of methyl orange are displayed in Fig. 4.21a, b, respectively.

4.6   Metal-Doped Metal Sulfides

In the past few decades, several metal sulfide photocatalysts have been investigated for water splitting. In case of metal sulfide photocatalyst, the CB consists of d and sp orbitals, whereas the VB consists of S3p orbitals. S3p is more negative than O 2p orbitals, which leads to CB positions negative enough to reduce H_2O to H_2 and narrow down the band gap with efficient reaction to the solar spectrum (Kudo and Miseki 2009). CdS and ZnS are the most studied metal sulfide semiconductors with excellent physical properties, wide band gap energy, and improved photocatalytic activity. To improve the catalytic activity of metal sulfides, researchers have developed various transition metal-doped sulfides. Doping of transition metal in sulfides reduces the band gap of the photocatalyst, improves the chemical stability by inducing holes into the acceptor energy level, and reduces the photocorrosion.

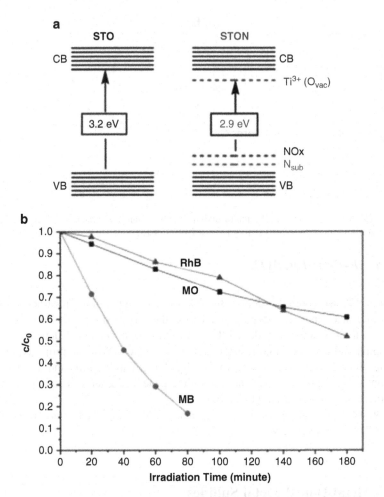

Fig. 4.20 (**a**) UV spectrum, (**b**) photocatalytic degradation. (This figure was obtained from https://doi.org/10.1039/c2cc33797e)

Mn-, Ni-, Cu-, etc., doped ZnS and Ni-, Co-, Sb-, Ce-, etc., doped CdS have been reported with improved photocatalytic activity.

4.6.1 Metal-Doped CdS

CdS belongs to II−VI semiconductor with a direct band gap of 2.4 eV and exhibits excellent photocatalytic activity with sensitivity to visible light. The reaction mechanism for photocatalysis in composite of cadmium sulfide is depicted by the succeeding equations (Xu et al. 2017):

$$\frac{\text{Cds}}{\text{g} - \text{C}_3\text{N}_4} \rightarrow \text{CdS} \, (\text{e}^- + \text{h}^+)$$

$$\frac{\text{CdS}(\text{e}^- + \text{h}^+)}{\text{g} - \text{C}_3\text{N}_4(\text{e}^- + \text{h}^+)} \rightarrow \frac{\text{CdS}(\text{e}^- + \text{e}^-)}{\text{g} - \text{C}_3\text{N}_4(\text{h}^+ + \text{h}^+)}$$

$$\text{e}^- + \text{O}_2 \rightarrow \dot{\text{O}}_2^-$$

$$2\text{e}^- + 2\text{H}^+ + \dot{\text{O}}_2^- \rightarrow \text{OH} + \text{OH}^-$$

$$\text{h}^- + \text{OH}^- \rightarrow \text{OH}$$

$$\text{h}^+ + \text{H}_2\text{O} \rightarrow \text{OH} + \text{H}^+$$

$$\dot{\text{O}}_2^- + \text{MO} \rightarrow \text{degradation product}$$

$$\text{OH} + \text{MO} \rightarrow \text{degradation product}$$

During the photocatalytic process, pure CdS faces some challenges like photocorrosion or photodissolution and lower stability upon light illumination, the photocatalytic activity of CdS is still good enough for various practical applications. To address earlier stated issues, doping of transition metal ions is an effective way to enhance the photocatalytic performance of CdS. Metal ion doping into CdS not only affects the photophysical nature of the semiconductor but also assists the

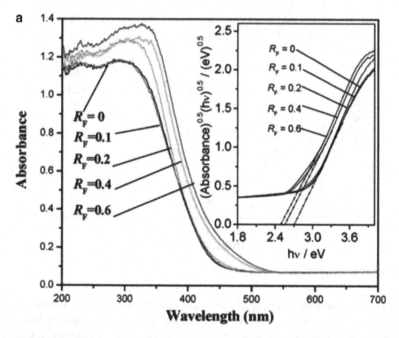

Fig. 4.21 (**a**) UV spectrum, (**b**) and (**c**) photocatalytic degradation of methyl orange with undoped and doped catalyst. (This figure was obtained from https://doi.org/10.1021/es702495w)

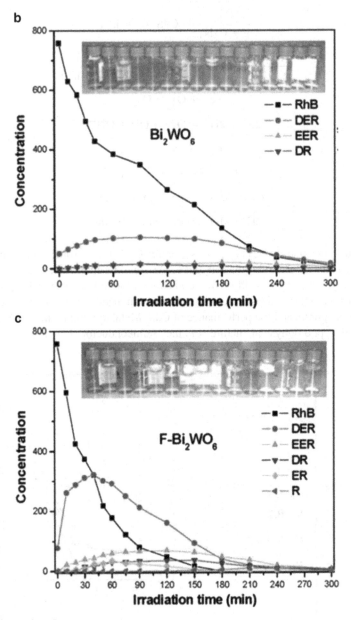

Fig. 4.21 (continued)

photochemical activity. Doping of metal ion in CdS can generate Schottky barrier and influence the surface property of the semiconductor. This Schottky barrier can capture e^- or h^+ and subsequently inhibit e^-/h^+ recombination. Additionally, several

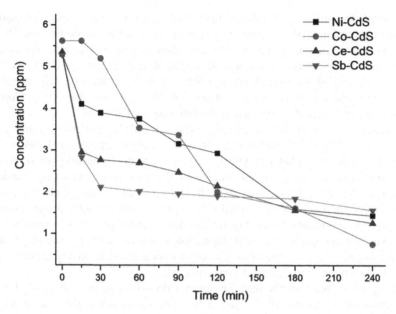

Fig. 4.22 Photodegradation of MB in the presence of Ni-, Co-, Ce-, and Sb-doped CdS catalysts under visible light. (This figure was obtained from https://doi.org/10.4236/mrc.2017.61001)

noble metals, such as Pt, Au, Ag, etc., lead to the separation of photogenerated electron and hole for the better photocatalytic activity.

Ertis et al. have synthesized Ni-, Co-, Sb-, Ce-doped CdS by chemical coprecipitation method and studied their photocatalytic activity for the degradation of methylene blue. Crystallite size of metal-doped CdS reduced in comparison to non-doped CdS, which increases the band gap of the materials by blue shifting the optical diffuse reflectance spectra and leads to higher photocatalytic activity. Among all metal-doped cases, Co-doped CdS showed the highest rate of methylene blue (MB) photodegradation than the other metal-doped CdS due to the lower crystallite size and increase in defect sites (Fig. 4.22). After 240 min of irradiation, Co–CdS showed 87% MB degradation, which is higher than undoped case because of the enhanced optical absorption in the visible region and increase in the lifetime of excited charge carriers. So, overall photocatalytic activity of Co-doped CdS will increase. When Co ion is replaced with other dopant ions, it causes some CdS crystal defects, which act as recombination sites and decrease the photocatalytic activity of the CdS compared to Co–CdS, but considerably more than CdS.

Ertis et al. explained the increase of photocatalytic activity of Sb doping in CdS with increase in the lifetime of excited charge carriers and increased photocatalytic degradation of methylene blue under visible light irradiation. They have synthesized Sb-doped CdS and Sb_2S_3 single and binary catalysts using CdS for methylene blue (MB) degradation under visible light irradiation. Sb-doped CdS catalyst, prepared by thioacetamide (TAA) by using ethylenediamine (EDA) as a chelating agent, shows enhanced photocatalytic activity. EDA exhibits better coordination between metal and sulfide by formation of stable complex with metal ion and serves as templating agent to control desired nucleation and growth rate of different compounds. Therefore, the photocatalytic activity for the degradation of methylene blue with single Sb_2S_3 and binary catalysts is lesser compared to Sb-doped CdS catalyst (84%). Sb–CdS–Na_2S and Sb–CdS–TAA showed 70% photodegradation of MB, which is higher than other catalysts (without doping of Sb) (Ertis and Boz 2017). Doping of Sb induces CdS to the degradation reaction without Cd^{2+} and S^{2-}.

Sasikala and coworkers have prepared indium-doped CdS dispersed on zirconium as a photocatalyst which exhibits enhanced photocatalytic activity for H_2 generation from H_2O. Improved photocatalytic activity is due to the increased photoresponse, high surface area, and amended lifetime of the photogenerated charge carriers. Among CdInS–Zr, CdS–Zr, and CdS, CdInS–Zr showed the highest photoresponse than CdS–Zr and CdS. Zr-supported CdInS provide more availability of electron compared to unsupported CdS. Figure 4.23 depicts the photocatalytic activity. It is observed that In-doped CdS exhibits higher photocatalytic activity, indicating that In doping improves the lifetime of the charge carriers and also relatively improves the optical absorption of the material (Sasikala et al. 2011).

Yang et al. have synthesized Zn-doped CdS with optimum level of doping concentration (molar ratio of Zn/Cd = 1:10), which has much higher photocatalytic activity compared with pure CdS. Doping of Zn^{2+} in CdS can act as charge trapper, which impedes electron–hole combination rate and increases the interfacial charge transfer for RhB degradation. When Zn/Cd ratio was increased or decreased from Zn/Cd = 1:10, it led to reduced photocatalytic activity. If Zn/Cd molar ratio is decreased from optimum concentration, the photocatalytic activity is low because Zn^{2+} is not able to act as interfacial charge transfer mediator. Furthermore, higher doping of Zn^{2+} decreases the distance between trapping sites in a particle, which increases the recombination rate and decreases the photocatalytic activity of the material. Reaction temperature and reaction time also play an efficient part in the photocatalytic activity of Zn-doped CdS (Yang et al. 2012). Reaction temperature and time at which sample obtained with high crystal quality and higher surface area give increased photocatalytic activity (Fig. 4.24).

4.6.2   Metal-Doped ZnS

ZnS is another most widely explored photocatalyst because of better energy conversion efficiency and rapid generation of electron–hole pairs under photoexcitation. It exhibits very negative potential for excited electrons and inactive to visible light due to its large band gap (~3.6 eV). Several researches have been done to extend the optical absorption of ZnS into the visible region by transition metal ions (Au, Ni, Cu) doping. However, doping can generate surface defects and improve the harvesting of light in photocatalytic materials, and increase the adsorption sites for charge migration to the adsorbed species, which preclude recombination of photogenerated electrons and holes. On the other hand, higher amount of defects play as traps for

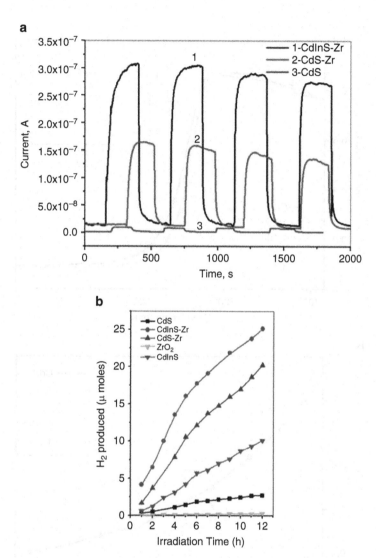

Fig. 4.23 (a) Photocurrent response of doped CdS to on–off cycles of light illumination. (b) Photocatalytic H_s generation as a function of irradiation time for ZrO_2, CdS, CdInS, CdS–Zr, CdInS–Zr. (This figure was obtained from https://doi.org/10.1039/C1JM12531A)

charge carriers, which decreases the photocatalytic activity of ZnS by increasing photogenerated electrons and holes recombination. Thus, control of defective sites plays significant role for photocatalytic activity.

Mn^{2+}-doped ZnS can incorporate new states within the band gap of host and can act as shallow traps to enhance the lifetime of photogenerated carriers. Mostly, Mn-doped ZnS exhibits cubic zinc blende structures and used as photocatalysts. The effect of Mn doping on photoluminescence and photocatalytic behavior of

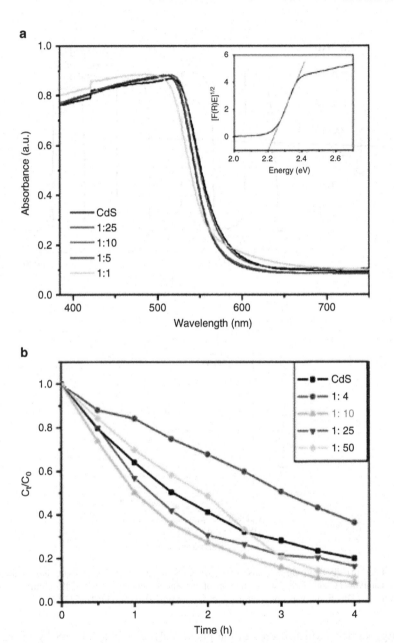

Fig. 4.24 (a) UV–Vis DRS of doped CdS. The inset is the plot of $[F(R)E]^{1/2}$ vs. photoenergy. (b) Photodegradation of RhB in the presence of pure CdS and doped CdS. (This figure was obtained from https://doi.org/10.1021/jp300939q)

$Zn_{1-x}Mn_xS$ was studied by Wang's group. They observed that Mn^{2+} ions increase the charge separation and led to excellent photocatalytic activity (Wang et al. 2017). $Mn_xZn_{1-x}S$ exhibits UV light absorption, so it is used as a sensitizer to modulate TiO_2.The produced $Mn_xZn_{1-x}S/TiO_2$ showed raised photodegradation of 4-chlorophenol under visible light (Manikandan et al. 2015). Wang et al. have synthesized the hexagonal wurtzite structure of Mn-doped ZnS which possesses more photocatalytic activity and photoreduction of Cr^{6+}under visible light irradiation than that of pristine ZnS. The photocatalytic activity rose with enhancement in Mn concentration up to 7% and decreased slowly beyond the optimum Mn concentration (Wang et al. 2017).

Wang et al. have explored the photocatalytic activity of ZnS upon doing metal ions (Fe, Ni, and Cu) in the reduction of fumaric acid to succinic acid. It was observed that 0.1 atom % of Fe and Ni doping decreases the photocatalytic efficiency of ZnS under irradiation containing UV components (λ>320 nm and 354 nm). Increased doping concentration of Fe or Ni continuously decreases the photocatalytic activity of ZnS, while doping of Cu in ZnS with 0.1 atomic % increases the yield of succinic acid 1.8 and 4.4 times compared to pristine ZnS, respectively. Doping of Cu^{2+}metal ions significantly enhanced the photocatalytic activity by altering the electronic band structure of ZnS inducing its optical absorption edges shift from UV band (344 nm) to longer wavelengths. Additionally, when the amount of Cu doping increased up to 0.3%, the photocatalytic activity of ZnS increased. This may be due to the color of the doped ZnS particles which acts as an optical filter and shortens the penetration distance of the incident light in the colloidal suspension. Further increment of Cu doping in ZnS decreases the photocatalytic activity because of the appearance of a separate CuS phase which may aggravate photoelectron–hole recombination in the volume of the photocatalysts as well as shielding the incident light (Wang 2016).

Rajabi et al. studied the photocatalytic property of transition metal (Mn^{2+}, Co^{2+}, and Ni^{2+} ions)-doped ZnS quantum dots (QDs) and explained that the smaller particle size and doping of Ni and Cu ions in ZnS QDs increases the band gap energy of ZnS due to the increase in the lifetime of electrons and holes. Hence, the overall photocatalytic activity of the metal ion-doped ZnS increases. Figure 4.25 indicates that by increasing the initial concentration of dye, discoloration efficiency (DE) % decreases. At higher concentration, more dye molecules are expected to be adsorbed on the photocatalyst surface, which decreases the active sites of the catalyst; hence, at the surface of photocatalyst, hydroxyl radical generation is reduced (Rajabi et al. 2015). It is clear from Fig. 4.25 that doped ZnS QDs is more active for MV dye decolorization as compared to pure ZnS QDs under UV light. Therefore, pure ZnS QDs show 95% of decolorization efficiency after 105 min, whereas doped samples, i.e., $Zn_{0.95}Mn_{0.05}S$, $Zn_{0.95}Ni_{0.05}S$, and $Zn0.95Co0.05S$, achieved decolorization in less than 60 min stirring time.

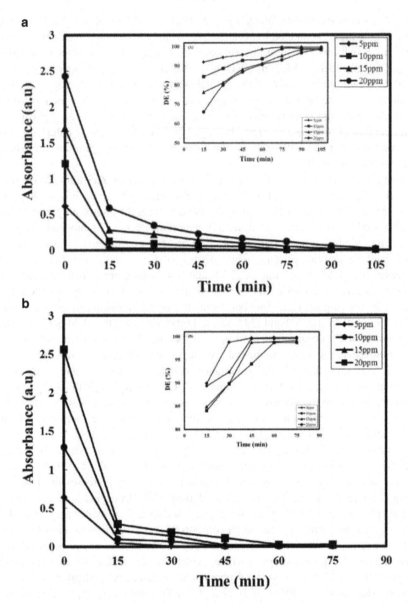

Fig. 4.25 The changes in the maximum wavelength of MV dye (max = 585 nm) at different initial concentration, in the presence of (**a**) ZnS, (**b**) $Zn_{0.95}Mn_{0.05}S$, (**c**) $Zn_{0.95}Ni_{0.05}S$, and (**d**) $Zn_{0.95}Co_{0.05}S$ QDs. (This figure was obtained from https://doi.org/10.1016/j.molcata.2015.01.029)

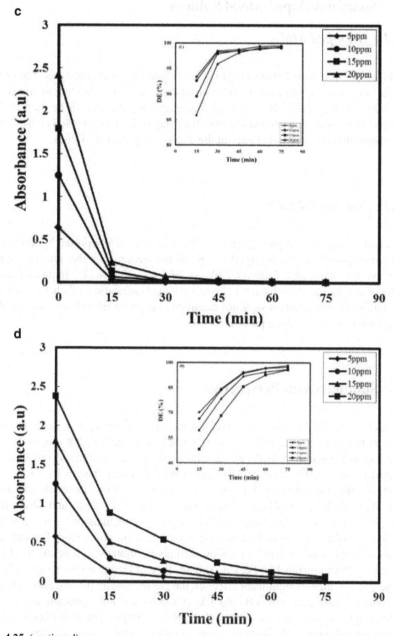

Fig. 4.25 (continued)

4.7 Nonmetal-Doped Metal Sulfides

4.7.1 N-Doped MoS₂

Liu et al. (2016) reported N-doped MoS_2 for visible light photocatalysis. The doped sample exhibits an enhanced light absorption in the wavelength range of 200–800 nm (Fig. 4.26). Band gap of the sample is estimated to be 2.08 eV for N-doped MoS_2 nanoflowers and a narrow band gap of 2.17 eV for MoS_2 nanosheets. The photocatalytic activity is better in the doped samples (Fig. 4.26c).

4.7.2 N-Doped ZnS

To eliminate the issue of photocorrosion, Zhou et al. (2015) reported N-doped ZnS for photocatalytic water splitting. It was predicted theoretically that after N doping, the band gap decreased from 3.67 eV to 2.67 eV (Fig. 4.27). Improvement in photocurrent in the doped case indicates efficient separation of charge carriers. The data from photocurrent signifies the improved photostability of N-doped ZnS at the time of photocatalytic process.

4.8 Summary and Perspective

Research in the field of photocatalysis has increased tremendously since its emergence in the early 1970s. TiO_2 is a widely used photocatalyst, but due to its wide band gap and absorption in the UV region of the spectrum, intense research was directed toward modification of TiO_2 like band gap tuning, heterostructuring, and search of other suitable materials. As a consequence, various materials like TiO_2, ZnO, WO_3, $FeVO_4$, $SrTiO_3$, and metal sulfides like CdS, ZnS, and MoS_2 have emerged for photocatalysis. Researchers have been trying to narrow down the band gap in the visible region to increase the solar spectrum absorption. Doping with transition metal and nonmetal ions is one promising strategy to tune the band gap. Doping of several metals like Fe, Cu, Mn, Sb, In, and La and nonmetal ions like N, S, and F has exhibited significant effect on the photocatalytic performance. Despite the suitable band gap, the VB and CB positions are not optimum for overall photocatalytic activity. So, there are still chances for improvement in photocatalytic activity by codoping with different metals or mixed metal–nonmetal doping.

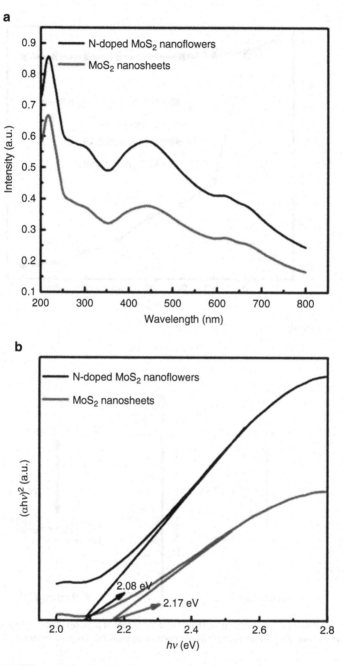

Fig. 4.26 (a) UV spectrum, (b) Kubelka–Munk plot, (c) photocatalytic dye degradation. (This figure was obtained from https://doi.org/10.1088/0957-4484/27/22/225403)

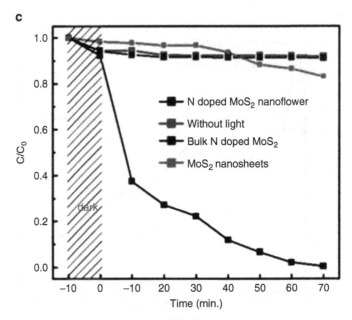

Fig. 4.26 (continued)

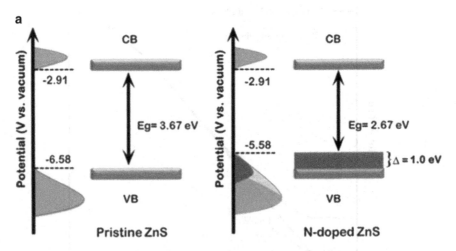

Fig. 4.27 (**a**) Potential energy diagram, (**b**) current vs. potential plot, (**c**) photocurrent generated in undoped and doped ZnS. (This figure was obtained from https://doi.org/10.1039/C4CP03736G)

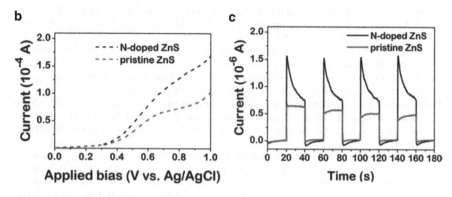

Fig. 4.27 (continued)

Acknowledgments We acknowledge the Department of Science and Technology, Science and Engineering Research Board, and Council of Scientific and Industrial Research, New Delhi, India, for providing financial support.

References

Al-Hamdi AM, Rinner U, Sillanpaa M (2017) Tin dioxide as a photocatalyst for water treatment: a review. Process Saf Environ Prot 107:190–205. https://doi.org/10.1016/j.psep.2017.01.022

Asahi R, Morikawa T, Ohwaki T, Aoki K, Taga Y (2001) Visible-light photocatalysis in nitrogen-doped titanium oxides. Science 293(5528):269 LP–269271. Retrieved from http://science.sciencemag.org/content/293/5528/269.abstract

Bousslama W, Elhouichet H, Férid M (2017) Enhanced photocatalytic activity of Fe doped ZnO nanocrystals under sunlight irradiation. Optik Int J Light Electron Opt 134:88–98. https://doi.org/10.1016/j.ijleo.2017.01.025

Di Valentin C, Pacchioni G, Selloni A, Livraghi S, Giamello E (2005) Characterization of paramagnetic species in N-doped TiO_2 powders by EPR spectroscopy and DFT calculations. J Phys Chem B 109(23):11414–11419. https://doi.org/10.1021/jp051756t

Dong P, Hou G, Xi X, Shao R, Dong F (2017) WO_3-based photocatalysts: morphology control, activity enhancement and multifunctional applications. Environ Sci Nano 4(3):539–557. https://doi.org/10.1039/C6EN00478D

Dutta DP, Ramakrishnan M, Roy M, Kumar A (2017) Effect of transition metal doping on the photocatalytic properties of $FeVO_4$ nanoparticles. J Photochem Photobiol A Chem 335:102–111. https://doi.org/10.1016/j.jphotochem.2016.11.022

Ertis IF, Boz I (2017) Synthesis and optical properties of Sb-doped CdS photocatalysts and their use in methylene blue (MB) degradation. Int J Chem React Eng. https://doi.org/10.1515/ijcre-2016-0102

Fan X, Fan J, Hu X, Liu E, Kang L, Tang C, Li Y (2014) Preparation and characterization of Ag deposited and Fe doped TiO_2 nanotube arrays for photocatalytic hydrogen production by water splitting. Ceram Int 40(10, Part A):15907–15917. https://doi.org/10.1016/j.ceramint.2014.07.119

Feng J, Wang Z, Zhao X, Yang G, Zhang B, Chen Z, Huang Y (2018) Probing the performance limitations in thin-film $FeVO_4$ photoanodes for solar water splitting. J Phys Chem C 122 (18):9773–9782. https://doi.org/10.1021/acs.jpcc.8b01330

Fu H, Zhang S, Xu T, Zhu Y, Chen J (2008) Photocatalytic degradation of RhB by fluorinated Bi_2WO_6 and distributions of the intermediate products. Environ Sci Technol 42(6):2085–2091. https://doi.org/10.1021/es702495w

Fujishima A, Honda K (1972) Electrochemical photolysis of water at a semiconductor electrode. Nature 238. https://doi.org/10.1038/238037a0

Gao F (2011) Effects of quantum confinement and shape on band gap of core/shell quantum dots and nanowires. Appl Phys Lett 98(19):193105. https://doi.org/10.1063/1.3590253

Gnanasekaran L, Hemamalini R, Saravanan R, Ravichandran K, Gracia F, Gupta VK (2016) Intermediate state created by dopant ions (Mn, Co and Zr) into TiO_2 nanoparticles for degradation of dyes under visible light. J Mol Liq 223:652–659. https://doi.org/10.1016/j.molliq.2016.08.105

Hoffmann MR, Martin ST, Choi W, Bahnemann DW (1995) Environmental applications of semiconductor photocatalysis. Chem Rev 95(1):69–96. https://doi.org/10.1021/cr00033a004

Kim S, Park H, Choi W (2004) Comparative study of homogeneous and heterogeneous photocatalytic redox reactions: PW12O403- vs TiO_2. J Phys Chem B 108(20):6402–6411. https://doi.org/10.1021/jp049789g

Kiriakidis G, Binas V (2014) Metal oxide semiconductors as visible light photocatalysts. J Korean Phys Soc 65(3):297–302. https://doi.org/10.3938/jkps.65.297

Klingshirn C, Hauschild R, Fallert J, Kalt H (2007) Room-temperature stimulated emission of ZnO: alternatives to excitonic lasing. Phys Rev B 75(11):115203. https://doi.org/10.1103/PhysRevB.75.115203

Kudo A, Miseki Y (2009) Heterogeneous photocatalyst materials for water splitting. Chem Soc Rev 38(1):253–278. https://doi.org/10.1039/B800489G

Kumar A, Sharma G, Naushad M et al (2014) Polyacrylamide/Ni0.02Zn0.98O nanocomposite with high solar light photocatalytic activity and efficient adsorption capacity for toxic dye removal. Ind Eng Chem Res 53:15549–15560. https://doi.org/10.1021/ie5018173

Kumar A, Kumar A, Sharma G et al (2018) Biochar-templated g-C3N4/Bi2O2CO3/CoFe2O4nano-assembly for visible and solar assisted photo-degradation of paraquat, nitrophenol reduction and CO2 conversion. Chem Eng J. https://doi.org/10.1016/j.cej.2018.01.105

Li Z-X, Shi F-B, Ding Y, Zhang T, Yan C-H (2011) Facile synthesis of highly ordered mesoporous ZnTiO3 with crystalline walls by self-adjusting method. Langmuir 27(23):14589–14593. https://doi.org/10.1021/la2034615

Li J, Zhao W, Guo Y, Wei Z, Han M, He H, Sun C (2015) Facile synthesis and high activity of novel $BiVO_4/FeVO_4$ heterojunction photocatalyst for degradation of metronidazole. Appl Surf Sci 351:270–279. https://doi.org/10.1016/j.apsusc.2015.05.134

Linsebigler AL, Lu G, Yates JT (1995) Photocatalysis on TiO_2 surfaces: principles, mechanisms, and selected results. Chem Rev 95(3):735–758. https://doi.org/10.1021/cr00035a013

Liu P, Liu Y, Ye W, Ma J, Gao D (2016) Flower-like N-doped MoS 2 for photocatalytic degradation of RhB by visible light irradiation. Nanotechnology 27(22):225403

Liu Y, Feng P, Wang Z, Jiao X, Akhtar F (2017) Novel fabrication and enhanced photocatalytic MB degradation of hierarchical porous monoliths of MoO_3 nanoplates. Sci Rep 7(1):1845. https://doi.org/10.1038/s41598-017-02025-3

Manikandan A, Manimegalai DK, Moortheswaran S, Arul Antony S (2015) Magneto-optical and photocatalytic properties of magnetically recyclable MnxZn1−xS (x = 0.0, 0.3, and 0.5) nanocatalysts. J Supercond Nov Magn 28. https://doi.org/10.1007/s10948-015-3089-3

Mishra M, Doo-Man C (2015) α-Fe_2O_3 as a photocatalytic material: a review. Appl Catal A Gen 498:126–141. https://doi.org/10.1016/j.apcata.2015.03.023

Montini T, Melchionna M, Monai M, Fornasiero P (2016) Fundamentals and catalytic applications of CeO_2-based materials. Chem Rev 116(10):5987–6041. https://doi.org/10.1021/acs.chemrev.5b00603

Neren Ökte A, Akalın Ş (2010) Iron (Fe3+) loaded TiO_2nanocatalysts: characterization and photoreactivity. React Kinet Mech Catal 100(1):55–70. https://doi.org/10.1007/s11144-010-0168-0

Ong CB, Ng LY, Mohammad AW (2018) A review of ZnO nanoparticles as solar photocatalysts: synthesis, mechanisms and applications. Renew Sust Energ Rev 81:536–551. https://doi.org/10.1016/j.rser.2017.08.020

Parida K, Pany S, Naik B (2013) Green synthesis of fibrous hierarchical meso-macroporous N doped TiO$_2$nanophotocatalyst with enhanced photocatalytic H$_2$ production. Int J Hydrog Energy 38:3545–3553. https://doi.org/10.1016/j.ijhydene.2012.12.118

Polisetti S, Deshpande PA, Madras G (2011) Photocatalytic activity of combustion synthesized ZrO$_2$ and ZrO$_2$–TiO$_2$ mixed oxides. Ind Eng Chem Res 50(23):12915–12924. https://doi.org/10.1021/ie200350f

Rumaiz AK, Woicik JC, Cockayne E, Lin HY, Jaffari GH, Shah SI (2009) Oxygen vacancies in N doped anatase TiO$_2$: experiment and first-principles calculations. Appl Phys Lett 95 (26):262111. https://doi.org/10.1063/1.3272272

Saravanan R, Prakash T, Gupta VK, Narayanan V, Stephen A (2013) Synthesis, characterization and photocatalytic activity of novel Hg doped ZnO nanorods prepared by thermal decomposition method. J Mol Liq 178:88–93. http://www.sciencedirect.com/science/article/pii/S0167732212004114

Saravanan R, Gupta VK, Edgar M, Gracia F (2014) Preparation and characterization of V$_2$O$_5$/ZnO nanocomposite system for photocatalytic application. J Mol Liq 198:409–412. http://www.sciencedirect.com/science/article/pii/S0167732214003432

Saravanan R, Khan MM, Gracia F, Qin J, Gupta VK, Stephen A (2016) Ce^{3+}-ion-induced visible-light photocatalytic degradation and electrochemical activity of ZnO/CeO2 nanocomposite. Nat Sci Rep 6:31641. http://www.nature.com/articles/srep31641

Saravanan R, Aviles J, Gracia F, Mosquera E, Vinod KG (2018) Crystallinity and lowering band gap induced visible light photocatalytic activity of TiO$_2$/CS (Chitosan) nanocomposites. Int J Biol Macromol 109:1239–1245. https://www.sciencedirect.com/science/article/pii/S0141813017323450

Sasikala R, Shirole AR, Sudarsan V, Girija KG, Rao R, Sudakar C, Bharadwaj SR (2011) Improved photocatalytic activity of indium doped cadmium sulfide dispersed on zirconia. J Mater Chem 21(41):16566–16573. https://doi.org/10.1039/C1JM12531A

Song H, Li Y, Lou Z, Xiao M, Hu L, Ye Z, Zhu L (2015) Synthesis of Fe-doped WO$_3$ nanostructures with high visible-light-driven photocatalytic activities. Appl Catal B Environ 166–167:112–120. https://doi.org/10.1016/j.apcatb.2014.11.020

Surendar T, Kumar S, Shanker V (2014) Influence of La-doping on phase transformation and photocatalytic properties of ZnTiO$_3$ nanoparticles synthesized via modified sol–gel method. Phys Chem Chem Phys 16(2):728–735. https://doi.org/10.1039/C3CP53855A

Tonda S, Kumar S, Anjaneyulu O, Shanker V (2014) Synthesis of Cr and La-codoped SrTiO$_3$ nanoparticles for enhanced photocatalytic performance under sunlight irradiation. Phys Chem Chem Phys 16(43):23819–23828. https://doi.org/10.1039/C4CP02963A

Urbach F (1953) The long-wavelength edge of photographic sensitivity and of the electronic absorption of solids. Phys Rev 92(5):1324. https://doi.org/10.1103/PhysRev.92.1324

Wang S, Lian JS, Zheng WT, Jiang Q (2012) Photocatalytic property of Fe doped anatase and rutile TiO$_2$ nanocrystal particles prepared by sol–gel technique. Appl Surf Sci 263:260–265. https://doi.org/10.1016/j.apsusc.2012.09.040

Wang L, Wang P, Huang B, Ma X, Wang G, Dai Y, Qin X (2017) Synthesis of Mn-doped ZnS microspheres with enhanced visible light photocatalytic activity. Appl Surf Sci 391:557–564. https://doi.org/10.1016/j.apsusc.2016.06.159

Wei F, Ni L, Cui P (2008) Preparation and characterization of N–S-codoped TiO$_2$ photocatalyst and its photocatalytic activity. J Hazard Mater 156(1):135–140. https://doi.org/10.1016/j.jhazmat.2007.12.018

Xu H, Wu L, Jin L, Wu K (2017) Combination mechanism and enhanced visible-light photocatalytic activity and stability of CdS/g-C3N4 heterojunctions. J Mater Sci Technol 33 (1):30–38. https://doi.org/10.1016/j.jmst.2016.04.008

Yang F, Yan N-N, Huang S, Sun Q, Zhang L-Z, Yu Y (2012) Zn-doped CdS nanoarchitectures prepared by hydrothermal synthesis: mechanism for enhanced photocatalytic activity and stability under visible light. J Phys Chem C 116(16):9078–9084. https://doi.org/10.1021/jp300939q

Yang X, Wang S, Sun H, Wang X, Lian J (2015) Preparation and photocatalytic performance of Cu-doped TiO_2 nanoparticles. Trans Nonferrous Metals Soc China 25(2):504–509. https://doi.org/10.1016/S1003-6326(15)63631-7

Yu J, Dai G, Huang B (2009) Fabrication and characterization of visible-light-driven plasmonic photocatalyst Ag/AgCl/TiO_2 nanotube arrays. J Phys Chem C 113(37):16394–16401. https://doi.org/10.1021/jp905247j

Zhou Y, Chen G, Yu Y, Feng Y, Zheng Y, He F, Han Z (2015) An efficient method to enhance the stability of sulphide semiconductor photocatalysts: a case study of N-doped ZnS. Phys Chem Chem Phys 17(3):1870–1876. https://doi.org/10.1039/C4CP03736G

Zou Z, Ye J, Sayama K, Arakawa H (2001) Direct splitting of water under visible light irradiation with an oxide semiconductor photocatalyst. Nature 414:625. https://doi.org/10.1038/414625a

Zou F, Jiang Z, Qin X, Zhao Y, Jiang L, Zhi J, Edwards P (2012) Template-free synthesis of mesoporous N-doped SrTiO3 perovskite with high visible-light-driven photocatalytic activity. Chem Commun 48:8514–8516. https://doi.org/10.1039/c2cc33797e

Chapter 5
Nanoparticles based Surface Plasmon Enhanced Photocatalysis

Mohammad Ehtisham Khan, Akbar Mohammad, and Moo Hwan Cho

Contents

Abstract Surface plasmonic resonance (SPR)-centered photocatalysts have helped to improve photocatalytic effectiveness for energy- and environmental-related applications, such as wastewater treatment and water splitting. The SPR phenomenon has been well studied in noble metal-based nanoparticles, such as Ag and Au, to achieve an efficient photocatalytic process, particularly for wastewater treatment. SPR-based nanostructures may involve plasmon-facilitated photocatalytic reaction mechanisms, such as Schottky junctions, through electron transfer, improved limited electric fields, and plasmon resonance-based energy transference. This chapter reviews the various factors that are involved in the photodegradation of dyes. A discussion on future directions within this field of investigation is also provided.

M. E. Khan (✉)
Department of Chemical Engineering and Technology, College of Applied Industrial Technology (CAIT), Jazan University, Jazan, Kingdom of Saudi Arabia

School of Chemical Engineering, Yeungnam University, Gyeongsan-si, Gyeongbuk, South Korea
e-mail: mekhan@jazanu.edu.sa

A. Mohammad · M. H. Cho (✉)
School of Chemical Engineering, Yeungnam University, Gyeongsan-si, Gyeongbuk, South Korea
e-mail: mhcho@ynu.ac.kr

© Springer Nature Switzerland AG 2020   133
M. Naushad et al. (eds.), *Green Photocatalysts*, Environmental Chemistry
for a Sustainable World 34, https://doi.org/10.1007/978-3-030-15608-4_5

Keywords Surface plasmon resonance · Metal nanoparticles · Au and Ag · Visible light · Photocatalysis · Wastewater treatment

5.1   Introduction

The combination of plasmon resonance-based noble metal nanostructures and semi-conductors for plasmon-boosted visible light-motivated dye degradation has attracted considerable attention from researchers (Hou and Cronin 2013). Photocatalysis is a novel technique that can be used in a variety of applications, including the degradation of organic dyes, antibacterial treatment, and fuel generation from water splitting. Several inorganic semiconducting nanostructured materials have been explored as photocatalysts, and the adaptability of these materials has been increased recently (Hou and Cronin 2013; Jiang et al. 2014). Special attention has been paid to the use of solar energy to facilitate the photocatalytic degradation of different kinds of organic dyes, which has been driven by issues concerning the global environment and water treatment (Fujishima and Honda 1972; Hou and Cronin 2013; Pathania et al. 2016; Sharma et al. 2017).

Plasmon-based nanostructures have been recognized as an effective solution to overcome the limitations of the process. They can act as subwavelength antennae to concentrate light; the resulting enhanced local electromagnetic field can promote the separation of electrons from holes (Hou and Cronin 2013). Plasmonic resonance effect-based photocatalysts have recently been investigated as a promising technology for the high-performance degradation of harmful dyes under visible light irradiation (Atwater and Polman 2010; Linic et al. 2011). This process comprises the distribution of metal nanoparticles (NPs; generally Au and Ag) into semiconductor photocatalysts, which greatly improves photoreactivity under visible light irradiation (Khan et al. 2015a; Khan et al. 2018; Linic et al. 2011; Wang et al. 2012; Saravanan et al. 2013, 2018). In this regard, the finding and the progress of visible-light encouraged photocatalysts is an authoritative, pleasing into deliberation that considerable fractions (44%) of the total solar spectrum is visible light, which could be appropriately consumed for this persistence of dye degradation (Khan et al. 2015a, 2018). Plasmon resonance-based photocatalysts have attracted significant attention because of their possible applications in a variety of fields, such as environmental remediation and the photodegradation of harmful substances and industrial pollutants in contaminated water (Khan et al. 2015b; Zhang et al. 2014).

Several efforts have been made to develop effective photocatalysts for the dye degradation of harmful dyes as model dyes. Many dyes are cast off in several fields, but their direct release into seas and rivers is a serious source of pollution (Jing et al. 2013; Zhang et al. 2014). These harmful dyes have important applications in many industries, including paper, furniture, textile, concrete, oil refining, and pharmaceuticals (Khan et al. 2017). In particular, the textile industry is the leading purchaser of toxic and harmful dyes (Naushad et al. 2016). Approximately 10% of the color dyes used in the textile and fabric industries are released directly into the atmosphere as an

Fig. 5.1 Schematic representation of wastewater treatment utsing plasmonic nanostructures as the photocatalysts

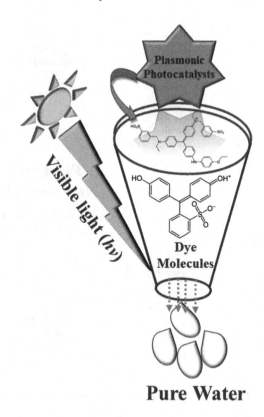

undesirable contaminant (Zhang et al. 2012). In this regard, plasmon resonance-based nanostructures are of interest for dye degradation to treat wastewater (Khan et al. 2015a; Wang and Xu 2012; Zhang et al. 2012) (Fig. 5.1).

5.2 Background on Metal Nanoparticles

The plasmon effects could be encouraged by noble metals. For example, Au and Ag are notable co-catalyst nanostructured materials for use in photocatalysis because of their plasmon effects (Camarero et al. 2003; Khan et al. 2018). Au and Ag NPs have been studied widely because of their unique optical and electronic properties for many applications in electronics, catalysis, and nano-biotechnology (Daneshvar et al. 2004; Khan et al. 2016, 2018; Maier 2007). Au NPs have also attracted remarkable attention for both heterogeneous and homogeneous visible light-induced catalysis (Daniel and Astruc 2004; Kreibig and Vollmer 1995; Maier 2007; Murray 2008). Recently, the confined surface plasmon resonance (LSPR) effect of nanomaterials, such as Ag and Au NPs, has been applied effectively in photocatalytic degradation under visible light irradiation. This chapter discusses

Fig. 5.2 Several possible fields of application using plasmonic nanoparticles/nanostructures (Khan et al. 2018)

the plasmonic behavior of nanogold photocatalysis with emphasis on the following aspects:

1. An introduction to the plasmonic effects of metal NPs
2. The easy fabrication process for plasmonic photocatalysts
3. Recent advances in effective plasmon-based photocatalysis
4. Applications of plasmonic photocatalysts for energy alterations by the visible-light irradiation
5. Reaction mechanisms involved in the plasmonic-based photocatalytic degradation of dye
6. Future research areas and potential applications

Figure 5.2 shows the different application categories of plasmonic nanoparticles/nanostructures. In photocatalytic progressions, the energy of the photons could be used to initiate several beneficial chemical reactions together with solar energy production in future fuel and water treatment applications. The photodegradation developments are potentially quite advantageous. Alterations from solar energy to chemical energy also have numerous advantages over conversions from solar energy to electric energy conversion using plasmonic nanostructures.

5.3   Basic Concepts and Valuable Effects of Plasmons

The basic concepts and advantages of special plasmonic effects are discussed in this section. Primarily, the collective oscillation of permitted electrons into the metal NPs is focused through the electromagnetic field of the incident light, in which the metallic NPs captivate visible and infrared light in specific sections. For example, Au and Ag NPs display a robust photo-absorption band of visible light because of their plasmonic resonance effects, which are maximized at approximately $\lambda = 530$ and $\lambda = 400$ nm, respectively (Guo and Wang 2011; Maier 2007). Ag NPs are oxidized easily, whereas Au NPs are more chemically stable in the presence of oxygen (Guo and Wang 2011; Rayalu et al. 2013). Plasmonic resonance effects in the absorption arise at the plasmon frequency region once the dielectric function drives to zero. Exposing metal NPs to light at their plasmon frequency creates intense electric fields at the surface of the NPs. The frequency of this resonance can be adjusted by varying the NPs' size, shape, material, and proximity to other NPs (Rayalu et al. 2013; Zhou et al. 2012). For example, the plasmon effect of Ag NPs, which lies in the UV region, could be moved into the visible array by reducing the NPs size. Correspondingly, it is possible to shift the plasmon effect of Au NPs from the visible range into the infrared wavelength array by reducing the NPs' size (Kim et al. 2010).

Figure 5.3 illustrates Mie's theory, which describes surface plasmon resonance (SPR) as the resonance photon-induced coherent oscillation of charges at the metal dielectric edge. SPR is predictable once the frequency of photons equals the usual frequency of the metal exterior electrons oscillating against the restoring energy of their positive nuclei (Su et al. 2003). In exceedingly conductive nanorange structures, permitted electrons are closely confined. When the nanostructures are exposed to electromagnetic energy at the plasma frequency, the spatial electron density redistributes and thus creates an electric field simultaneously—a columbic reinstating energy of the positively charged surface; the nucleus brings joint alternations of the charges in the particles, comparable to an oscillating mechanism after stretching and releasing (Sarina et al. 2013; Su et al. 2003). The beneficial aspects of plasmon-based catalytic effects are seen under visible light irradiation. Plasmon-based photocatalysis has attracted recent attention as a promising technology for the high-performance photocatalytic degradation of harmful and toxic dyes (Warren and Thimsen 2012).

Fig. 5.3 Graphic representation of surface plasmon resonance in small spherical metallic NPs below visible light illumination

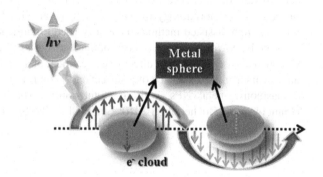

Plasmonic effect-based photocatalysis has two separate features: the Schottky junction and a localized surface plasmon effect; both features benefit photocatalysis in a dissimilar manner. For example, the Schottky junction occurs from the interaction of a metal NP surface and other semiconductor nanostructure, forming an inner electric field in a section of the photocatalyst that is close to the metal or semiconductor edge. This could be possibly provide strength for the electrons and holes to move in different directions, once they are formed inside or near the Schottky junction (Bai et al. 2015; Warren and Thimsen 2012). Upon accumulation, the metal fraction delivers a quick path for charge allocation; its surface acts as a charge trapping center to more active sites for photo-based reactions (Bai et al. 2015).

5.4 Methods for the Fabrication of Plasmonic Metal NPs as Photocatalysts

A number of chemical, physical, and biological methods have been investigated for the fabrication of plasmonic photocatalysts (Warren and Thimsen 2012; Zhang et al. 2013), including a chemical reduction process, physical vapor deposition method, hydrothermal method, and electrochemically active biofilm-assisted synthesis method (Sun and Xia 2002; Warren and Thimsen 2012). Plasmonic effect-based metal NPs are reported to have a high abundance of surface plasmon excitation. For example, Au and Ag NPs have attracted significant attention for their spatial SPR effects in the visible light region, which could be used in several environmental remediation applications.

Basically, the surface plasmon effect depends on the controlled size, well-ordered shape, and precise morphology. The precise morphology of plasmonic metal NPs is a remarkable concern because of their distinct plasmonic behavior (Jiang et al. 2014; Sun and Xia 2002). Numerous approaches have been reported for the creation of plasmonic metal NPs with precise shapes and sizes to improve their photocatalytic results for the degradation of pollutant dyes. Likewise, a precursor of $AgNO_3$ (silver nitrate) could be a useful candidate because it is less expensive and more abundant in nature. A few reducing mediators, such as sodium borohydride and micro-organism refereed biofilms, are being used to reduce the metal ions that exist in solutions of metal oxides. In other cases, stabilizing or capping agents are introduced to achieve and control the morphology of the resultant metal NPs. In photo-based chemical synthesis, light-assisted methods have been used to fabricate plasmonic metal NPs. Light-mediated fabrication has been applied to the synthesis of plasmonic metal NPs, such as laser ablation or direct laser treatment of an aqueous solution of a metal salt precursor in the existence of a surfactant to control the precise shapes and sizes of plasmonic metal NPs, with the light source working as a reducing agent (Gramotnev and Bozhevolnyi 2010; Lim et al. 2006; Sharma et al. 2009).

5.5 Effects and Mechanisms of Plasmonic Resonance in Photocatalysis

Plasmon resonance is the joint swinging of a permitted charge in a conducting nanostructured material. The light under the plasma incidence is reflected so that the electrons in the metallic NPs shelter in the electric field near the light. The light directly above the plasma frequency region is transmitted so that the electrons cannot respond quickly or sufficiently enough to monitor it. The plasmonic oscillations are confined to the surfaces of conducting nanostructured materials and act together in the presence of light (Khan et al. 2016; Xia and Halas 2005).

A group of plasmon resonance-based metallic NPs display SPR in the visible range with varied bandgap semiconducting and carbon-based nanostructured materials for the visible light-induced degradation of dyes by a hot electron mechanism. Basically, the metallic NPs are able to perform as sensitizers, captivating incident photons and shifting the excited and higher energy electrons (i.e., hot electrons) to close semiconducting materials or carbon-based nanostructured materials (Xia and Halas 2005; Zarazua et al. 2016). The constant amplification of plasmon states could be helpful for metal NPs in photo irradiation, which is subsequent to a hot electron dose into the CB (Conduction Band) (Conduction Band) of semiconducting materials. These excited electrons could be cast-off for the enhanced photodegradation of different kinds of dyes (Su et al. 2007; Zarazua et al. 2016).

Fundamentally, the improved light-scattering increases the usual photon track length in plasmon resonance-based nanostructured materials, thus improving the development of charge carriers in the semiconducting materials that take part in photocatalytic reactions. This effect is contingent on the light-scattering ability of plasmon resonance-based nanostructures, which could be assessed through the intended light scattering to absorption ratios (Hou and Cronin 2013; Su et al. 2007). A high scattering-to-absorption ratio is a significant factor in applying photons efficiently to improve the performance of plasmon photocatalysis. The sizes, shape, and water-splitting filling amount of metal nanostructures also regulate the improvement of water-splitting efficiency through scattering. For example, the photo reaction rate was reported for silver nanocubes compared with silver nanostructure and nanowires, primarily owing to the differences in their scattering efficiency (Mock et al. 2002).

5.6 Mechanisms of Photocatalytic Dye Decomposition

The photocatalysis process could be used for the degradation of harmful and toxic dyes, the photocatalytic hydrogen process, and the photosynthesis of beneficial chemicals. The elementary mechanisms of nano-semiconductor-based photodegradation of dyes contain photo-based chemical progressions of visible light absorption, electron hole-pair generation, separation, and free charge carrier-

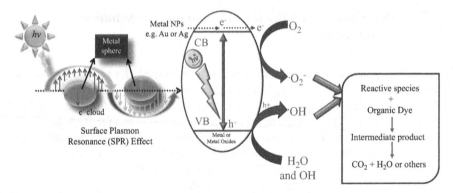

Fig. 5.4 A possible general mechanistic path using localized SPR for the photocatalytic degradation of organic dyes (Khan et al. 2018)

persuaded redox reactions. This process should be helpful in an extensive range of applications, such as wastewater treatment, water splitting, and the self-cleaning of exteriors (Le et al. 2008; Mock et al. 2002; Qu et al. 2013).

In Fig. 5.4, the important steps in the photocatalytic degradation of organic dyes or pollutants are shown. Once the metal NP catalysts are illuminated through photons with an energy greater than or equivalent to their band gap energy (E_g), an electron (e^--cb) is excited from the valence band to the conduction band, which leaves a hole (h^+-vb) (Linic et al. 2011; Xu and Schoonen 2000). In this photocatalytic process, the hot electron-injected excited electrons and holes shift to the surface in an alternative state of nanostructured materials. The charge recombination rate is suppressed repeatedly by a scavenger or another doped nanostructured material, which could easily trap the electrons or holes. Accordingly, the improved crystalline nanostructure materials with fewer vacancies or defects are able to lessen the trapping states and charge recombination sites, which results in an enhanced capability for the use of the photogenerated transporters in the anticipated photo reactions. For the advanced photocatalytic efficiency, the electron–hole pairs must be well defined, and the charges should shift quickly across the surface or interface to hinder the recombination rate of the nanostructured material.

Ag and Au are the most commonly used noble metals NPs for contact with several kinds of semiconducting nanostructure materials due to their strong plasmonic effects. These effects are primarily attributed to their unique morphology, size, and composition, as well as the advantageous properties of the surrounding medium (Armelao et al. 2006; García 2011; Moores and Goettmann 2006).

5.7   Conclusion

This chapter provided information on plasmon-enhanced photocatalysis, particularly on three important aspects of plasmon-based effects: (i) light absorption, (ii) hot electron injection, and (iii) near-field enhancement. The plasmon-based reaction mechanisms and their photo-based effects indicate that the plasmon effect is a very promising approach to improve the photocatalytic performance of metals and metal oxides. Significant advances have been made to extend the technology for practical applications.

Acknowledgment This study was supported by the Priority Research Centers Program (Grant No. 2014R1A6A1031189) through the National Research Foundation of Korea (NRF) funded by the Korean Ministry of Education.

References

Armelao L et al (2006) Recent trends on nanocomposites based on Cu, Ag and Au clusters: a closer look. Coord Chem Rev 250(11–12):1294–1314. https://doi.org/10.1016/j.ccr.2005.12.003

Atwater HA, Polman A (2010) Plasmonics for improved photovoltaic devices. Nat Mater 9(3):205. https://doi.org/10.1038/nmat2866

Bai S et al (2015) Toward enhanced photocatalytic oxygen evolution: synergetic utilization of plasmonic effect and schottky junction via interfacing facet selection. Adv Mater 27 (22):3444–3452. https://doi.org/10.1002/adma.201501200

Camarero L et al (2003) Photo-assisted oxidation of indigocarmine in an acid medium. Environ Eng Sci 20(4):281–287. https://doi.org/10.1089/109287503322148555

Daneshvar N, Salari D, Khataee AR (2004) Photocatalytic degradation of azo dye acid red 14 in water on ZnO as an alternative catalyst to TiO_2. J Photochem Photobiol A Chem 162 (2–3):317–322. https://doi.org/10.1016/S1010-6030(03)00378-2

Daniel M-C, Astruc D (2004) Gold nanoparticles: assembly, supramolecular chemistry, quantum-size-related properties, and applications toward biology, catalysis, and nanotechnology. Chem Rev 104(1):293–346. https://doi.org/10.1021/cr030698

Fujishima A, Honda K (1972) Electrochemical photolysis of water at a semiconductor electrode. Nature 238(5358):37. https://doi.org/10.1038/238037a0

García MA (2011) Surface plasmons in metallic nanoparticles: fundamentals and applications. J Phys D Appl Phys 44(28):283001. https://doi.org/10.1088/0022-3727/44/28/283001

Gramotnev DK, Bozhevolnyi SI (2010) Plasmonics beyond the diffraction limit. Nat Photonics 4 (2):83. https://doi.org/10.1038/nphoton.2009.282

Guo S, Wang E (2011) Noble metal nanomaterials: controllable synthesis and application in fuel cells and analytical sensors. Nano Today 6(3):240–264. https://doi.org/10.1016/j.nantod.2011. 04.007

Hou W, Cronin SB (2013) A review of surface plasmon resonance-enhanced photocatalysis. Adv Funct Mater 23(13):1612–1619. https://doi.org/10.1002/adfm.201202148

Jiang R et al (2014) Metal/semiconductor hybrid nanostructures for plasmon-enhanced applications. Adv Mater 26(31):5274–5309. https://doi.org/10.1002/adma.201400203

Jing L et al (2013) Surface tuning for oxide-based nanomaterials as efficient photocatalysts. Chem Soc Rev 42(24):9509–9549. https://doi.org/10.1039/C3CS60176E

Khan ME, Khan MM, Cho MH (2015a) Biogenic synthesis of a Ag–graphene nanocomposite with efficient photocatalytic degradation, electrical conductivity and photoelectrochemical performance. New J Chem 39(10):8121–8129. https://doi.org/10.1039/C5NJ01320H

Khan ME, Khan MM, Cho MH (2015b) Green synthesis, photocatalytic and photoelectrochemical performance of an Au–Graphene nanocomposite. RSC Adv 5(34):26897–26904. https://doi. org/10.1039/C5RA01864A

Khan ME, Khan MM, Cho MH (2016) CdS-graphene nanocomposite for efficient visible-light-driven photocatalytic andphotoelectrochemical applications. J Colloid Interface Sci 482:221–232. https://doi.org/10.1016/j.jcis.2016.07.070

Khan ME, Khan MM, Cho MH (2017) Ce^{3+}-ion, surface oxygen vacancy, and visible light-induced photocatalytic dye degradation and photocapacitive performance of CeO$_2$-graphene nanostructures. Sci Rep 7(1):5928. https://doi.org/10.1038/s41598-017-06139-6

Khan ME, Khan MM, Cho MH (2018) Recent progress of metal-graphene nanostructures in photocatalysis. Nanoscale 10:9427–9440. https://doi.org/10.1039/c8nr03500h

Kim K-H, Husakou A, Herrmann J (2010) Linear and nonlinear optical characteristics of composites containing metal nanoparticles with different sizes and shapes. Opt Express 18 (7):7488–7496. https://doi.org/10.1364/OE.18.007488

Kreibig U, Vollmer M (1995) Theoretical considerations. In: Optical properties of metal clusters. Springer, Berlin, pp 13–201. https://doi.org/10.1007/978-3-662-09109-8_2

Le F et al (2008) Metallic nanoparticle arrays: a common substrate for both surface-enhanced Raman scattering and surface-enhanced infrared absorption. ACS Nano 2(4):707–718. https://doi.org/10.1021/nn800047e

Lim PY et al (2006) Synthesis of ag nanospheres particles in ethylene glycol by electrochemical-assisted polyol process. Chem Phys Lett 420(4–6):304–308. https://doi.org/10.1016/j.cplett.2005.12.075

Linic S, Christopher P, Ingram DB (2011) Plasmonic-metal nanostructures for efficient conversion of solar to chemical energy. Nat Mater 10(12):911. https://doi.org/10.1038/nmat3151

Maier, SA (2007). Plasmonics: fundamentals and applications. Springer. New York. ISBN: 9780387331508.

Mock JJ et al (2002) Shape effects in plasmon resonance of individual colloidal silver nanoparticles. J Chem Phys 116(15):6755–6759. https://doi.org/10.1063/1.1462610

Moores A, Goettmann F (2006) The plasmon band in noble metal nanoparticles: an introduction to theory and applications. New J Chem 30(8):1121–1132. https://doi.org/10.1039/B604038C

Murray RW (2008) Nanoelectrochemistry: metal nanoparticles, nanoelectrodes, and nanopores. Chem Rev 108(7):2688–2720. https://doi.org/10.1021/cr068077e

Naushad M, Abdullah ALOthman Z, Rabiul Awual M et al (2016) Adsorption of rose Bengal dye from aqueous solution by amberlite Ira-938 resin: kinetics, isotherms, and thermodynamic studies. Desalin Water Treat 57:13527–13533. https://doi.org/10.1080/19443994.2015.1060169

Pathania D, Gupta D, Al-Muhtaseb AH et al (2016) Photocatalytic degradation of highly toxic dyes using chitosan-g-poly(acrylamide)/ZnS in presence of solar irradiation. J Photochem Photobiol A Chem 329:61–68. https://doi.org/10.1016/j.jphotochem.2016.06.019

Qu X, Alvarez PJJ, Li Q (2013) Applications of nanotechnology in water and wastewater treatment. Water Res 47(12):3931–3946. https://doi.org/10.1016/j.watres.2012.09.058

Rayalu SS et al (2013) Photocatalytic water splitting on Au/TiO$_2$ nanocomposites synthesized through various routes: enhancement in photocatalytic activity due to SPR effect. Appl Catal B Environ 142:684–693. https://doi.org/10.1016/j.apcatb.2013.05.057

Saravanan R, Karthikeyan N, Gupta VK, Thangadurai P, Narayanan V, Stephen A (2013) ZnO/Ag nanocomposite: an efficient catalyst for degradation studies of textile effluents under visible light. Mater Sci Eng C 33:2235–2244. http://www.sciencedirect.com/science/article/pii/S0928493113000593

Saravanan R, Manoj D, Qin J, Naushad M, Gracia F, Lee AF, MansoobKhan MM, Gracia-Pinilla MA (2018) Mechanothermal synthesis of Ag/TiO$_2$ for photocatalytic methyl orange degradation and hydrogen production. Process Saf Environ Prot 120:339–347. https://doi.org/10.1016/j.psep.2018.09.015

Sarina S, Waclawik ER, Zhu H (2013) Photocatalysis on supported gold and silver nanoparticles under ultraviolet and visible light irradiation. Green Chem 15(7):1814–1833. https://doi.org/10.1039/C3GC40450A

Sharma VK, Yngard RA, Lin Y (2009) Silver nanoparticles: green synthesis and their antimicrobial activities. Adv Colloid Interf Sci 145(1–2):83–96. https://doi.org/10.1016/j.cis.2008.09.002

Sharma G, Kumar A, Naushad M et al (2017) Photoremediation of toxic dye from aqueous environment using monometallic and bimetallic quantum dots based nanocomposites. J Clean Prod. https://doi.org/10.1016/j.jclepro.2017.11.122

Su K-H et al (2003) Interparticle coupling effects on plasmon resonances of nanogold particles. Nano Lett 3(8):1087–1090. https://doi.org/10.1021/nl034197f

Su YH et al (2007) Layer-by-layer au nanoparticles as a Schottky barrier in a water-based dye-sensitized solar cell. Appl Phys A 88(1):173–178. https://doi.org/10.1007/s00339-007-3988-7

Sun Y, Xia Y (2002) Shape-controlled synthesis of gold and silver nanoparticles. Science 298 (5601):2176–2179. https://doi.org/10.1126/science.1077229

Wang JL, Xu LJ (2012) Advanced oxidation processes for wastewater treatment: formation of hydroxyl radical and application. Crit Rev Environ Sci Technol 42(3):251–325. https://doi.org/10.1080/10643389.2010.507698

Wang P et al (2012) Plasmonic photocatalysts: harvesting visible light with noble metal nanoparticles. Phys Chem Chem Phys 14(28):9813–9825. https://doi.org/10.1039/C2CP40823F

Warren SC, Thimsen E (2012) Plasmonic solar water splitting. Energy Environ Sci 5 (1):5133–5146. https://doi.org/10.1039/C1EE02875H

Xia Y, Halas NJ (2005) Shape-controlled synthesis and surface plasmonic properties of metallic nanostructures. MRS Bull 30(5):338–348. https://doi.org/10.1557/mrs2005.96

Xu Y, Schoonen MAA (2000) The absolute energy positions of conduction and valence bands of selected semiconducting minerals. Am Mineral 85(3–4):543–556. https://doi.org/10.2138/am-2000-0416

Zarazua I et al (2016) Effect of the electrophoretic deposition of Au NPs in the performance CdS QDs sensitized solar cells. Electrochim Acta 188:710–717. https://doi.org/10.1016/j.electacta.2015.11.127

Zhang N, Liu S, Xu Y-J (2012) Recent progress on metal core@semiconductor shell nanocomposites as a promising type of photocatalyst. Nanoscale 4(7):2227–2238. https://doi.org/10.1039/C2NR00009A

Zhang X et al (2013) Plasmonic photocatalysis. Rep Prog Phys 76(4):046401. https://doi.org/10.1088/0034-4885/76/4/046401

Zhang J et al (2014) Self-assembly of a Ag nanoparticle-modified and graphene-wrapped TiO$_2$ nanobelt ternary heterostructure: surface charge tuning toward efficient photocatalysis. Nanoscale 6(19):11293–11302. https://doi.org/10.1039/C4NR03115F

Zhou X et al (2012) Surface plasmon resonance-mediated photocatalysis by noble metal-based composites under visible light. J Mater Chem 22(40):21337–21354. https://doi.org/10.1039/C2JM31902K

Chapter 6
Reduced Graphene Oxide-Based Photocatalysis

R. Suresh, R. V. Mangalaraja, Héctor D. Mansilla, Paola Santander, and Jorge Yáñez

Contents

Abstract Advanced oxidation processes (AOPs), especially photocatalysis, have arisen as a promising solution for the decomposition of harmful organic pollutants in contaminated water. Among various photocatalytic materials, nanocomposites based on reduced graphene oxide (RGO) coupled with semiconductor(s) and/or metal nanoparticles are of significant relevance since they have suitable physicochemical and optical properties such as excellent adsorption of pollutants from aqueous solutions, solar light harvesting tendencies, high photocharge separation efficiency, and excellent stability. Hence, RGO-based composites can have superior

R. Suresh (✉) · J. Yáñez (✉)
Department of Analytical and Inorganic Chemistry, Faculty of Chemical Sciences, University of Concepción, Concepción, Chile
e-mail: jyanez@udec.cl

R. V. Mangalaraja
Department of Materials Engineering, Faculty of Engineering, University of Concepción, Concepción, Chile

Technological Development Unit (UDT), University of Concepcion, Coronel Industrial Park, Coronel, Chile

H. D. Mansilla
Department of Organic Chemistry, Faculty of Chemical Sciences, University of Concepción, Concepción, Chile

P. Santander
Center of Biotechnology, University of Concepción, Concepción, Chile

Millenium Nuclei on Catalytic Processes towards Sustainable Chemistry (CSC), Concepción, Chile

© Springer Nature Switzerland AG 2020   145
M. Naushad et al. (eds.), *Green Photocatalysts*, Environmental Chemistry for a Sustainable World 34, https://doi.org/10.1007/978-3-030-15608-4_6

photocatalytic activity regarding the photodegradation of a broad range of organic pollutants. This chapter mainly focuses on the photocatalytic performances of recently developed RGO-based binary, ternary, and quaternary nanocomposites. The role of RGO in the improvement of photocatalysts is explained taking into account the most recent literature. An improvement strategy like metal and non-metal doping of RGO or semiconductors is also outlined. The photocatalytic activities of RGO-based ternary nanocomposites such as RGO/dual semiconductors, RGO/metal/semiconductor, and RGO/dual metal-free semiconductors are also briefly explained. The photocatalytic performance of RGO-based ternary nanocomposites is described. Furthermore, the plausible photocatalytic pathway for generation of free radicals by RGO-based composites is also explained in detail. This book chapter will be useful to researchers in the field of material science for developing new RGO-based photocatalysts with superior activity and low production costs.

Keywords Reduced graphene oxide · Photocatalyst · Composites · Pollutants · Adsorption · Environmental remediation · Binary composite · Ternary composite · Reducing graphene

6.1 Introduction

Carbon has various allotropes such as graphite, diamond, fullerene (C60), carbon nanotubes, graphene, and carbon quantum dots. These carbon allotropes have been extensively used as support materials for semiconductor photocatalysts which in turn find applications in the degradation of organic pollutants in contaminated water. For instance, graphite deposited TiO_2 (Zhang et al. 2011) was used as photocatalyst for photodegradation of methyl orange under visible light illumination. Boron-doped diamond containing ZnO composites (Gao et al. 2015) has been used in the photocatalytic degradation of methyl orange. Fullerene decorated TiO_2 (Cho et al. 2015) exhibits significant photocatalytic activity toward degradation of methylene blue under visible light illumination. Carbon nanotubes/TiO_2 (Jiang et al. 2013) nanocomposites have been utilized for photocatalytic degradation of methyl orange under ultraviolet light irradiation. Visible light-driven carbon quantum dots modified BiOBr (Liang et al. 2018) photocatalysts have been used for the removal of rhodamine B and ciprofloxacin. Among them, graphene is of pivotal importance for material scientists (Li et al. 2016). This is because of its unique structure and large surface area, extremely hydrophobic nature, and good electronic conducting properties. Therefore, its composites with inorganic/organic semiconductor nanoparticles, dyes, metallic nanoparticles, and metal organic frameworks exhibit superior photocatalytic activity toward the degradation of various organic pollutants (Hui et al. 2016; Sharma et al. 2016). Nevertheless, graphene has low dispersibility in water, i.e., it can easily undergo agglomeration irreversibly in aqueous solution to form graphite. This is caused by the strong van der Waals interactions and π–π

stacking between graphene sheets. Hence, its surface area decreases, which limits its photocatalytic applications. This major drawback can be eradicated by introducing hydrophilic functional groups on the graphene sheets, which can be achieved either by oxidation of graphene sheets or reduction of graphene oxide. In general, the reduction of graphene oxide is a widely used method to introduce functional groups on graphene nanosheets (Pei and Cheng 2012). The main advantages of this process are as follows: (i) the reduction of graphene oxide not only eliminates oxygen-containing functional groups but also develops atomic scale lattice defects along with conjugated graphitic network; (ii) the method is relatively simple, cheap, and faster than its alternatives; and (iii) this method gives high yields with controllable properties.

Graphene oxide contains one or few layers of graphitic oxide (Kanishka et al. 2018). The most acceptable chemical structure of graphene oxide was proposed by Lerf-Klinowski. In this model, basal planes have hydroxyl and epoxide groups, whereas the edges have carboxyl and carbonyl groups. This results in mixed sp^2-sp^3 carbon-containing sheets. Graphene oxide has more sp^3 domains and high oxygen-containing functional groups; thus it acquires high chemical reactivity and hydrophilicity. Therefore, graphene oxide itself is widely used as an outstanding material in many fields including environmental remediation. Nevertheless, the application of graphene oxide to produce large scale functionalized graphene is highly desirable for photocatalysis.

The chemical reduction process can eliminate more functional groups from graphene oxide and form graphene sheets with least functional groups, most accurately known as reduced graphene oxide (RGO). Although the RGO sheets have comparatively lower electronic conductivity than those of graphene, RGO is considered as a versatile material for photocatalytic applications. Moreover, RGO can readily form stable aqueous dispersion through electrostatic stabilization, and it shows better adsorption capacity for aromatic pollutants due to its lower oxygen content, π–π interactions, higher hydrophobicity, greater surface area, and more defect sites. When RGO is composited with semiconductor photocatalysts, the RGO may promote efficient charge transfer through its conjugated structure and thus inhibit the recombination of photogenerated electron–hole pairs. Furthermore, RGO can provide larger adsorptive centers due to its π-conjugated two-dimensional planar structure, which makes the absorption of organic molecules onto its surface more effective. For these reasons, much interest was focused on RGO-based composites with photocatalytic applications.

In this book chapter, photocatalytic performance of RGO-based binary, ternary, and quaternary nanocomposites, fabricated with nanostructured inorganic semiconductors, plasmonic metal nanoparticles, and semiconducting polymers, are outlined.

6.2 RGO-Based Binary Photocatalysts

In general, pure semiconductor photocatalysts exhibit relatively low photocatalytic performance, due to the insufficient utilization of light energy and fast recombination of photogenerated charge carriers. These demerits could be eliminated by coupling RGO with suitable photocatalytic materials like metal oxides, metal chalcogenides, noble metals, and even organic semiconductors. The RGO-based binary composites can be activated by low-powered and low-cost light sources such as light-emitting diodes or fluorescent ultraviolet (UV) lamps. In this sense, this provides an opportunity to develop cost-effective photocatalysts.

Nanostructured metal oxide photocatalysts are immensely used in the photocatalytic degradation of various organic pollutants. Widely studied metal oxide photocatalysts are TiO_2, ZnO, Fe_2O_3, Cu_2O, NiO, WO_3, MoO_3, V_2O_5, and CeO_2. Both TiO_2 and ZnO are UV active photocatalysts while Fe_2O_3, Cu_2O, NiO, WO_3, MoO_3, V_2O_5, and CeO_2 are visible active photocatalysts. Furthermore, they have good electron transport properties, they are easy to synthesize and have low production costs. Their crystal structure, band gap, morphology, chemical stability, and reusability also favor their photocatalytic performance. They are also relatively biocompatible with the environment.

Like metal oxides, metal vanadates, a significant class of inorganic semiconductor nanomaterials, have also received much attention in photocatalysis. Different metal vanadates such as bismuth vanadate ($BiVO_4$), silver vanadate ($AgVO_3$), copper vanadate ($Cu_2V_2O_7$), iron vanadate ($FeVO_4$), and cerium vanadate ($CeVO_4$) are reported for their great photocatalytic activity. Metal vanadates also have suitable properties such as low band gap, large surface area, and different morphology. Moreover, vanadium in vanadate moiety has multiple valence states leading to rich structures of its compounds. They also have good dispersibility, non-toxicity, and good electronic conductivity.

Metal tungstates nanostructures are another important class of semiconducting materials. In recent years, they also grabbed considerable attention in photocatalysis. In general, metal tungstates are classified either as scheelite or as wolframite, depending on their structure type. Bivalent metal cations with large size possess scheelite-type tetragonal structure, whereas cations with lesser size are categorized as wolframite-type monoclinic structure. In scheelite type, four oxygen atoms are bonded with each tungsten atom, whereas in wolframite type every six oxygen atoms are bonded with each tungsten atom. The metal tungstates such as iron tungstate ($FeWO_4$), nickel tungstate ($NiWO_4$), bismuth tungstate (Bi_2WO_6), and zinc tungstate ($ZnWO_4$) are used as photocatalyst for photodegradation of various organic pollutants.

Similarly, metal phosphates, metal niobates, metal sulfides, metal chalcogenides, metal halides, oxyhalides, and hydroxides have also received special attention as potential visible light-driven photocatalysts.

However, it has been well established that these semiconductor photocatalysts alone cannot be utilized as an efficient photocatalyst for photodecomposition of organic pollutants, since they own some particular drawbacks. They are as follows:

(a) Widely studied semiconductors such as TiO_2 and ZnO are only active in ultraviolet region. Solar light comprises about 40% visible light energy. Therefore, visible light-driven semiconductors are highly desirable.
(b) Visible light active semiconductors like Fe_2O_3 and $BiVO_4$ have fast recombination rate of photo-charges.
(c) Semiconductor photocatalysts such as CdS, Fe_2O_3, and ZnO undergo photocorrosion easily.
(d) The conduction and valence band potentials of semiconductors may not be suitable for consumption of photogenerated electrons during oxygen reduction reactions, and thus obstacle oxidative degradation of pollutants by holes.

To overcome these demerits, researchers aimed to synthesize RGO nanosheets containing different kinds of semiconductors. Recently reported RGO/semiconductor binary composites with their photocatalytic performance and experimental conditions have been given as Table 6.1 and Fig. 6.1. These binary composites exhibit enhanced photocatalytic performance due to the following reasons:

(a) An effective photogenerated electron transfer from the conduction band level of semiconductor to RGO.
(b) Greater specific surface area of RGO nanosheets.
(c) RGO prevents agglomeration of semiconductor nanoparticles during its synthesis and thus highly dispersed nanoparticles on RGO can be obtained. Semiconductor nanoparticles also have high surface to volume ratio and they act as active centers.
(d) RGO reduces the band gap energy of ultraviolet active semiconductor nanoparticles. For instance, ZnO/RGO nanocomposite with visible light photocatalytic activity has been reported by Prabhu et al. (2018).
(e) RGO can affect the formation of crystal structure of semiconductors. Thus, it intensively affects their photocatalytic activity.
(f) RGO sheets can inhibit photocorrosion of the semiconductor nanoparticles.

Moreover, the photocatalytic mechanism of RGO/semiconductor has also been well studied by various research groups. For example, Rahimi et al. (2018) have proposed a photocatalytic mechanism of RGO-NiO nanowires on photodegradation of methyl orange molecules. They found that both superoxide and hydroxide radicals are involved in the photodegradation of methyl orange molecules. From the literatures, the general photocatalytic mechanism of RGO/semiconductor composite can be summarized by the following way (Fig. 6.2): Upon illumination of light with sufficient energy on semiconductor/RGO, electron–hole pairs will be created within semiconductor structure. The photoexcited electrons are located in the conduction band and the holes remain in the valence band (Eq. 6.1). Then, the photogenerated electrons are transferred to the RGO surfaces (Eq. 6.2) where

Table 6.1 The photocatalytic performances of RGO-based binary composites

Photocatalyst/ mass	Light source	Pollutant/conc./ volume	Degradation time/ efficiency (%)	References
TiO_2/RGO	150 W medium-pressure Hg lamp	C.I. basic blue 9 and C.I. basic red 1/5 mgL^{-1}	240 min/ 91.2 and 88.3	Stefanska et al. (2018)
ZnO/RGO/ 0.05 g	500 W halogen lamp	Methylene blue/ 10 mgL^{-1}/200 mL	180 min/ 93%	Prabhu et al. (2018)
α-Fe_2O_3/ RGO/10 mg +0.5 mL of H_2O_2	300 W Xe lamp	Rhodamine-B/15 mg/ L/	10 min/ 100%	Zou et al. (2018)
Cu_2O/RGO/ 100 mg	500 W Xe lamp	Rhodamine B/10 mgL^{-1}/100 mL	150 min/ 98.9	Wang et al. (2014)
RGO-NiO/ 20 mg	250 W metal halide lamp	Methyl orange/ 2.5×10^{-5} M/50 mL	300 min/N/ A	Rahimi et al. (2018)
CeO_2/RGO/ 50 mg	Direct sunlight	Methylene blue/ 10^{-5} M/50 mL	50 min/72%	Kumar and Kumar (2017)
RGO-CdO/ 50 mg	UV light/254 nm	Methylene blue/ 10^{-5} M/50 mL	110 min/ 80%	Kumar et al. (2016)
SnO_2@RGO/ 20 mg	500 W Hg lamp	Methyl orange/ 20 mgL^{-1}/50 mL,	60 min	Shen et al. (2017)
RGO/Mn_3O_4/ 10 mg	Natural sunlight	Methylene blue/ 12 mgL^{-1}/50 mL	60 min/ 95.80	Ghosh et al. (2017)
WO_3–RGO/ 30 mg	150 W Xe lamp	Methylene blue/ 15 mg mL^{-1}/20 mL	120 min/100	Fu et al. (2015)
CdS–rGO/ 30 mg	500 W Xe lamp	Congo red/60 mg L^{-1}/60 mL	240 min/ 94.8	Zou et al. (2016)
10RGO-Bi_2S_3/100 mg	500 W Xe lamp	Rhodamine B/10 mg L^{-1}/50 mL	180 min/ 92.2	Zhang et al. (2016)
CuS-rGO/ 300 mg/L	Sunlight (intensity 585 W/m^2)	Congo red/300 mg L^{-1}/ 50 mL	450 min/ 98.76	Borthakur et al. (2016)
Cu_2SnS_3/ RGO/100 mg	500 W Xe lamp	Rhodamine B/10 mg L^{-1}/100 mL	210 min/ 87.0	Yao et al. (2017)
RGO-ZnS/ 25 mg	Sunlight (intensity-800 $\pm$ 10 lx)	Reactive black 5/20 mg L^{-1}/100 mL	300 min/ 90.0	Shamsabadi et al. (2018)
MoS_2-RGO/ 30 mg	300 W high-pressure Hg arc lamp	Methylene blue/ 12 mg mL^{-1}/50 mL	30 min/80.0	Wang et al. (2017a)
MnS_2/RGO/ 50 mg	Six 15 W Hg-vapor lamps	Carbophenothion/ 10 mg L^{-1}/50 mL	450 min/	Fakhri and Kahi (2017)
Cu_3Se_2/rGO (15%)/10 mg	Solar light and visible light	Methylene blue/ 10 mg L^{-1}/30 mL	60 min/91% and 73%	Nouri et al. (2017)
RGO-Ag_2Se/ 0.05 g	8 W visible light	Rhodamine B/200 mmol/50 mL	150 min/ 70.3	Zhu et al. (2015)
		Texbrite BA-L	150 min/ 74.3	

(continued)

Table 6.1 (continued)

Photocatalyst/ mass	Light source	Pollutant/conc./ volume	Degradation time/ efficiency (%)	References
ZnSe/RGO/ 10 mg	Sunlight	Methylene blue/ 10 mg L^{-1}/30 mL	120 min/ 89.5	Yousefi et al. (2018a)
BiVO$_4$/ RGO30/ 100 mg	500 W Xe lamp	$K_2Cr_2O_7$/10 mg L^{-1}/ 100 mL	180 min/ 90.3	Zhang et al. (2018a)
β-SnWO$_4$- RGO	W-halogen lamp	Methylene orange/ 1×10^{-4} M/100 mL	25 min/90.0	Thangavel et al. (2014)
		Rhodamine B/1 $\times 10^{-4}$ M/ 100 mL	20 min/91.0	
RGO– ZnWO$_4$–2/ 10 mg	Visible light	Methyl orange/ 10 mgL^{-1}/500 mL	25 min	Wang et al. (2013)
		RhB/10 mg/L/ 500 mL	25 min	
RGO– Bi$_2$WO$_6$–3/ 0.1 g	Halogen lamp	Methylene blue/ 15 mg L^{-1}/200 mL	90 min/93.5	Xu et al. (2013)
RGO-BiPO$_4$/ 0.04 g	3 W UV lamp	Rhodamine B/10^{-5} M/40 mL	2 h/87.5	Peng et al. (2017)
RGO/SrTiO$_3$/ 0.01 g	Sunlight (intensity = 1.2×10^5 lux)	Rhodamine B/0.04 mg/100 mL	120 min/ 94.590.0	Rosy and Kalpana (2018)
		Rose bengal/ 0.02 mg RB in 100 mL of water		
MgFe$_2$O$_4$– RGO/	Sunlight	Malachite green	120 min/ 98.0	Shetty et al. (2017)
RGO–Ag (1:0.25)/ 22 mg	250 W Hg vapor lamp	Phenol, bisphenol A, and atrazine/100 mg/ L/50 mL.	480 min	Bhunia and Jana (2014)
Ag$_2$CO$_3$/ RGO/0.2 g	250 W Xe arc lamp	Rhodamine B/1 $\times 10^{-5}$ mol L^{-1}/ 50 mL	40 min/100	Wang et al. (2017b)
RGO-BiOBr/ 50 mg	300 W Xe lamp	Rhodamine-B/ 10 ppm/100 mL	70 min/ ≈100%	Alansi et al. (2018)
BiOI–RGO (4 wt%)/ 80 mg	250 W Xenon lamp	Methylorange/10 mg/ L/80 mL	60 min/ ~85%	Vinoth et al. (2017)
RGO/gra-phitic-C$_3$N$_4$/ 100 mg	250 W xenon lamp	Rhodamine B, tetra-cycline/10 mg L^{-1}/ 100 mL	20 min/ 60 min/ 96.1%/ 89.5%	Song et al. (2018)

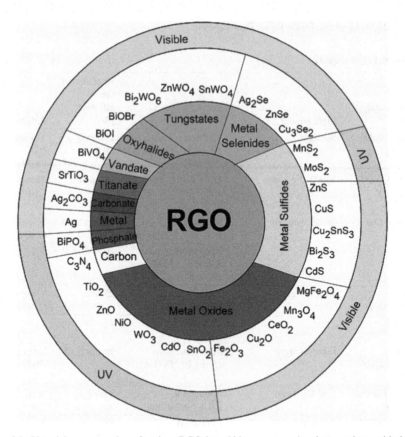

Fig. 6.1 Pictorial representation of various RGO-based binary composite photocatalysts with their suitable light energy

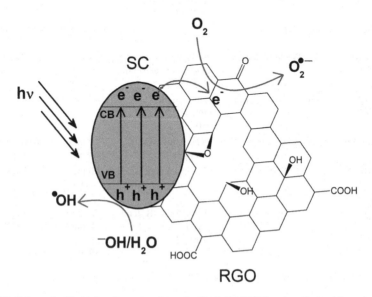

Fig. 6.2 Schematic illustration for generation of radicals by RGO/semiconductor binary composites. *CB* conduction band, *VB* valence band, *SC* semiconductor, *RGO* reduced graphene oxide

oxygen molecules absorb these electrons and form superoxide radicals (Eq. 6.3). According to Eqs. 6.4 and 6.5, the superoxide radicals may also convert into hydroxide radicals (•OH). In this way, RGO decreases recombination rate of photogenerated electron–hole pairs. The holes interact with water molecules to form hydroxyl radicals through an oxidative reaction (Eq. 6.6). The produced hydroxyl radicals are having high oxidizing power and hence degrade targeted pollutants in water (Eq. 6.7).

$$RGO - SC + h\nu \rightarrow RGO - SC \ (e^-_{CB} + h^+_{VB}) \tag{6.1}$$

$$e^-_{CB}(SC) \rightarrow e^-_{trap}(RGO) \tag{6.2}$$

$$e^-_{trap}(RGO) + O_2 \rightarrow O_2^- \tag{6.3}$$

$$O_2^- + 2HO^\bullet + H^+ \rightarrow H_2O_2 + O_2 \tag{6.4}$$

$$H_2O_2 \rightarrow 2OH^\bullet \tag{6.5}$$

$$h^+_{VB} + OH/H_2O \rightarrow {}^\bullet OH + H^+ \tag{6.6}$$

$${}^\bullet OH + Pollutant - Degradaded \ product \tag{6.7}$$

The photocatalytic performance of RGO-based binary composites can be further improved by doping processes either in the semiconductor or RGO. Generally, dopants are metal ions and non-metal ions. Transition metal ions such as Co^{2+}, V^{4+}, Fe^{3+}, etc. and alkaline earth metal ions like Mg^{2+}, Ca^{2+}, Sr^{2+}, etc. have been used as dopants in semiconductors. Usually non-metals, such as nitrogen, sulfur, boron, and fluorine, are used as dopant for both semiconductors and RGO.

Dopants can significantly alter the band gap energy of semiconductors and thus shift their absorption from ultraviolet region to the visible region. For example, colorless ZnO nanorods have a band gap of 3.2 eV, whereas Co-doping via an electrochemical method in ZnO produces green colored Co-doped ZnO powder with band gap of 1.5 eV (Miao et al. 2017). Such a Co-doped ZnO sample effectively catalyzes photodegradation of methylene blue under visible light irradiation. Similarly, TiO_2 synthesized by hydrothermal methods absorbs in the ultraviolet region (3.16 eV) while nitrogen-doped TiO_2 extends its absorption up to the (2.96 eV) visible region (Zhang et al. 2018b; Miranda et al. 2013; Qin et al. 2017). Dopants can also act as efficient charge trapping centers within semiconductor band gap energy. Therefore, recombination of photocharges is effectively reduced. Hence, electrons and holes are available for generation of hydroxyl and superoxide radicals.

Non-metal doping induces more defects and vacancies in RGO, resulting in remarkable improvements in their photocatalytic properties (Raghavan et al. 2018). For example, sulfur-doped RGO/TiO_2 composites exhibit greater photocatalytic efficiency toward methylene blue degradation than that of undoped RGO/TiO_2 composites. Some examples for metal- and non-metal-doped semiconductor/RGO composites with their photocatalytic performance have been compared in Table 6.2. The general scheme for cationic doped semiconductor/RGO composite is shown in Fig. 6.3.

Table 6.2 The improvement strategy in RGO-based binary composites as photocatalysts

Photocatalyst/mass	Light source	Pollutant/concentration/volume	Degradation time (min)	Efficiency (%)	References
N-doped TiO$_2$/RGO/10 mg	Solar simulator (Sun 2000, ABET technology)	Rhodamine B/10 ppm/50 mL	120 min	90%	Zhang et al. (2018b)
Fe$_{2.5}$Cr$_{0.2}$Ce$_{0.3}$O$_4$/RGO/0.2–0.8 g/L	Visible light (125 W)	Methylene blue/30 mg L^{-1}/100 mL	120 min	66.3%	Rad et al. (2018)
Cobalt-doped ZnO/RGO/100 mg	300 W Xe lamp	Methylene blue/ 2×10^{-5} mol L^{-1}/100 mL	180 min	–	Miao et al. (2017)
In,S-TiO$_2$@RGO/0.8 g/L	300 W tungsten xenon lamp	Atrazine/20 mg L^{-1}/25 mL	18 min	100%	Khavar et al. (2018)
Zn$_{(1-x)}$Mg$_x$O/RGO	Sunlight	Methylene blue	60 min	91%	Yousefi et al. (2018b)
Alkaline earth metalions (Mg^{2+}, Ca^{2+}, Sr^{2+} & Ba^{2+}) doped CdSe/RGO/0.05 g	350 W Xenon lamp	Tetracycline hydrochloride/15 mg L^{-1}/100 mL	60 min	85.6%	Zhou et al. (2015)
BiOI/nitrogen-doped RGO/0.1 g	300 W Xe arc lamp	Rhodamine B/50 mg L^{-1}/100 mL	90 min	92%	Lu et al. (2018)
Sulfur-based RGO-TiO$_2$/0.01 g	300 W Xenon lamp	Methylene blue/0.3 mg L^{-1}/100 ml	120 min	93%	Raghavan et al. (2018)
N-[RGO-TiO$_2$]/50 mg	350 W tungsten	Methylene blue, congo red/3.12×10^{-5} M/100 mL	180/70 min	88/96	Appavu et al. (2016)

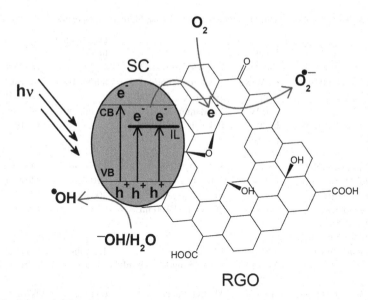

Fig. 6.3 Schematic illustration for generation of radicals by RGO-based binary composites. *CB* conduction band, *VB* valence band, *IL* impurity (dopant) level, *SC* semiconductor, *RGO* reduced graphene oxide

6.3 RGO-Based Ternary Photocatalysts

RGO containing two different photoactive materials are known as RGO-based ternary composites. The photoactive material may be inorganic semiconductors, conducting polymers, metal organic frameworks, and noble metal nanoparticles. Some examples for RGO-based ternary photocatalysts with their performance are given in Table 6.3. The major goals for fabricating RGO-based ternary composites are as follow:

1. To achieve solar light active photocatalysts. For example, addition of NiO into TiO$_2$/RGO (Sharma and Lee 2016) induces a bathochromic shift for the band gap transition of the composite. This is due to the interaction of 3d level of Ni^{2+} ions with that of Ti^{4+} in the conduction band. The prepared NiO-TiO$_2$/RGO shows 88.4% photocatalytic efficiency for photodegradation of o-chlorophenol under visible light illumination.

2. To retard the fast recombination rate of photogenerated electron–hole pairs. For instance, Zn/Fe layered hydroxide/RGO composites with improved photocatalytic degradation of paracetamol have been reported by Zhu et al. (2018). The observed improvement in photocatalytic performance is caused by an improvement in electron–hole pair separation efficiency or a lowering of recombination rate. The added RGO acts as an electron sink and improve charge separation efficiency.

Table 6.3 The photocatalytic performances of RGO-based ternary composites

Photocatalyst/mass	Light source	Pollutant/conc./volume	Degradation time/ efficiency	References
TiO_2/NiO-RGO/0.01% H_2O_2/30 mg	60 W Xe lamp	o-chlorophenol/ 100 mg L^{-1}/100 mL	8 h/88.4%	Sharma and Lee (2016)
RGO-$ZnFe_2O_4$-polyaniline/0.5 mL of H_2O_2/20 mg	250 W high-pressure Hg lamp	Rhodamine B/15 mg L^{-1}/50 mL	10 min/90%	Feng et al. (2016)
Polyaniline/SnS_2/N-doped RGO/300 mg	Visible light	$K_2Cr_2O_7$/50 mg L^{-1}/ 300 mL	120 min/ ~100%	Zhang et al. (2018c)
TiO_2/chitosan/RGO/ cylindrical pieces (diameter of 5 mm, length of 10 mm)	UV lamp	Methyl orange/ 250 mgL^{-1}/100 mL	300 min/ 97%	Chen et al. (2017)
ZnO/RGO/$CuInS_2$/ 10 mg	300 W Xenon lamp	Rhodamine B/5.625 mg L^{-1}/40 mL	120 min/ ~100%	Xu et al. (2018)
ZnO/RGO/polyaniline/ 25 mg	High-pressure Hg lamp	Methyl orange/ 10 mg L^{-1}/50 mL	60 min/ ~100%	Wu et al. (2016)
Graphitic-C_3N_4/ graphene oxide/RGO/ 80 mg	Halogen lamp	Phenol/10 mg L^{-1}/ 160 mL	72 h/28%	Aleksandrzak et al. (2017)
Zn/Fe layered double hydroxides/RGO/25 mg	500 W Xe lamp	Paracetamol/5 mg L^{-1}/ 50 mL	420 min/ 95%	Zhu et al. (2018)
SnS_2-$BiFeO_3$/RGO/ 0.2–0.5 g L^{-1}	250 W Hg Lamp	Methylene blue/methyl orange/5–15 mg L^{-1}/ 150 mL	90 min/ >95%	Bagherzadeh and Kaveh (2018)
Ag/$LaMnO_3$/RGO/ 0.02 g	300 W Xe lamp	Direct green BE/0.02 g L^{-1}/50 mL	2 h/99.8%	Hu et al. (2016)
Ag/Ag_2S/RGO/0.05 g	300 W Xe lamp	Ciprofloxacin/ 10 mg L^{-1}/100 mL	60 min/ 87.6%	Huo et al. (2018)
Ag-AgBr-RGO	300 W Xe lamp	Rhodamine B (10 mg L^{-1})/p-nitrophenol (5 mg L^{-1})/ 100 mL	30 min/ 180 min/ 100%/99.3%	Yang et al. (2018)
TiO_2/Ag/RGO	150 W high-pressure Hg lamp	Methylene blue/ 20 mg L^{-1}	45 min/ ~100%	Vasilaki et al. (2015)
Ag-RGO-$BiVO_4$/0.20 g	500 W Xe lamp	Rhodamine B/5 mg L^{-1}/200 mL	10 h/80.2%	Du et al. (2016)
Pd/MIL-101/RGO/ 50 mg	500 W Xe lamp	Brilliant green/acid fuchsin/25 mg L^{-1}/ 200 mL	15 min/ 20 min/ 100%/100%	Wu et al. (2015)
$NiWO_4$-ZnO-nitrogen-doped RGO/20 mg	250 W Hg lamp	Methylene blue/ 10 mg L^{-1}/200 mL	120 min/ 99.5%	Sadiq et al. (2017a, b)

(continued)

Table 6.3 (continued)

Photocatalyst/mass	Light source	Pollutant/conc./volume	Degradation time/ efficiency	References
RGO/CoFe$_2$O$_4$/Ag/ 0.01 g	500 W Xe lamp	Chlorinated paraffins/ 0.01 mL	12 h/91.9%	Chen et al. (2016)
CeVO$_4$/RGO/BiVO$_4$/ 50 mg	500 W Xe lamp	Tetracycline/ 20 mg L^{-1}/100 mL	60 min/ ~100%	Liu et al. (2018a)
BiVO$_4$/SiO$_2$/RGO	1000 W Xe lamp	C.I. reactive blue 19/50 μg mL^{-1} + 5 mmol L^{-1} H$_2$O$_2$ and 1 mL 3% HCl	45 min/98%	Liu et al. (2018b)
RGO/BiVO$_4$/CdS/ 0.3 g L^{-1}	350 W Xe lamp	Isoniazid/1,4-dioxane/ 50 mg L^{-1}/100 mL	300 min/ 88.0% and 93.0%	Selvam et al. (2018)
Au-doped WO$_3$/TiO$_2$/ RGO/1.0 g L^{-1}	9 W UV lamp	2,4-dichlorophenoxy acetic acid/ 3.2 × 10^{-3} mol L^{-1}/ 350 mL	150 min	Iliev et al. (2018)
N-doped RGO-FeWO$_4$/ Fe$_3$O$_4$/0.02 g	250 W Hg lamp	Methylene blue/ 10 mg L^{-1}/200 mL	100 min/ ~100%	Sadiq et al. (2017a, b)
RGO/WS$_2$/Mg-doped ZnO	UV light	Rhodamine B	20 min/ ~100%	Yu et al. (2017)
AgCl@Ag/N-RGO/ 1.6 g L^{-1}	150 W Xe arc lamp	2-chlorophenol/ 10 mg L^{-1}/60 mL	150 min/ ~100%	Wang et al. (2017c)

3. To achieve stability of certain type of photocatalyst. For example, visible light active Ag-AgBr/RGO (Yang et al. 2018) photocatalysts have been reported for the photodegradation of rhodamine B and p-nitrophenol. In general, visible light illumination causes photoreduction of Ag$^+$ ions into Ag^0. Therefore, during photocatalysis, holes cause oxidation of bromide and leaching of Ag$^+$ ions from AgBr (Eq. 6.8), while electrons reduce Ag$^+$ ions as Ag^0 (Eq. 6.9). Nevertheless, bromine oxidizes pollutant molecule and again converted as bromide ion (Eq. 6.10), whereas electron derived by Ag^0 will be immediately transferred to RGO and hence photoreduction of Ag$^+$ in AgBr was hindered (Eq. 6.11). Again, bromide and Ag$^+$ ions combine to form AgBr (Eq. 6.12). In this way, RGO ensures the stability of Ag-AgBr photocatalyst during photocatalytic experiments.

$$AgBr + h^+{}_{VB} \rightarrow Ag^+ + Br^o \tag{6.8}$$

$$Ag^+ + e^-{}_{CB} \rightarrow Ag^o \tag{6.9}$$

$$Br^o + Pollutant \rightarrow Br^- + Degradaded\ product \tag{6.10}$$

$$Ag^o \rightarrow Ag^+ + e^-_{trap}(RGO) \tag{6.11}$$

$$Ag^+ + Br^- \rightarrow AgBr \tag{6.12}$$

4. To produce a biocompatible photocatalyst. For instance, TiO_2/chitosan/RGO composites (Chen et al. 2017) with enhanced photocatalytic decolorization of methyl orange have been reported. The chitosan is a natural polymer which shows excellent adsorption property due to its hydroxyl and amino groups. This biocompatible polymer also provides reaction sites on the TiO_2 surface. However, to provide more dispersibility, RGO is added with TiO_2/chitosan composite.

5. To produce metal-free photocatalysts. For example, graphitic C_3N_4/graphene oxide/RGO composites (Aleksandrzak et al. 2017) with greater photocatalytic activity toward phenol degradation under visible light have been reported. The graphitic C_3N_4 with band gap of 2.7 eV is relevant for visible light-assisted photocatalysis. However, it was found that C_3N_4/graphene oxide/RGO shows better photocatalytic activity than that of pure C_3N_4. This improvement is credited to the enhanced emitted upconverted photoluminescence of RGO, resulting in improved generation of electron–hole pairs.

The photocatalytic mechanism of RGO-based ternary composites was well explored in literatures. For examples, photocatalytic degradation mechanisms of $CeVO_4$/RGO/$BiVO_4$ photocatalyst, proposed by Liu et al. (2018a), are discussed below. On the basis of free radicals scavenging experiments, they have proposed Z-scheme-type photocatalytic mechanism for $CeVO_4$/RGO/$BiVO_4$ composite on photodegradation of tetracycline. In photocatalytic process, photogenerated electrons from conduction band of $BiVO_4$ recombine with holes in the valence band of $CeVO_4$ (Eq. 6.13). Here, RGO sheets act as the bridge for the transport of charges from $BiVO_4$ to $CeVO_4$. Due to the position of band potentials, superoxide radicals and hydroxyl radicals are produced at the conduction band of $CeVO_4$ and valence band of $BiVO_4$, respectively. By this way, the electron–hole separation occurs effectively in $CeVO_4$/RGO/$BiVO_4$. Hence, the as-produced radicals degraded tetracycline molecules in water.

$$BiVO_4(e^-_{CB} + h^+_{VB}) + CeVO_4(e^-_{CB} + h^+_{VB}) \xrightarrow{RGO} BiVO_4(h^+_{VB}) + CeVO_4(e^-_{CB}) \tag{6.13}$$

Similarly, electron–hole pair separation occurred within ZnO/polyaniline/RGO composite also (Wu et al. 2016). During UV irradiation, electronic excitation from HOMO to LUMO occurs within polyaniline structure. Due to energy level, the holes in valence band of ZnO are transferred to HOMO of polyaniline, while electrons in LUMO of polyaniline are migrated to conduction band of ZnO. As a result, an effective electron–hole pair separation could be achieved, which significantly enhances the photocatalytic activity of ZnO/polyaniline/RGO.

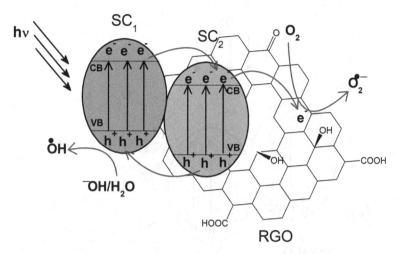

Fig. 6.4 Schematic illustration for generation of radicals by RGO-based ternary composites. *CB* conduction band, *VB* valence band, *SC₁* and *SC₂* semiconductors, *RGO* reduced graphene oxide

From the above examples, the general photocatalytic mechanism of RGO-based ternary composites can be outlined by the following manner (Fig. 6.4): In a typical photocatalytic experiment, the photogenerated electrons–holes migrate to the surface of the photocatalyst. The interface between two different photocatalytic materials could inhibit the recombination of electron–hole pair. On the other hand, electrons and holes will migrate to various suitable directions. Generally, electrons from the conduction band will move to that having the least potential. The holes will migrate from more positive valence band potential to less positive valence band potential. Then, photogenerated electrons are transferred to the RGO surfaces which further slow down recombination rate of photo-charges. Holes interact with water molecules to produce hydroxyl radicals (•OH), while electrons reduce oxygen molecules to give superoxide radicals. Both superoxide species and hydroxyl radicals are highly reactive and effectively decompose organic pollutant molecules.

6.4 RGO-Based Quaternary Composites

The RGO containing three different photoactive materials are known as RGO-based quaternary composites. Photoactive materials may be semiconductors and plasmonic metals. The main goal when fabricating RGO quaternary composites is to reduce charge recombination and thus improve photocatalytic performance. For example, in order to photo-oxidize 2,4-dichlorophenoxyacetic acid, Au-doped WO_3/TiO_2/RGO quaternary composites have been used (Iliev et al. 2018). Similarly, RGO/TiO_2/Pd-Ag (Khojasteh et al. 2018) and TiO_2-Bi_2O_3/(BiO)$_2$CO$_3$-RGO (Žerjav et al. 2018)

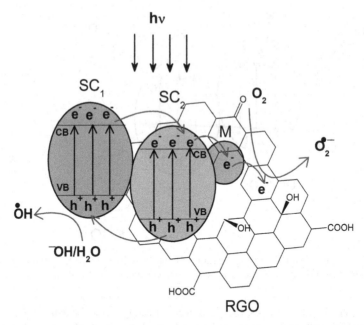

Fig. 6.5 Schematic illustration for generation of radicals by RGO-based ternary composites. *CB* conduction band, *VB* valence band, *SC₁* and *SC₂* semiconductors, *M* metal nanoparticles, *RGO* reduced graphene oxide

have been utilized as effective photocatalysts for photodegradation of rhodamine B and bisphenol A, respectively.

The general photocatalytic pathway for production of free radicals by RGO-based quaternary composite photocatalysts can be explained in the following way (Fig. 6.5): Electron–hole pairs will be generated within semiconductors when they are irradiated with light of suitable energy. The electrons are transferred to the conduction band and the holes remain in the valence band. The photogenerated electrons are transferred to the RGO surfaces through semiconductor materials, and hence efficient photo-charges separation could be achieved. The holes interact with water molecules to form hydroxyl radicals, while the electrons generate superoxide radicals via the reduction of oxygen molecules. These superoxide and hydroxyl radicals can decompose harmful organic pollutants in water.

6.5   Conclusions

An overview of the developments of RGO-based composites as photocatalysts is described. As a support material for photocatalysts, RGO has several advantages such as greater adsorption, the fact that it decreases recombination rate of photocharges, the narrowing of the band gap of semiconductors, easy synthesis, and the fact that it decreases photocorrosion. These properties make it a great potential in the field of visible light-assisted photocatalytic technology. This chapter is mainly focused on the photocatalytic performance of RGO-based binary, ternary, and quaternary nanocomposites toward degradation of organic pollutants in waste water. The summary of this chapter is as follows:

1. The reasons for selection of RGO as support for photocatalytic materials are systematically introduced.
2. RGO-based binary composites such as RGO/metal oxides, RGO/metal vanadates, RGO/metal tungstates, RGO/metal phosphates, RGO/metal chalcogenides, and RGO/organic semiconductors with appropriate examples have been described. From these reviews, the vital roles of RGO have been pointed out. They are the following: (a) RGO effectively transfers electrons and thus increases photogenerated charge separation efficiency, (b) provides large surface area for adsorption of pollutants, (c) decreases size of the semiconductors during synthesis step, (d) narrows the band gap of semiconductor, (e) affects the crystal structure of certain metal oxides, and (f) inhibits photocorrosion of semiconductors. The strategies for enhancing the photocatalytic activity of RGO-based binary composites such as doping with metal and non-metal in semiconductor as well as RGO were also described in detail. The doping process on the semiconductor and RGO has been confirmed to be a successful method to achieve superior photocatalysts under solar light illumination.
3. The development of RGO-based ternary composites such as RGO/dual semiconductors, RGO/semiconductor/metal, and RGO/dual metal-free semiconductors with enhanced visible light-driven photocatalytic activity has also been outlined. The main goals for fabrication of RGO-based ternary composites are (a) to achieve solar light active photocatalysts, (b) to achieve stability of such photocatalysts, (c) to achieve biocompatible photocatalysts, (d) to achieve metal-free photocatalysts, and (d) to ensure the stability of certain type of photocatalytic materials.
4. RGO-based quaternary composites are also presented at the end of this chapter. Moreover, the generalized pathway for production of superoxide and hydroxyl free radicals by RGO-based binary, ternary, and quaternary composites has also been described in detail. RGO could be a potential support for photocatalytic materials for water treatment and environmental remediation in the coming decades.

Acknowledgments RS acknowledges the National Commission for Scientific and Technological Research (CONICYT), Santiago, Chile, for the financial assistance in the form of postdoctoral fellowship (Fondecyt Project No: 3160499). The authors thank the support of the projects FONDECYT 1151296; Center Optics and Photonics, Grant CONICYT-PFB-0824 and CONICYT/FONDAP/15110019.

References

Alansi AM, Qunaibit MA, Alade IO, Qahtan TF, Saleh TA (2018) Visible-light responsive BiOBr nanoparticles loaded on reduced grapheme oxide for photocatalytic degradation of dye. J Mol Liq 253:297–304. https://doi.org/10.1016/j.molliq.2018.01.034

Aleksandrzak M, Kukulka W, Mijowska E (2017) Graphitic carbon nitride/graphene oxide/reduced graphene oxide nanocomposites for photoluminescence and photocatalysis. Appl Surf Sci 398:56–62. https://doi.org/10.1016/j.apsusc.2016.12.023

Appavu B, Kannan K, Thiripuranthagan S (2016) Enhanced visible light photocatalytic activities of template free mesoporous nitrogen doped reduced graphene oxide/titania composite catalysts. J Ind Eng Chem 36:184–193. https://doi.org/10.1016/j.jiec.2016.01.042

Bagherzadeh M, Kaveh R (2018) A new SnS_2-$BiFeO_3$/reduced graphene oxide photocatalyst with superior photocatalytic capability under visible light irradiation. J Photochem Photobiol A 359:11–22. https://doi.org/10.1016/j.jphotochem.2018.03.031

Bhunia SK, Jana NR (2014) Reduced graphene oxide-silver nanoparticle composite as visible light photocatalyst for degradation of colorless endocrine disruptors. ACS Appl Mater Interfaces 6:20085–20092. https://doi.org/10.1021/am505677x

Borthakur P, Boruah PK, Darabdhara G, Sengupta P, Das MR, Boronin AI, Kibis LS, Kozlova MN, Fedorov VE (2016) Microwave assisted synthesis of CuS-reduced graphene oxide nanocomposite with efficient photocatalytic activity towards azo dye degradation. J Environ Chem Eng 4:4600–4611. https://doi.org/10.1016/j.jece.2016.10.023

Chen X, Zhao Q, Li X, Wang D (2016) Enhanced photocatalytic activity of degrading short chain chlorinated paraffins over reduced graphene oxide/$CoFe_2O_4$/Ag nanocomposite. J Colloid Interface Sci 479:89–97. https://doi.org/10.1016/j.jcis.2016.06.053

Chen C, Zhang Y, Zeng J, Zhang F, Zhou K, Bowen CR, Zhang D (2017) Aligned macroporous TiO_2/chitosan/reduced graphene oxide (rGO) composites for photocatalytic applications. Appl Surf Sci 424:170–176. https://doi.org/10.1016/j.apsusc.2017.02.137

Cho EC, Ciou JH, Zheng JH, Pan J, Hsiao YS, Lee KC, Huang JH (2015) Fullerene C70 decorated TiO_2 nanowires for visible-light-responsive photocatalyst. Appl Surf Sci 355:536–546. https://doi.org/10.1016/j.apsusc.2015.07.062

Du M, Xiong S, Wu T, Zhao D, Zhang Q, Fan Z, Zeng Y, Ji F, He Q, Xu X (2016) Preparation of a microspherical silver-reduced graphene oxide-bismuth vanadate composite and evaluation of its photocatalytic activity. Materials 9:160(14 pages). https://doi.org/10.3390/ma9030160

Fakhri A, Kahi DS (2017) Synthesis and characterization of MnS_2/reduced graphene oxide nanohybrids for with photocatalytic and antibacterial activity. J Photochem Photobiol B 166:259–263. https://doi.org/10.1016/j.jphotobiol.2016.12.017

Feng J, Hou Y, Wang X, Quan W, Zhang J, Wang Y, Li L (2016) In-depth study on adsorption and photocatalytic performance of novel reduced graphene oxide-$ZnFe_2O_4$-polyaniline composites. J Alloys Compd 681:157–166. https://doi.org/10.1016/j.jallcom.2016.04.146

Fu L, Xia T, Zheng Y, Yang J, Wang A, Wang Z (2015) Preparation of WO_3-reduced graphene oxide nanocomposites with enhanced photocatalytic property. Ceram Int 41:5903–5908. https://doi.org/10.1016/j.ceramint.2015.01.022

Gao S, Jiao S, Lei B, Li H, Wang J, Yu Q, Wang D, Guo F, Zhao L (2015) Efficient photocatalyst based on ZnO nanorod arrays/p-type boron-doped-diamond heterojunction. J Mater Sci Mater Electron 26:1018–1022. https://doi.org/10.1007/s10854-014-2498-6

Ghosh S, Basu S, Baskey(Sen) M (2017) Decorating mechanism of Mn_3O_4 nanoparticles on reduced graphene oxide surface through reflux condensation method to improve photocatalytic performance. J Mater Sci Mater Electron 28:17860–17870. https://doi.org/10.1007/s10854-017-7727-3

Hu J, Liu Y, Men J, Zhang L, Huang H (2016) Ag modified $LaMnO_3$ nanorods-reduced graphene oxide composite applied in the photocatalytic discoloration of direct green. Solid State Sci 61:239–245. https://doi.org/10.1016/j.solidstatesciences.2016.10.008

Hui KN, Hui KS, Zhang XL et al (2016) Photosensitization of ZnO nanowire-based electrodes using one-step hydrothermally synthesized CdSe/CdS (core/shell) sensitizer. Sol Energy 125:125–134. https://doi.org/10.1016/j.solener.2015.12.002

Huo P, Liu C, Wu D, Guan J, Li J, Wang H, Tang Q, Li X, Yan Y, Yuand S (2018) Fabricated Ag/Ag_2S/reduced graphene oxide composite photocatalysts for enhancing visible light photocatalytic and antibacterial activity. J Ind Eng Chem 57:125–133. https://doi.org/10.1016/j.jiec.2017.08.015

Iliev V, Tomova D, Bilyarska L (2018) Promoting the oxidative removal rate of 2,4-dichlorophenoxyacetic acid on gold-doped WO_3/TiO_2/reduced graphene oxide photocatalysts under UV light irradiation. J Photochem Photobiol A 351:69–77. https://doi.org/10.1016/j.jphotochem.2017.10.022

Jiang T, Zhang L, Ji M, Wang Q, Zhao Q, Fu X, Yin H (2013) Carbon nanotubes/TiO_2 nanotubes composite photocatalysts for efficient degradation of methyl orange dye. Particuology 11:737–742. https://doi.org/10.1016/j.partic.2012.07.008

Kanishka K, Silva HD, Huang HH, Yoshimura M (2018) Progress of reduction of graphene oxide by ascorbic acid. Appl Surf Sci 447:338–346. https://doi.org/10.1016/j.apsusc.2018.03.243

Khavar AHC, Moussavi G, Mahjoub AR, Satari M, Abdolmaleki P (2018) Synthesis and visible-light photocatalytic activity of In,S-TiO_2@rGO nanocomposite for degradation and detoxification of pesticide atrazine in water. Chem Eng J 345:300–311. https://doi.org/10.1016/j.cej.2018.03.095

Khojasteh H, Niasari MS, Sangsefidi FS (2018) Photocatalytic evaluation of RGO/TiO_2NWs/Pd-Ag nanocomposite as an improved catalyst for efficient dye degradation. J Alloys Compd 746:611–618. https://doi.org/10.1016/j.jallcom.2018.02.345

Kumar S, Kumar A (2017) Enhanced photocatalytic activity of rGO-CeO_2 nanocomposites driven by sunlight. Mater Sci Eng B 223:98–108. https://doi.org/10.1016/j.mseb.2017.06.006

Kumar S, Ojha AK, Walkenfort B (2016) Cadmium oxide nanoparticles grown in situ on reduced graphene oxide for enhanced photocatalytic degradation of methylene blue dye under ultraviolet irradiation. J Photochem Photobiol B 159:111–119. https://doi.org/10.1016/j.jphotobiol.2016.03.025

Li X, Yu J, Wageh S, Ghamdi AAA, Xie J (2016) Graphene in photocatalysis: a review. Small 12:6640–6696. https://doi.org/10.1002/smll.201600382

Liang Z, Yang J, Zhou C, Mo Q, Zhang Y (2018) Carbon quantum dots modified BiOBr microspheres with enhanced visible light photocatalytic performance. Inorg Chem Commun 90:97–100. https://doi.org/10.1016/j.inoche.2018.02.013

Liu Q, Shen J, Yang X, Zhang T, Tang H (2018a) 3D reduced graphene oxide aerogel-mediated Z-scheme photocatalytic system for highly efficient solar-driven water oxidation and removal of antibiotics. Appl Catal B-Environ 232:562–573. https://doi.org/10.1016/j.apcatb.2018.03.100

Liu B, Lin L, Yu D, Sun J, Zhu Z, Gao P, Wang W (2018b) Construction of fiber-based $BiVO$ /SiO /reduced graphene oxide (RGO) with efficient visible light photocatalytic activity. Cellulose 25:1089–1101. https://doi.org/10.1007/s10570-017-1628-8

Lu B, Zeng S, Li C, Wang Y, Pan X, Zhang L, Mao H, Lu Y, Ye Z (2018) Nanoscale p-n heterojunctions of BiOI/nitrogen-doped reduced graphene oxide as a high performance photocatalyst. Carbon 132:191–198. https://doi.org/10.1016/j.carbon.2018.02.038

Miao Y, Wang X, Wang W, Zhou C, Feng G, Cai J, Zhang R (2017) Synthesis of cobalt-doped ZnO/rGO nanoparticles with visible-light photocatalytic activity through a cobalt-induced electrochemical method. J Energy Chem 26:549–555. https://doi.org/10.1016/j.jechem.2016.10.017

Miranda C, Mansilla H, Yánez J, Obregón S, Colón G (2013) Improved photocatalytic activity of g-C_3N_4/TiO_2 composites prepared by a simple impregnation method. J Photochem Photobiol A 253:16–21. https://doi.org/10.1016/j.jphotochem.2012.12.014

Nouri M, Saray AM, Azimi HR, Yousefi R (2017) High solar-light photocatalytic activity of using Cu_3Se_2/rGO nanocomposites synthesized by a green co-precipitation method. Solid State Sci 73:7–12. https://doi.org/10.1016/j.solidstatesciences.2017.09.001

Pei S, Cheng HM (2012) The reduction of graphene oxide. Carbon 50:3210–3228. https://doi.org/10.1016/j.carbon.2011.11.010

Peng H, Jing N, Miao Y, Yan LS (2017) Facile solvothermal synthesis of reduced graphene oxide-BiPO$_4$ nanocomposite with enhanced photocatalytic activity. Chin J Anal Chem 45(3):357–362. https://doi.org/10.1016/S1872-2040(17)61000-4

Prabhu S, Pudukudy M, Sohila S, Harish S, Navaneethan M, Navaneethan D, Ramesh R, Hayakawa Y (2018) Synthesis, structural and optical properties of ZnO spindle/reduced graphene oxide composites with enhanced photocatalytic activity under visible light irradiation. Opt Mater 79:186–195. https://doi.org/10.1016/j.optmat.2018.02.061

Qin J, Yang C, Cao M, Zhang X, Saravanan R, Limpanart S, Mab M, Liu R (2017) Two-dimensional porous sheet-like carbon-doped ZnO/g-C$_3$N$_4$nanocomposite with high visible-light photocatalytic performance. Mater Lett 189:156–159. https://doi.org/10.1016/j.matlet.2016.12.007

Rad TS, Khataee A, Pouran SR (2018) Synergistic enhancement in photocatalytic performance of Ce(IV) and Cr(III) co-substituted magnetite nanoparticles loaded on reduced graphene oxide sheets. J Colloid Interface Sci 528:248–262. https://doi.org/10.1016/j.jcis.2018.05.087

Raghavan N, Thangavel S, Sivalingam Y, Venugopal G (2018) Investigation of photocatalytic performances of sulfur based reduced graphene oxide-TiO$_2$ nanohybrids. Appl Surf Sci 449:712–718. https://doi.org/10.1016/j.apsusc.2018.01.043

Rahimi K, Zafarkish H, Yazdani A (2018) Reduced graphene oxide can activate the sunlight-inducedphotocatalytic effect of NiO nanowires. Mater Design 144:214–221. https://doi.org/10.1016/j.matdes.2018.02.030

Rosy A, Kalpana G (2018) Reduced graphene oxide/strontium titanate heterostructured nanocomposite as sunlight driven photocatalyst for degradation of organic dye pollutants. Curr Appl Phys 18:1026–1033. https://doi.org/10.1016/j.cap.2018.05.019

Sadiq MMJ, Shenoy US, Bhat DK (2017a) Enhanced photocatalytic performance of N-doped RGO-FeWO$_4$/Fe$_3$O$_4$ ternary nanocomposite in environmental applications. Mater Today Chem 4:133–141. https://doi.org/10.1016/j.mtchem.2017.04.003

Sadiq MMJ, Shenoy US, Bhat DK (2017b) NiWO$_4$-ZnO-NRGO ternary nanocomposite as an efficient photocatalyst for degradation of methylene blue and reduction of 4-nitro phenol. J Phys Chem Solids 109:124–133. https://doi.org/10.1016/j.jpcs.2017.05.023

Selvam NCS, Kim YG, Kim DJ, Hong WH, Kim W, Park SH, Jo WK (2018) Reduced graphene oxide-mediated Z-scheme BiVO$_4$/CdS nanocomposites for boosted photocatalytic decomposition of harmful organic pollutants. Sci Total Environ 635:741–749. https://doi.org/10.1016/j.scitotenv.2018.04.169

Shamsabadi TM, Goharshadi EK, Shafaee M, Niazi Z (2018) ZnS@reduced graphene oxide nanocomposite as an effective sunlight driven photocatalyst for degradation of reactive black 5: a mechanistic approach. Sep Purif Technol 202:326–334. https://doi.org/10.1016/j.seppur.2018.04.001

Sharma A, Lee BK (2016) Integrated ternary nanocomposite of TiO$_2$/NiO/reduced grapheme oxide as a visible light photocatalyst for efficient degradation of o-chlorophenol. J Environ Manag 181:563–573. https://doi.org/10.1016/j.jenvman.2016.07.016

Sharma G, Naushad M, Pathania D, Kumar A (2016) A multifunctional nanocomposite pectin thorium(IV) tungstomolybdate for heavy metal separation and photoremediation of malachite green. Desalin Water Treat 57:19443–19455. https://doi.org/10.1080/19443994.2015.1096834

Shen H, Zhao X, Duan L, Liu R, Wu H, Hou T, Jiang X, Gao H (2017) Influence of interface combination of RGO-photosensitized SnO$_2$@RGO core-shell structures on their photocatalytic performance. Appl Surf Sci 391:627–634. https://doi.org/10.1016/j.apsusc.2016.06.031

Shetty K, Lokesh SV, Rangappa D, Nagaswarupa HP, Nagabhushana H, Anantharaju KS, Prashantha SC, Vidya YS, Sharma SC (2017) Designing MgFe$_2$O$_4$ decorated on green mediated reduced graphene oxide sheets showing photocatalytic performance and luminescence property. Physica B 507:67–75. https://doi.org/10.1016/j.physb.2016.11.021

Song C, Fan M, Shi W, Wang W (2018) High-performance for hydrogen evolution and pollutant degradation of reduced graphene oxide/two-phase g-C$_3$N$_4$ heterojunction photocatalysts. Environ Sci Pollut R 25:14486–14498. https://doi.org/10.1007/s11356-018-1502-8

Stefanska KS, Fluder M, Tylus W, Jesionowski T (2018) Investigation of amino-grafted TiO_2/reduced graphene oxide hybridsas a novel photocatalyst used for decomposition of selected organic dyes. J Environ Manag 212:395–404. https://doi.org/10.1016/j.jenvman.2018.02.030

Thangavel S, Venugopal G, Kim SJ (2014) Enhanced photocatalytic efficacy of organic dyes using β-tin tungstate reduced graphene oxide nanocomposites. Mater Chem Phys 145:108–115. https://doi.org/10.1016/j.matchemphys.2014.01.046

Vasilaki E, Georgaki I, Vernardou D, Vamvakaki M, Katsarakis N (2015) Ag-loaded TiO_2/reduced graphene oxide nanocomposites forenhanced visible-light photocatalytic activity. Appl Surf Sci 353:865–872. https://doi.org/10.1016/j.apsusc.2015.07.056

Vinoth R, Babu SG, Ramachandran R, Neppolian B (2017) Bismuth oxyiodide incorporated reduced graphene oxide nanocomposite material as an efficient photocatalyst for visible light assisted degradation of organic pollutants. Appl Surf Sci 418:163–170. https://doi.org/10.1016/j.apsusc.2017.01.278

Wang W, Shen J, Li N, Ye M (2013) Synthesis of novel photocatalytic RGO–$ZnWO_4$ nanocomposites with visible light photoactivity. Mater Lett 106:284–286. https://doi.org/10.1016/j.matlet.2013.05.042

Wang A, Li X, Zhao Y, Wu W, Chen J, Meng H (2014) Preparation and characterizations of Cu_2O/reduced graphene oxide nanocomposites with high photo-catalytic performances. Powder Technol 261:42–48. https://doi.org/10.1016/j.powtec.2014.04.004

Wang C, Wang AW, Feng J, Li Z, Chen B, Wu QH, Jiang J, Lu J, Li YY (2017a) Hydrothermal preparation of hierarchical MoS_2-reduced graphene oxide nanocomposites towards remarkable enhanced visible-light photocatalytic activity. Ceram Int 43:2384–2388. https://doi.org/10.1016/j.ceramint.2016.11.026

Wang W, Liu Y, Zhang H, Qian Y, Guo Z (2017b) Re-investigation on reduced graphene oxide/AG_2CO_3 composite photocatalyst: an insight into the double-edged sword role of RGO. Appl Surf Sci 396:102–109. https://doi.org/10.1016/j.apsusc.2016.11.030

Wang L, Shi Y, Wang T, Zhang L (2017c) Silver chloride enwrapped silver grafted on nitrogen-doped reduced graphene oxide as a highly efficient visible-light-driven photocatalyst. J Colloid Interface Sci 505:421–429. https://doi.org/10.1016/j.jcis.2017.06.037

Wu Y, Luo H, Zhang L (2015) Pd nanoparticles supported on MIL-101/reduced graphene oxide photocatalyst: an efficient and recyclable photocatalyst for triphenylmethane dye degradation. Environ Sci Pollut Res 22:17238–17243. https://doi.org/10.1007/s11356-015-5364-z

Wu H, Lin S, Chen C, Liang W, Liu X, Yang H (2016) A new ZnO/rGO/polyaniline ternary nanocomposite as photocatalyst with improved photocatalytic activity. Mater Res Bull 83:434–441. https://doi.org/10.1016/j.materresbull.2016.06.036

Xu J, Ao Y, Chen M (2013) A simple method for the preparation of Bi_2WO_6-reduced grapheme oxide with enhanced photocatalytic activity under visible light irradiation. Mater Lett 92:126–128. https://doi.org/10.1016/j.matlet.2012.10.038

Xu T, Hu J, Yang Y, Que W, Yin X, Wu H, Chen L (2018) Ternary system of ZnO nanorods/reduced graphene oxide/$CuInS_2$ quantum dots for enhanced photocatalytic performance. J Alloys Compd 734:196–203. https://doi.org/10.1016/j.jallcom.2017.10.275

Yang Y, Zhang W, Liu R, Cui J, Deng C (2018) Preparation and photocatalytic properties of visible light driven Ag-AgBr-RGO composite. Sep Purif Technol 190:278–287. https://doi.org/10.1016/j.seppur.2017.09.003

Yao S, Xu L, Gao Q, Wang X, Kong N, Li W, Wang J, Li G, Pu X (2017) Enhanced photocatalytic degradation of Rhodamine B by reduced graphene oxides wrapped-Cu_2SnS_3 flower-like architectures. J Alloys Compd 704:469–477. https://doi.org/10.1016/j.jallcom.2017.02.069

Yousefi R, Azimi HR, Mahmoudian MR, Basirun WJ (2018a) The effect of defect emissions on enhancement photocatalytic performance of ZnSe QDs and ZnSe/rGO nanocomposites. Appl Surf Sci 435:886–893. https://doi.org/10.1016/j.apsusc.2017.11.183

Yousefi R, Azimi HR, Mahmoudian MR, Cheraghizade M (2018b) Highly enhanced photocatalytic performance of $Zn_{(1-x)}Mg_xO$/rGO nanostars under sunlight irradiation synthesized by one-pot refluxing method. Adv Powder Technol 29:78–85. https://doi.org/10.1016/j.apt.2017.10.014

Yu W, Chen X, Mei W, Chen C, Tsang Y (2017) Photocatalytic and electrochemical performance of three-dimensional reduced graphene oxide/WS$_2$/Mg-doped ZnO composites. Appl Surf Sci 400:129–138. https://doi.org/10.1016/j.apsusc.2016.12.138

Žerjav G, Djinović P, Pintar A (2018) TiO$_2$-Bi$_2$O$_3$/(BiO)$_2$CO$_3$-reduced graphene oxide composite as an effective visible light photocatalyst for degradation of aqueous bisphenol A solutions. Catal Today 315:237–246. https://doi.org/10.1016/j.cattod.2018.02.039

Zhang D, Yang X, Zhu J, Zhang Y, Zhang P, Li G (2011) Graphite-like carbon deposited anatase TiO$_2$ single crystals as efficient visible-light photocatalysts. J Sol-Gel Sci Technol 58:594–601. https://doi.org/10.1007/s10971-011-2433-8

Zhang L, Li N, Jiu H, Zhang Q (2016) Solvothermal synthesis of reduced graphene oxide-Bi$_2$S$_3$ nanorod composites with enhanced photocatalytic activity under visible light irradiation. J Mater Sci Mater Electron 27:2748–2753. https://doi.org/10.1007/s10854-015-4086-9

Zhang L, Wang A, Zhu N, Sun B, Liang Y, Wu W (2018a) Synthesis of butterfly-like BiVO$_4$/RGO nanocomposites and their photocatalytic activities. Chin J Chem Eng 26:667–674. https://doi.org/10.1016/j.cjche.2017.09.007

Zhang Y, Yang HM, Park SJ (2018b) Synthesis and characterization of nitrogen-doped TiO$_2$ coatings on reduced graphene oxide for enhancing the visible light photocatalytic activity. Curr Appl Phys 18:163–169. https://doi.org/10.1016/j.cap.2017.12.001

Zhang F, Zhang Y, Zhang G, Yang Z, Dionysiou DD, Zhu A (2018c) Exceptional synergistic enhancement of the photocatalytic activity of SnS$_2$ by coupling with polyaniline and N-doped reduced graphene oxide. Appl Catal B Environ 236:53–63. https://doi.org/10.1016/j.apcatb.2018.05.002

Zhou M, Han D, Liu X, Ma C, Wang H, Tang Y, Huo P, Shi W, Yan Y, Yang J (2015) Enhanced visible light photocatalytic activity of alkaline earth metal ions-doped CdSe/rGO photocatalysts synthesized by hydrothermal method. Appl Catal B-Environ 172–173:174–184. https://doi.org/10.1016/j.apcatb.2015.01.004

Zhu L, Ye S, Ali A, Ulla K, Cho KY, Oh WC (2015) Modified hydrothermal synthesis and characterization of reduced graphene oxide-silver selenide nanocomposites with enhanced reactive oxygen species generation. Chin J Catal 36:603–611. https://doi.org/10.1016/S1872-2067(14)60275-8

Zhu J, Zhu Z, Zhang H, Lu H, Zhang W, Qiu Y, Zhu L, Küppers S (2018) Calcined layered double hydroxides/reduced graphene oxide composites with improved photocatalytic degradation of paracetamol and efficient oxidation-adsorption of As(III). Appl Catal B-Environ 225:550–562. https://doi.org/10.1016/j.apcatb.2017.12.003

Zou L, Wangn X, Xu X, Wang H (2016) Reduced grapheme oxide wrapped CdS composites with enhanced photocatalytic performance and high stability. Ceram Int 42:372–378. https://doi.org/10.1016/j.ceramint.2015.08.119

Zou X, Zhou Y, Wang Z, Chen S, Xiang B, Qiang Y, Zhu S (2018) Facile synthesis of α-Fe$_2$O$_3$ pyramid on reduced graphene oxide for supercapacitor and photo-degradation. J Alloys Compd 744:412–420. https://doi.org/10.1016/j.jallcom.2018.02.126

Chapter 7
Functionalized Polymer-Based Composite Photocatalysts

Sebastian Raja and Luiz H. C. Mattoso

Contents

Abstract The tremendous risk of climate change and environmental problems has drastically endangered the survival and development of human society, which certainly demand modern technologies that require high-performance materials with superior properties. Photocatalysis is a promising, greener, and eco-friendly technology for pollution remediation, energy conversion, and chemical synthesis. Moreover, this technology has emerged as a powerful tool to design and in the development of engineered materials. In this association, functionalized polymers and their (nano) composites are not only used as fillers in polymer matrices but also play a significant role as hosts or supporting materials for inorganic metals and semiconductors (TiO_2, ZnO, Fe_2O_3, and so on). As efficient supporting materials in photocatalysis, functionalized polymeric composites derived either from natural or synthetic polymers are gaining much attention from governments, industries, and academia on the account of their low cost, environmental compatibility, and replacement capability for petroleum-derived products. This chapter provides a comprehensive understanding for obtaining photocatalytic organic/inorganic hybrid (nano)

S. Raja (✉) · L. H. C. Mattoso
National Nanotechnology Laboratory for Agribusiness, Embrapa Instrumentação, São Carlos, Brazil
e-mail: sebastianrajaorg@gmail.com; luiz.mattoso@embrapa.br

© Springer Nature Switzerland AG 2020
M. Naushad et al. (eds.), *Green Photocatalysts*, Environmental Chemistry for a Sustainable World 34, https://doi.org/10.1007/978-3-030-15608-4_7

composites from natural and synthetic functionalized polymers and their wastewater treatment applications. Two different categories of functionalized polymers, namely, (i) natural polymers and (ii) synthetic polymers, and their composite materials along with their recent advancement in wastewater treatment by light-responsive photocatalysis are figured out.

Keywords Heterogeneous photocatalyst · Polymer composite · Semiconductor · Natural polymer · Synthetic polymer · Wastewater treatment · Organic pollutant · Dye degradation and environment

7.1 Introduction

Environment-related problems are the key challenges for all living species; these are hampering the development of modern society. Water is one of the basic needs and vital for all known forms of life. Worldwide, 3.2 million people die per year due to the consumption of contaminated water, poor sanitation, and lack of hygiene. Moreover, industrial discharge of chemicals, agricultural movements, and environmental changes are the key factors for water contamination (Shahat et al. 2015; Awual et al. 2016). Water pollutants exist in many hazardous wastes including pharmaceutical wastes, pesticides, herbicides, textile dyes, resins, and phenolic compounds (Saravanan et al. 2017). Consequently, effective methods and techniques are required to tackle the problems caused by the increasing level of pollutants and toxic compounds in both air and water.

Photocatalysis is a promising, greener, and eco-friendly technology for pollution remediation, energy conversion, and chemical synthesis. Principally, photocatalysis is the acceleration or speed up of a chemical reaction in the presence of light and catalysts. It can be classified into two categories, namely, (i) homogeneous photocatalysis and (ii) heterogeneous photocatalysis. Nevertheless, heterogeneous photocatalysis is considered as one of the best tools for wastewater treatment due to its great potential and high performance for removing various organic pollutants and harmful bacteria by utilizing sunlight with photocatalyst (Kou et al. 2017; Dong et al. 2015). It is well known that the light source and the nature of the photocatalyst are the key factors for obtaining better efficiency of photocatalyst.

Among various inorganic metal semiconductors (TiO_2, ZnO, Fe_2O_3, and so on), TiO_2 (also known as titania) is the most widely studied photocatalyst for wastewater treatment till date (Hamayun et al. 2018; Ajmal et al. 2014; Gnanasekaran et al. 2015, 2016). Moreover, the promising features such as low cost, nontoxicity, high stability, and unique photocatalytic efficiency make TiO_2 a potential candidate in organic pollutant removal through photocatalytic reaction. The general mechanism of photocatalysis is depicted in Fig. 7.1.

Further, the mechanism mainly involves four steps (Singh et al. 2013).

i) Generation of photo-induced holes (h+VB) and electrons (e−CB)

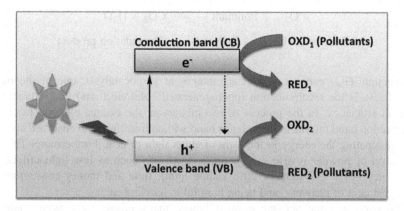

Fig. 7.1 General mechanism of photocatalysis

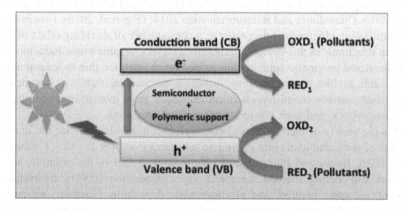

Fig. 7.2 General mechanism of pollutant removal by polymer-supported photocatalysis

$$TiO_2 + h\nu \rightarrow h^+{}_{VB} + e^-{}_{CB}$$

ii) Formation of hydroxide radicals ($^\bullet$OH) by photogenerated holes

$$H_2O + h^+{}_{VB} \rightarrow H^+ + {}^\bullet OH$$

iii) Formation of superoxide anion radical ($O_2{}^{\cdot-}$)

$$O_{2\,(ads)} + e^-{}_{CB} \rightarrow O_2{}^{\cdot-} \left(O_{2(ads)}:\ \text{Adsorbed oxygen}\right)$$

iv) Redox reaction of organic pollutant (Fig. 7.2)

$$\cdot OH + \text{Pollutant} \rightarrow\rightarrow\rightarrow CO_2 + H_2O$$

$$O_2^{\cdot -} + \text{Pollutant} \rightarrow \rightarrow \rightarrow CO_2 + H_2O$$

$$\text{Pollutant} + h^+{}_{VB} \rightarrow \text{Pollutant}^{\cdot +} \rightarrow \text{Degradation product}$$

Though TiO_2 endows several advantages in photocatalysis, one of the main drawbacks is the recombination (photogenerated holes/electrons) that reduces the overall efficiency. In the process of recombination, the excited electron from the conduction band returns to the valence band without reacting with adsorbed species and dissipating the energy in the form of either light or heat. Furthermore, TiO_2 in the form of powder is also giving some drawbacks such as low light utilization efficiency (<1% UV light or 20% visible light), time and money-consuming in posttreatment of recovery, and being harmful to human health.

To overcome the abovementioned issues, many efforts have been made to immobilize inorganic semiconductors on various substrates, such as silica, activated carbon, graphene oxide (GO), and carbon nitride (C_3N_4) (Colmenares et al. 2016; Li et al. 2015; Chowdhury and Balasubramanian 2014; Ong et al. 2016). However, the light utilization efficiency is decreased as a consequence of shielding effect of these opaque structures. In this context, immobilization of inorganic semiconductors with functionalized polymeric supports has gained much attention due to several advantages such as low cost, flexibility, attractive mechanical stability, low density, controllable surface chemistry, chemical resistance, more availability, high durability, transparency, and high UV resistance without oxidation.

A wide variety of methods have been explored in the literature for the immobilization of semiconductors into polymeric substrates (Singh et al. 2013; Colmenares et al. 2016; Boury and Plumejeau 2015; Singh et al. 2013), for example, sol-gel method, dip-coating method, chemical vapor deposition (CVD), hydrothermal method, sol-spray method, and electrophoretic deposition. However, sol-gel and sputtering methods have been considered as the main low-temperature deposition techniques for incorporating semiconductors into various polymer substrates. In this context, this chapter provides insight into various functionalized polymeric composites (synthetic and natural) as support for metal oxide semiconductors along with recent advancements in organic pollutant degradation by light-responsive heterogeneous photocatalysis. Figure 7.2 shows the general mechanism of pollutant removal by polymer-supported photocatalysis.

7.2 Composites from Synthetic Polymers

Man-made polymers are generally known as synthetic polymers. They are classified into four main categories, namely, thermoplastics, thermosets, elastomers, and synthetic fibers. They can be commonly found in many consumer products. They are further categorized by variation in main chain as well as side chains. For instance, backbones of common synthetic polymers are made of carbon-carbon (C-C) bond, whereas the heterochain polymers possess other elements (e.g., O, S, and N) along the backbone. The various types of synthetic polymers and their significance in photocatalysis are discussed below.

7.2.1 Common Synthetic Polymers

The most common synthetic polymers are high-density polyethylene (HDPE), low-density polyethylene (LDPE), polypropylene (PP), polyethylene terephthalate (PET), polystyrene (PS), polyvinyl alcohol (PVA), polyvinyl chloride (PVC), poly (vinylidenedifluoride) (PVDF), and so on. They play a pivotal role in heterogeneous photocatalysis as supporting material in order to enhance their catalytic performance. In this association, the first polymeric hybrid material based on synthetic polymer was made from polyethylene (PE) film with titanium oxide for photocatalytic degradation of phenol (Tennakone et al. 1995). The catalytic efficiency of about 50% was achieved after 2.5 h of illumination under UV light. Since then, much effort has been made along this direction. Some of the recent achievements are discussed below.

Singh et al. (2014) achieved polystyrene (PS)/Ag$^+$-doped TiO_2 composite photocatalyst by solvent-cast techniques (impregnation and strewing). The reported photocatalyst showed 94% and 83% catalytic efficiencies for the degradation of methylene blue (MB) in aqueous solutions under ultraviolet and solar irradiation, respectively. Further, the photocatalytic activity was found to be sustained after three cycles.

Photocatalytic hybrid materials with excellent catalytic activity and high stability were achieved from polypropylene (PP) supports with zinc oxide (ZnO) nanorods (Colmenares et al. 2015). The resulting PP/ZnO composite exhibited 76% of photocatalytic efficiency after three cycles in comparison with commercial ZnO (53%). It reveals that the improvement of catalytic activity could be due to the wettability and adsorptive properties of the composite.

Recently, Sayed et al. (2018) reported the solar light-responsive polyvinyl alcohol (PVA)-assisted TiO_2/Ti film via a facile one-pot hydrothermal approach. In photocatalytic reaction, the highest catalytic performance of 98.3% was achieved in 60 min for the degradation of ciprofloxacin (CIP) by the addition of peroxymonosulfate (PMS) into PVA-assisted TiO_2/Ti film as a function of pH.

Hegedűs et al. (2016) reported polyvinyl alcohol (PVA)-TiO_2 composite foil by a simple solution-casting method. From this investigation, the PVA-TiO_2 composites performed as suitable candidates for photocatalytic pretreatment to remove Triton X-100, a widely used nonionic detergent. The highest degradation efficiency of the contaminant about 65% was achieved after two cycles.

Low-density polyethylene (LDPE) and high-density polyethylene (HDPE) immobilized TiO_2 (LDPE, HDPE/TiO_2) composite were achieved by Romero-Sáez et al. (2017). The highest photocatalytic efficiency of ~ 90% was achieved after three cycles during the degradation of methyl orange (MO) under visible light irradiation.

Table 7.1 Photocatalytic performances of common synthetic polymer-based supports into various inorganic semiconductors for organic pollutant degradation

Composite	Contaminant	Light source	Photocatalytic performance	Reference
PS/Ag$^+$-doped TiO$_2$ composite	MB	UV and solar light	Degradation of MB with 94%. The catalytic efficiency was to be sustained after three cycles	Singh et al. (2014)
PP/ZnO composite	Phenol	UV light	Phenol degradation performance reached 76% after three cycles. This efficiency is higher than that of commercial ZnO (53%)	Colmenares et al. (2015)
PVA-assisted TiO$_2$/Ti composite	CIP	Solar light	Photodegradation efficiency of CIP achieved 98.3% in 60 min	Sayed et al. (2018)
PVA/TiO$_2$ composite foil	Triton X-100	UV light	Degradation efficiency of Triton X-100 reached 65% after two cycles	Hegedűs et al. (2016)
LDPE, HDPE/TiO$_2$ composite	MO	Visible light	Degradation of MO was achieved 90% after three cycles	Romero-Sáez et al. (2017)
PVDF/ZnIn$_2$S$_4$	MO	UV-visible	The catalytic efficiency was achieved 92% for the removal of MO	Peng et al. (2012)
PVDF/TiO2, BiOI, ZnO	MB	UV-visible	The catalytic performance was exhibited 75% for the removal of MB	Zanrosso et al. (2018)

Poly(vinylidenefluoride) (PVDF)/ZnIn$_2$S$_4$ composite was reported by Peng et al. (2012) for the degradation of MO under UV-visible light illumination. In photocatalysis, the photocatalytic performance of ~92% was exhibited for the removal of the dye.

Similarly, Zanrosso et al. (2018) demonstrated the PVDF/TiO2, BiOI, and ZnO composites for the removal of MB under UV and visible light irradiation. The catalytic efficiency of ~75% was achieved during the photocatalytic reaction. The photocatalytic performances based on synthetic polymers and their composites are figured out in Table 7.1.

7.2.2 Conducting or Conjugated Polymers (CPs)

CPs are another category of synthetic polymers that have become a fascinating area of research because of their unique feature, i.e., electrical conductivity (π-conjugated electronic system). According to the literature (Hall 2003 and Swager 2017), poly (sulfur nitride) is reported as the first conducting polymer followed by polyacetylene, polyheterocycles (N-, and S-), and polyaniline (PANI). In this association, combination of CPs with metal oxide semiconductors (e.g., TiO$_2$, ZnO, Fe$_2$O$_3$, and so on) has received much attention nowadays due to their desirable properties such as high absorption coefficient in the visible region, high mobility of charge carriers, excellent environmental stability, low cost, and noticeable optical and photophysical properties. Moreover, hybrid composites of CPs/metal oxide semiconductors offer a wide range of applications such as conductive coating, energy storage, electrocatalysis, photovoltaics, and photocatalysis (Ibanez et al. 2018; Ansari et al. 2015). Basically, CPs are synthesized either from electrochemical polymerization or from chemical methods such as cross coupling, oxidative polymerization, and vapor phase polymerization. In photocatalysis, CPs play a crucial role in order to tune the physical properties (conductivity) during their addition process with semiconductor metal oxides. Moreover, they avoid the release of harmful metal oxide nanoparticle (e.g., TiO_2) into the environment. Thus, the modification of CPs provides alternative pathways for pollutant degradation. In this way, a PANI /TiO_2-based composite photocatalyst was reported by Yu et al. (2012) via a simple impregnation-hydrothermal method for the degradation of organic dyes such as anthraquinone, reactive brilliant blue, and KN-R under visible light irradiation.

Gilja et al. (2017) developed PANI/TiO_2 composite photocatalyst by chemical oxidation method. In this investigation, composites with different amounts of PANI were synthesized, and the highest photocatalytic efficiency of 98% was achieved for the degradation of the reactive red azo dye (RR45) in 60 min under UV irradiation, which is higher than that of pure TiO_2 (92%). In solar irradiation, the complete discoloration of the dye was observed in 90 min.

A novel plasmonic photocatalyst of PANI-Ag/AgCl was achieved successfully by deposition-precipitation reaction (Ghaly et al. 2017). The highest catalytic efficiency (98%) was observed for the degradation of MB under simulated solar light. Further, the stability test revealed that the reduction in efficiency was only less (98–84%), indicating that PANI not only improves the performance of Ag/AgCl but also enhances its stability.

Xu et al. (2017) fabricated a polypyrrole (PPy)/bismuth iodide oxide (BiOI) nanocomposite through an in situ precipitation strategy at room temperature. The photodegradation of RhB was achieved about 83% under visible light irradiation. It is stated that the enhancement of catalytic amount reduces the performances after a few cycles. The photocatalyst also achieved 61% of efficiency for bisphenol degradation under illumination of visible light.

A poly(3,4-ethylenedioxythiophene (PEDOT)/ZnO photocatalyst was developed via in situ synthesis (Katančić et al. 2017). A synthesized composite photocatalyst was evaluated for the degradation of the model compound RR45 dye under visible-light irradiation. The highest performance of 75% was exhibited after 90 min for the degradation of RR45 dye in wastewater.

Ghosh et al. (2015) introduced a photocatalyst based on PEDOT/$FeCl_3$ nanospindle through template-mediated controlled synthesis. The catalyst was employed for the degradation of MO and phenol under UV and visible light irradiation. The catalytic performance was achieved about 100% after 15 min of irradiation under UV light; this efficiency was much higher than that of P25 TiO_2, which is recognized as one of the best photocatalysts under UV light. Under visible light irradiation, the performance displayed 100% degradation of MO in 180 min. More importantly, the catalytic activity of PEDOT spindles was retained at over 98% after successful recycling.

A composite of PEDOT/TiO$_2$ nanofibers was fabricated (Liu et al. 2016) via electrospinning and calcination method. In this investigation, the highest catalytic performance was achieved 125% within 30 min under UV irradiation for the degradation of phenazopyridine (PAP).

Ghoreishi et al. (2016) demonstrated composites based on PANI and polypyrrole (PPy)-anchored WO$_3$/TiO$_2$. The reported catalyst achieved excellent degradation efficiency with PANI (100%) and PPy (95%) during the degradation of methylene blue (MB) under UV light irradiation. It is revealed that the enhanced performance is due to high porosity, better charge separation, and synergistic effect of the composite.

Similarly, poly(aniline-*co-o*-aminophenol) [poly(ANI-*co-o*AP)]-immobilized ZnO composite (Sivakumar et al. 2014) reached a catalytic efficiency up to 88% for the degradation of methylene blue (MB) under UV light illumination.

Based on the high charge carrier mobility of CPs, a composite based on PPy/Bi$_2$WO$_6$ was designed by Zhang et al. (2014) in order to promote efficient charge separation for the degradation of phenol. The photocatalytic performance of the composite was achieved about 100% after 120 min under simulated sunlight illumination for the degradation of phenol, whereas only 72.4% was achieved for pure Bi$_2$WO$_6$ within the same time period. Further, it is revealed that the photogenerated holes in the valence band of Bi$_2$WO$_6$ transferred to the highest occupied molecular orbital (HOMO) of PPy, leading to rapid photoinduced charge separation in photocatalytic reaction. Table 7.2 summarizes the most recent achievements of CPs in wastewater treatment.

7.2.3 Coordination Polymers (COPs)

Coordination polymers (COPs) are a class of crystalline materials that are constructed by organic ligands with metals or metal clusters in an extended network (0D, 1D, 2D, and 3D) and have attracted increasing attention in catalysis. Moreover, the characteristic features such as high dispersion of active sites, pore size and topology, and tunable adsorption properties along with their intrinsic hybrid nature endow COPs as potential candidates in photocatalysis (Colmenares and Kuna 2017; Kuila et al. 2017; Zhang et al. 2016; Kan et al. 2012). In addition, the modification of metal cations and organic linker composition offer effective solar light absorption properties, thus allowing solar energy to chemical energy via photosynthesis. These crystalline materials are classified into two categories, namely metal organic complex (MOC) and metal organic framework (MOF) in the literature.

In addition, COPs could afford the synergistic effect and the fundamental principles of light harvesting and energy transfer phenomena. Several numbers of COPs have been investigated as supporting material in photocatalysis. In this context, MOFs are a special group of COPs that possess crystalline porous networks involving strong metal-ligand interaction, whereas MOCs are constructed by metal ions coordinated by organic ligands in zero-dimensional or mononuclear configurations

Table 7.2 Photocatalytic performances based on CPs/inorganic semiconductor in wastewater treatment

Composite	Contaminant	Light source	Photocatalytic performance	Reference
PANI/TiO$_2$	RR45 dye	UV and solar	Degradation of RR45 with 92% of efficiency in 60 min under UV irradiation A 100% discoloration of dye in 90 min under solar irradiation	Gilja et al. (2017)
PANI/Ag-AgCl composite	MB	UV	MB degradation performance was reached 98%	Ghaly et al. (2017)
PPy/BiOI composite	RhB and Bisphenol	Solar light	RhB degradation efficiency of 83%, whereas for bisphenol, 63%	Xu et al. (2017)
PEDOT/ZnO composite	RR45 dye	Visible light	Degradation efficiency of RR45 reached 75% in 90 min	Katančić et al. (2017)
PEDOT/FeCl$_3$ composite	MO	UV and visible light	Degradation of MO was achieved 100% in 15 min under UV, whereas 100% in 180 min under visible light	Ghosh et al. (2015)
PEDOT/TiO$_2$	PAP	UV	The catalytic efficiency was achieved 125% in 30 min	Liu et al. (2016)
PANI, PPy-WO$_3$/ TiO$_2$ composite	MB	UV	The catalytic performance was exhibited 100% with PANI, whereas 95% with PPy	Ghoreishi et al. (2016)
[poly(ANI-co-oAP)]/ZnO composite	MB	UV	The catalytic efficiency of 88% achieved for the degradation of dye	Sivakumar et al. (2014)
PPy/Bi$_2$WO$_6$ composite	Phenol	Solar light	Photocatalytic efficiency was achieved about 100% after 120 min, whereas only 72.4% was achieved for pure Bi$_2$WO$_6$ within the same time period	Zhang et al. (2014)

Table 7.3 Classification and basic features of COPs

Coordination polymers (COPs)	
Metal organic complex (MOC)	Metal organic frameworks (MOFs)
Two-dimensional (2D) or three-dimensional (3D) structure	Zero-dimensional (0D) or one-dimensional (1D) structure
Strong metal-ligand interaction	Mononuclear configuration
Tunable pore size	

(Table 7.3). On the other hand, the basic characteristic (i.e., metal-organic coordination) is identical for both MOF and MOC. However, the recent spectacular development of MOFs has drawn enormous attention in many fields offering a wide variety of applications such as gas storage and separation, drug delivery, and electronics (Mon et al. 2018; Yuan et al. 2018; Li et al. 2016). Furthermore, high metal-containing nodes, organic linkers, controlled and easy synthesis, and light-absorbing capacity are the significant characteristics of MOFs which could facilitate photocatalytic reaction for specific applications. Also, the remarkable functionalities of MOFs have attracted much attention in heterogeneous photocatalysis (Zhu et al. 2017; Dias and Petit 2015). Thus, this section reviews the recent progress of MOFs in wastewater treatment by light irradiation.

Xamena et al. (2007) introduced the first photocatalyst based on MOF5 (Zn_4O $(BDC)_6$ for the degradation of phenol. Since then, several numbers of photocatalytic MOFs have been reported in the literature. For instance, Co, Ni, and ZnMOFs were subjected for the degradation of various organic pollutants such as RhB, Orange G, Remazol Brilliant Blue R, and MB dyes (Mahata et al. 2006).

Recently, Xia et al. (2017) reported a photocatalyst based on Ln-MOFs (Ce, Tb, Dy) from anthracene carboxylate ligand by solvothermal synthesis. It is demonstrated that the broadband visible-light absorption and efficient photoinduced charge generation facilitate efficient degradation of RhB. The highest catalytic performance of about 99% was achieved in 12 min under visible-light irradiation.

Zhao et al. (2017) demonstrated the pillared-layer MOF (NNU-36) as an efficient photocatalyst for Cr(IV) reduction and Rhodamine B (RhB) dye degradation under visible light. The photocatalytic reaction reveals that 46.6% of RhB dyes degraded in 70 min. This slower degradation rate is due to the fast recombination of photogenerated electrons and holes, as a consequence of Cr(IV) reduction.

Copper (Cu)-based 3D-MOF $[Cu(4,4'-bipy)Cl]_n$ and $[Co(4,4'-bipy)\cdot(HCOO)_2]_n$ were developed by Zhang et al. (2018a, b) through a hydrothermal method for photocatalytic degradation of MB under visible light. It is found that the highest degradation efficiency of 93.93% was observed after 150 min.

Islam et al. (2017) achieved hierarchical BiOI/MOF photocatalyst via a simple precipitation method. In photocatalytic reaction, the highest catalytic efficiency of 88% was observed for the composite BiOI/MOF during the degradation of RhB, while 28% was observed for bare BiOI. The final photocatalyst exhibited excellent reusability for up to five consecutive cycles.

Sha et al. (2015) developed AgI/UiO-66 (zirconium-based MOF) photocatalyst by facile solution method. The photocatalytic test revealed that the ratio of Ag/Zr could effectively influence the catalytic performance during the degradation of RhB under visible light. Further, the 1:1 ratio of Ag/Zr afforded a high photocatalytic efficiency.

A photocatalyst was derived (Meng et al. 2014) from cluster-based coordination complexes of cadmium (Cd) with multidentate carboxylic acid and N-donor ligands for the degradation of MO. The highest catalytic efficiency was achieved about ~ 97% in the presence of H_2O_2 under high-pressure mercury lamp irradiation.

Peng et al. (2018) demonstrated a photocatalyst based on copper (II) coordination polymers by hydrothermal method. The highest photocatalytic performances of 97.1% and 99.2% were achieved for the degradation of MB and MO, respectively, in the presence of H_2O_2 under UV-light illumination after 150 min. Table 7.4 summarizes the catalytic performances of MOFs from the recent findings in the literature.

Table 7.4 Recent advancements in COP-based photocatalysts in organic pollutant degradation

Composite	Contaminant	Light source	Photocatalytic performance	Reference
Ln-MOFs (Ce, Tb, Dy)	RhB	Visible	The broadband visible light absorption and efficient photoinduced charge generation achieved the degradation efficiency of 99% in 10 min	Xia et al. (2017)
NNU-36 MOF	RhB	Visible	RhB degradation efficiency was obtained as 46.6% This low efficiency was due to the fast recombination of photogenerated electrons/holes	Zhao et al. (2017)
Cu-MOF	MB	Visible light	MB degradation efficiency was achieved about 93.93%, in 150 min	Zhang et al. (2018a, b)
Hierarchical BiOI/MOF	RhB	Visible light	Degradation efficiency was reached 88% with excellent reusability after five cycles	Islam et al. (2017)
AgI/UiO-66 MOF	RhB	Visible	Ratio of Ag/Zr could effectively influence the catalytic performance The high performance was achieved for the ratio of 1:1	Wu et al. (2015)
Cd-MOF	MO	Mercury lamp	The highest catalytic efficiency was achieved about ~ 97% in the presence of H_2O_2	Meng et al. (2014)
Cu-MOF	MB and MO	UV	The photocatalytic performances of 97.1% and 99.2% for the degradation of MB and MO, respectively, in the presence of H_2O_2 after 150 min	Peng et al. (2018)

7.3 Composites from Natural Polymers (NPs)

Polymers derived from renewable resources or agro-industrial wastes are called natural polymers (NPs). Like synthetic polymers, NPs can also serve as a desirable support for inorganic semiconductors in photocatalysis (Colmenares et al. 2016; Boury and Plumejeau 2015; Singh et al. 2013) due to the presence of a large number of functional groups (such as -OH, -NH$_2$, and COOH) and significant features including excellent sustainability, environment-friendly, biodegradability, and more abundance. These characteristic features make natural polymers as important feedstock to produce photocatalytic composites. Moreover, the immobilization of NPs with inorganic semiconductors offers high surface area, superior adsorption properties, high hydroxyl (-OH) groups, and reduced recombination in the photocatalytic process. Though there are several numbers of natural polymers reported in the literature, this chapter mainly describes the importance of most significant natural polymers, namely, cellulose and chitin or chitosan, and their composites along with their recent progress in heterogeneous photocatalysis for organic pollutant degradation.

7.3.1 Cellulose-Derived Composites

Cellulose is the most abundant renewable biopolymer resource on earth with a total annual production of about 10^{11}–10^{12} tons (Shaghaleh et al. 2018). It is obtained from various renewable resources from fibrous crops to agro-industrial wastes (Inamuddin et al. 2015; Mohammad et al. 2012). Moreover, it is considered as the strongest candidate for replacing petroleum-derived polymers due to its abundance and eco-friendly features such as renewability, biocompatibility, and biodegradability. Cellulose condenses β-(1 → 4)-glycosidic bonds with cellobiose repeating units (Fig. 7.3), which displays unbranched straight chain with high hydrophilicity and chirality. It allows large capacity of chemical modification due to the presence of large number of hydroxyl groups on the surface. Moreover, it possesses an ordered structure, high thermal stability, and high crystallinity. Further, the structure of cellulose comprises cell walls and bundles of cellulose molecules ($C_6H_{10}O_5$) as microfibrils in the form of stabilized or elongated by hydrogen bonds. Cotton and wood contain the high content of cellulose (> 95%) among various sources such as plants, algae, marine creatures, and bacteria (Menon et al. 2017).

The nanometer dimension of cellulose is commonly described as nanocellulose (NC). It is classified into three categories, namely, cellulose nanofibrils (CNF), cellulose nanocrystals (CNC), and bacterial cellulose (BC). It possesses remarkable physical properties, flexible surface chemistry, and low thermal expansion in addition to transparency, high elasticity, and anisotropy. Furthermore, many attractive features of NC such as their inherent renewability, sustainability, high strength, large specific surface area, and nanoscale dimension offer a wide range of applications such as polymer reinforcement, nanocomposite formulations, biosensing, food and nutraceuticals, and tissue engineering scaffolds (France et al. 2017; Golmahammadi et al. 2017; Rajinipriya et al. 2018). Moreover, the high density of hydroxyl groups on the surface of NC endows for easy modifications with various functional groups (namely, carboxyl, sulfate ester, amine, aldehyde groups, etc.), small organic molecules, nanoparticles, and grafting with polymers (Espino-Pérez et al. 2014; Chen et al. 2015; Lin and Dufresne 2014; Nechyporchuk et al. 2016; Habibi 2014). By utilizing its structural and chemical features, cellulose can be assembled into various substrates for several applications including energy, environmental, biomedical, and catalytic applications (Mohammad et al. 2018; Hoeng et al. 2016; Chen et al. 2018; Julkapli and Bagheri 2017; Wang et al. 2017b). Moreover, composites derived

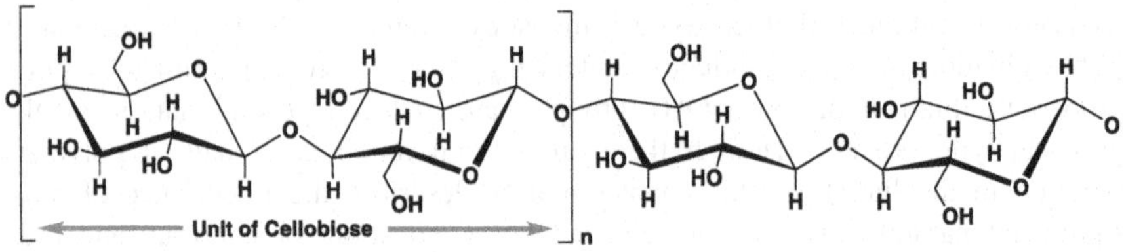

Fig. 7.3 Structural representation of cellulose with cellobiose repeating units

from cellulose and their derivatives are an attractive area of research that has attained a high degree of momentum in heterogeneous catalysis for improving the catalytic efficiencies. They could also serve as a desirable support for semiconductor metal oxides to improve photocatalyst performances. Thus, this section provides the recent findings of cellulose and their composites in wastewater treatment by photocatalysis.

A photocatalyst comprising TiO_2@Ag nanoparticles on cellulose microfiber matrices (CMF) was constructed by Zhang et al. 2018a, b. The resulted CMF/TiO_2@Ag composite served as an efficient photocatalyst in order to remove 4-chlorophenol (4-CP) into small molecules such as hydroquinone (HQ), benzoquinone (BQ), and quinhydrone. Moreover, the photocatalyst was considered as a suitable robust substrate for plasmon-enhanced molecular spectroscopy. This dual functionality provides unique opportunities to precisely monitor the detailed photocatalytic molecular transformations occurring at molecule-catalyst interfaces.

Rathod et al. (2018) developed a nanocellulose (NC)-supported TiO_2 photocatalyst through an ultrasonic impregnation method. During the photocatalytic degradation of mefenamic acid (anthranilic acid derivative drug), the catalytic efficiency reached up to 89% after 160 min under ultraviolet (UV) light irradiation. It also revealed that the catalytic efficiency was increased with increased content of nanosized TiO_2 (up to 10 wt%), which could be due to the rise in the number of active sites, which resulted in more effective interactions and more hydroxyl (OH^-) moieties.

A novel cellulose-derived photocatalyst CCNF/BiOBr from carbon nanofibers (CCNF) and bismuth oxybromide (BiOBr) was prepared through electrospinning followed by hydrothermal process (Geng et al. 2018). The reported composite was employed for the degradation of RhB under continuous visible-light irradiation. The photocatalytic performance of the composite catalyst was better than that of pure BiOBr. This investigation provides a promising way to design new photocatalyst from biomass residues for the effective treatment of wastewater.

Su et al. (2017) demonstrated a photocatalyst based on Cu_2O nanoparticle (NP)-functionalized cellulose-based aerogel with a three-dimensional (3D) porous structure and abundant active sites for visible-light photocatalysis. The photocatalytic performance of about 95.79% was achieved for the degradation of a model compound MB; this efficiency was higher than that of pure Cu_2O. The improved catalytic activity could be attributed to the enhanced separation of photogenerated electrons and holes in in situ deposition of Cu_2O in the cellulose matrix.

A green and portable photocatalyst based on cellulose nanofiber (CNF)/TiO_2 aerogel was employed (Li et al. 2018) for the photocatalytic degradation of MB under UV light and sunlight irradiation. The as-prepared hybrid photocatalyst exhibited excellent photocatalytic activity (98%), which is attributed to the pollutant adsorption capability of CNF along with electron hole blocking from TiO_2 recombination process. The recent achievements based on cellulose derivatives with inorganic semiconductor and their photocatalytic performances are summarized in Table 7.5.

Table 7.5 Photocatalytic performances of cellulose-derived composites for pollutant degradation in water

Composite	Contaminant	Light source	Photocatalytic performance	Reference
CMF-supported TiO_2 @ Ag NPs	4-CP	UV (365 nm)	Degradation of 4-CP yields the intermediates such as hydroquinone (HQ), benzoquinone (BQ), and quinhydrone, which were further confirmed from SERS	Zhang et al. (2018a, b)
NC/TiO_2	Mefenamic acid (MEF)	UV	The highest catalytic performance was reached up to 89% after 160 min	Rathod et al. (2018)
CCNF/ BiOBr fiber	RhB	Visible light	Degradation of RhB achieved 100% in 90 min In weakly acidic condition (pH 5) At a higher pH (11), the degradation efficiency is decreased by less than 25%	Geng et al. (2018)
Cellulose/ Cu_2O NPs aerogel	MB	Visible-light irradiation ($\lambda > 400$ nm; 350-W xenon lamp)	Photodegradation of MB reached 95.79%, much higher than the value of pure Cu_2O (73.59%)	Su et al. (2017)
NC/TiO_2 aerogel	MB	UV and sunlight irradiation	Degradation of dye reached beyond 98% after 4 cycles After 7th cycle, the degradation is reduced to 74% due to the mass loss (19%) of the composite	Li et al. (2018)

7.3.2 Chitin or Chitosan-Derived Composites

Chitin or chitosan (deacetylated), a linear polysaccharide composed of (1–4)-linked 2-acetamido-2-deoxy-b-D-glucopyranose units (Zargar et al. 2015), is a natural polymer, first discovered by Henri Braconnot in 1811 (Figs. 7.4 and 7.5). It is the most abundant natural polymer after cellulose with an annual production of 10^{10}–10^{11} tons (Goodday 1994). They are mainly extracted from various living organisms such as lobsters, shells or crabs, insects, shrimps, and so on via fermentation approaches. They are found naturally in the form of crystalline microfibrils, which form the structural components of many organisms. Structurally, chitin is similar to cellulose, but C2 position has an acetamide group ($-NHCOCH_3$).

Nanostructures of chitin possesses in the form of nanofibrils or nanowhiskers that can be isolated by various chemical treatments. They possess many significant features such as nontoxicity, biocompatibility, biodegradability, and environmental-friendliness, which serve as chelating agents for many applications including water treatment, drug delivery, wound-healing agents, pressure-sensitive adhesive taps, and so on (Rameshthangam et al. 2017; Zargar et al. 2015).

Like cellulose, chitin and chitosan have also been utilized as supports for semiconductor metal oxides in heterogeneous photocatalysis (Lee et al. 2015). Some of the recent findings are discussed below.

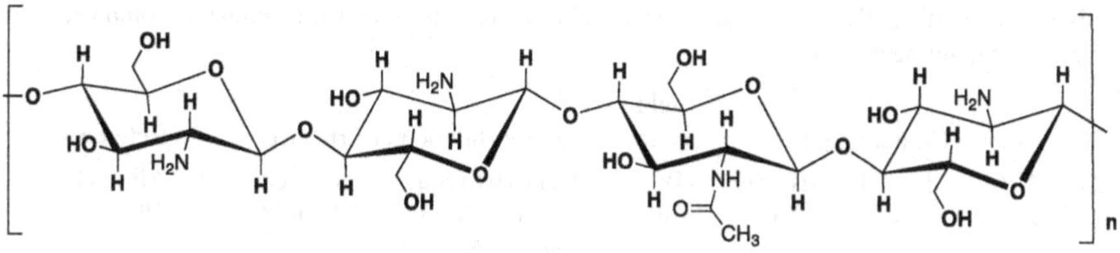

Fig. 7.4 Chemical structure of chitin polysaccharide

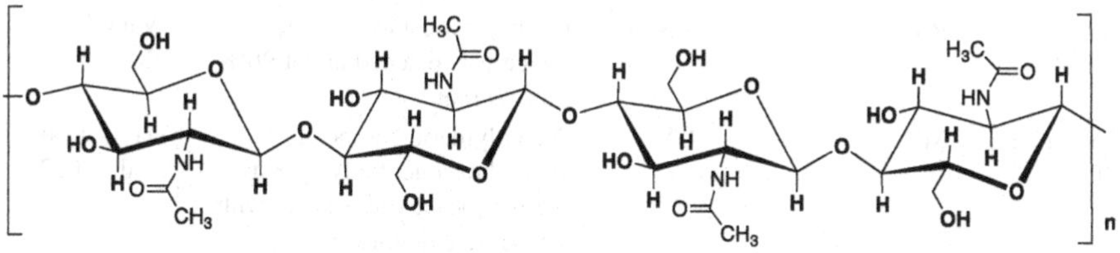

Fig. 7.5 Chemical structure of chitosan polysaccharide

Ali et al. (2018) developed a series of novel photocatalysts based on chitosan-titanium oxide (CS/TiO_2) fibers templated with zero-valent metal nanoparticles (ZV-MNPs) for the degradation of various organic model dyes such as MO, MB, acridine orange (AR), Congo red (CR), and (di) nitrophenols (4-NP, 2-NP, 3-NP, and 2,6-DNP). The high catalytic efficiency was achieved in 4 min for MO, CR, and MB. Moreover, degradation of AO and nitrophenols was achieved in less than 15 min. Besides, the reusability of the catalyst was also tested against MO and MB dyes where more than 97% of the dye reduction was achieved in less than 14 min in four cycles. It concluded that the reported photocatalyst provides not only high performance but also a promising reusability.

Wang et al. (2017a, b) reported a photocatalyst based on CS/Ag/AgCl composite from a simple one-step method. The composites exhibited efficient photocatalytic activity for the degradation of RhB under visible-light irradiation. The result indicated that the rate of degradation was of about 96% after 40 min at 20% mass ratio.

A photocatalyst from aerochitin-TiO_2 composite was explored by Dassanayake et al. (2017) for the degradation of MB. The efficiency of photocatalyst exhibited about 98% under UV-light irradiation (6 W) for MB degradation in 200 min.

A composite photocatalyst ChW/Ag_3PO_4 was developed by Zhang and Liu (2017) via a simple chemical deposition of Ag_3PO_4 nanoparticles into the surface of chitin whisker (ChW). The catalytic activity and stability of the photocatalyst were tested for the degradation of RhB under visible light. As a result, the RhB dye degraded up to 94.5% in 15 min after five cycles, while pure Ag_3PO_4 achieved only about 41.5%.

Shahabuddin et al. (2015) achieved chitosan (CS)-grafted polyaniline/Co_3O_4 nanocube composites (CS-PANI/Co_3O_4) for photocatalytic degradation of MB dye. The photocatalytic reaction resulted in a high degradation efficiency of about 88% for MB under UV-light irradiation after 180 min. The resulted CS-PANI/Co_3O_4

Table 7.6 Photocatalytic performances of chitin and chitosan/semiconductor composites in organic pollutant degradation

Composite	Contaminant	Light source	Photocatalytic performance	Reference
ZV-MNPs/ CS-TiO$_2$	MO, CR, MB, AO and 4-NP, 2-NP, 3-NP, and 2,6-DNP	UV-visible	Catalytic reduction was achieved in 4 min for MO, CR, and MB and in 15 min for AO and all nitrophenols The catalyst exhibited excellent catalytic activity with reusability	Ali et al. (2018)
CS/Ag/ AgCl	RhB	Visible	The highest catalytic performance was reached about 96% after 40 min	Wang et al. (2017a)
Aerochitin-TiO$_2$	MB	UV	The polymeric composite showed excellent adsorptivity and photocatalytic activity with 98% of dye degradation	Dassanayake et al. (2017)
ChW/ Ag$_3$PO$_4$	RhB	Visible light	Photodegradation of RhB reached up to 94.5%, while pure Ag$_3$PO$_4$ exhibited only 41.5%	Zhang and Liu (2017)
CS-PANI/ Co$_3$O nanocube	MB	UV	Degradation efficiency of composite reached 88% after 180 min The photocatalyst is capable of generating AOS, leading to the dye degradation	Shahabuddin et al. (2015)

nanocomposite is capable of generating advanced oxidation species (AOS), leading to the degradation of MB dye. The photocatalytic performances of chitin or chitosan-based composites with semiconductor metal oxides are summarized in Table 7.6.

7.4 Conclusion and Future Perspectives

Light-responsive heterogeneous photocatalysis for wastewater treatment has gained enormous attention during the past few decades. This chapter extensively demonstrated the utilization of various functionalized polymers (natural and synthetic) and their significance in photocatalysis along with current progress in the light-responsive photocatalysis in the context of pollutant degradation. These organic-inorganic hybrid composite photocatalysts offer better photocatalytic performances than the individual components due to the synergistic effect from the intrinsic properties of polymers and semiconductors. Moreover, polymeric support offers advantages such as high surface area for adsorbing more amount of pollutants, improved photocatalytic performance by the reduction of charge carrier recombination, and longer photoelectron lifetime (Luo et al. 2015; Ohtani Tonoi 2014).

In the upcoming years, the global market will rely on heterogeneous photocatalysis, especially for environmental remediation. However, achieving high performances of photocatalysts is a challenging task for the scientific community. As a consequence, design and development of novel composite materials with improved properties will be a promising and innovative research to be addressed in future investigations. The high performance of photocatalyst can be achieved by engineering band-gap structures of the semiconductors. In addition to the degradation efficiencies, other factors such as rate of reaction, quantum yield, and multi-activity assessment should also be taken into account in order to achieve high photocatalytic performance more precisely.

However, there are some concerns about the development of synthetic polymers derived from nonrenewable resources due to the limited accessibility and low sustainability along with environmental concerns. Thus, it is highly desirable to look for other alternatives that should address the above issues. In this association, natural polymers are of great importance owing to low cost, sustainability, more abundance, environment-friendliness, biodegradability, and nontoxicity. However, low thermal stability, physiochemical resistance, production cost, and scalability are the few issues that should be addressed extensively. Thus, there has been intense investigation focusing toward these materials to address the above issues and enable them as potential candidates for heterogeneous photocatalysis.

Acknowledgments The authors thank Embrapa, FAPESP (Proc. No. 2015/00094-0; Proc. No. 2017/22017-3), MCTI/SISNANO, REDEAGRONANO, DEMa/UFSCar, and CNPq for the financial support.

References

Ajmal A, Majeed I, Malik RN, Idriss H, Nadeem MA (2014) Principles and mechanisms of photocatalytic dye degradation on TiO_2 based photocatalysts: a comparative overview. RSC Adv 4:37003–37026. https://doi.org/10.1039/c4ra06658h

Ali F, Khan SB, Kamal T, Alamry KA, Asiri AM (2018) Chitosan-titanium oxide fibers supported zero-valent nanoparticles: highly efficient and easily retrievable catalyst for the removal of organic pollutants. Sci Rep 8:6260. https://doi.org/10.1038/s41598-018-24311-4

Ansari MO, Khan MM, Ansari SA, Cho MH (2015) Polythiophene nanocomposites for photodegradation applications: past, present and future. J Saudi Chem Soc 19(5):494–504. https://doi.org/10.1016/j.jscs.2015.06.004

Awual MR, Hasan MM, Eldesoky GE et al (2016) Facile mercury detection and removal from aqueous media involving ligand impregnated conjugate nanomaterials. Chem Eng J 290:243–251. https://doi.org/10.1016/j.cej.2016.01.038

Boury B, Plumejeau S (2015) Metal oxides and polysaccharides: an efficient hybrid association for materials chemistry. Green Chem 17:72–88. https://doi.org/10.1039/c4gc00957f

Chen J, Lin N, Huang J, Dufresne A (2015) Highly alkynyl-functionalization of cellulose nanocrystals and advanced nanocomposites thereof *via* click chemistry. Polym Chem 6 (24):4385–4395. https://doi.org/10.1039/C5PY00367A

Chen W, Yu H, Lee S, Wei T, Li J, Fan Z (2018) Nanocellulose, a promising nanomaterial for advanced electrochemical energy storage. Chem Soc Rev 47(8):2837–2872. https://doi.org/10.1039/c7cs00790f

Chowdhury S, Balasubramanian R (2014) Graphene/semiconductor nanocomposites (GSNs) for heterogeneous photocatalytic decolorization of wastewaters contaminated with synthetic dyes: a review. Appl Catal B Environ 160–161:307–324. https://doi.org/10.1016/j.apcatb.2014.05.035

Colmenares JC, Kuna E (2017) Photoactive hybrid catalysts based on natural and synthetic polymers: a comparative overview. Molecules 22(5):790. https://doi.org/10.3390/molecules22050790

Colmenares JC, Kuna E, Jakubiak S, Michalski J, Kurzydłowski K (2015) Polypropylene nonwoven filter with nanosized ZnO rods: promising hybrid photocatalyst for water purification. Appl Catal B Environ 170-171:273–282. https://doi.org/10.1016/j.apcatb.2015.01.031

Colmenares JC, Varma RS, Lisowski P (2016) Sustainable hybrid photocatalysts: titania immobilized on carbon materials derived from renewable and biodegradable resources. Green Chem 18:5736–5750. https://doi.org/10.1039/c6gc02477g

Dassanayake RS, Rajakaruna E, Abidi N (2017) Preparation of aerochitin-TiO$_2$ composite for efficient photocatalytic degradation of methylene blue. J Appl Polym Sci 135:45908. https://doi.org/10.1002/app.45908

Dias EM, Petit C (2015) Towards the use of metal–organic frameworks for water reuse: a review of the recent advances in the field of organic pollutants removal and degradation and the next steps in the field. J Mater Chem A 3:22484–22506. https://doi.org/10.1039/c5ta05440k

Dong S, Feng J, Fan M, Pi Y, Hu L, Han X, Liu M, Sun J, Sun J (2015) Recent developments in heterogeneous photocatalytic water treatment using visible light- responsive photocatalysts: a review. RSC Adv 5(19):14610–14630. https://doi.org/10.1039/c4ra13734e

Espino-Pérez E, Domenek S, Belgacem N, Sillard C, Bras J (2014) Green process for chemical functionalization of Nanocellulose with carboxylic acids. Biomacromolecules 15 (12):4551–4560. https://doi.org/10.1021/bm5013458

France KJD, Hoare T, Cranston ED (2017) Review of hydrogels and aerogels containing Nanocellulose. Chem Mater 29(11):4609–4631. https://doi.org/10.1021/acs.chemmater.7b00531

Geng A, Meng L, Han J, Zhong Q, Li M, Han S, Mei C, Xu L, Tan L, Gan L (2018) Highly efficient visible-light photocatalyst based on cellulose derived carbon nanofiber/BiOBr composites. Cellulose 25:4133–4144. https://doi.org/10.1007/s10570-018-1851-y

Ghaly HA, El-Kalliny AS, Gad-Allah TA, Abd El-Sattar NEA, Souaya ER (2017) Stable plasmonic Ag/AgCl–polyaniline photoactive composite for degradation of organic contaminants under solar light. RSC Adv 7(21):12726–12736. https://doi.org/10.1039/c6ra27957k

Ghoreishi KB, Asim N, Che Ramli ZA, Emdadi Z, Yarmo MA (2016) Highly efficient Photocatalytic degradation of methylene blue using carbonaceous WO$_3$/TiO$_2$ composites. J Porous Mater 23(3):629–637. https://doi.org/10.1007/s10934-015-0117-4

Ghosh S, Kouame NA, Remita S, Ramos L, Goubard F, Aubert P, Dazzi A, Deniset-Besseau A, Remita H (2015) Visible-light active conducting polymer nanostructures with superior photocatalytic activity. Sci Rep 5(18002):1–9. https://doi.org/10.1038/srep18002

Gilja V, Novaković K, Travas-Sejdic J, Hrnjak-Murgić Z, Kraljić MK, Žic M (2017) Stability and synergistic effect of Polyaniline/TiO$_2$ Photocatalysts in degradation of Azo dye in wastewater. Nano 7(12):412. https://doi.org/10.3390/nano7120412

Gnanasekaran L, Hemamalini R, Ravichandran K (2015) Synthesis and characterization of TiO$_2$ quantum dots for photocatalytic application. J Saudi Chem Soc 19:589–594. https://doi.org/10.1016/j.jscs.2015.05.002

Gnanasekaran L, Hemamalini R, Saravanan R, Ravichandran K, Gracia F, Gupta VK (2016) Intermediate state created by dopant ions (Mn, Co and Zr) into TiO$_2$ nanoparticles for degradation of dyes under visible light. J Mol Liq 223:652–659. https://doi.org/10.1016/j.molliq.2016.08.105

Golmohammadi H, Morales-Narváz E, Naghdi T, Merkoçi A (2017) Nanocellulose in sensing and biosensing. Chem Mater 29(13):5426–5446. https://doi.org/10.1021/acs.chemmater.7b01170

Gooday GW (1994) Physiology of microbial degradation of chitin and chitosan. Biochem Microb Degrad:279–312. Springer Netherlands. https://doi.org/10.1007/978-94-011-1687-9-9

Habibi Y (2014) Key advances in the chemical modification of nanocelluloses. Chem Soc Rev 43 (5):1519–1542. https://doi.org/10.1039/c3cs60204d

Hall N (2003) Twenty-five years of conducting polymers. Chem Commun 0(1):1–4. https://doi.org/10.1039/B210718J

Hegedűs P, Szabó-Bárdos E, Horváth O, Szabó P, Horváth K (2016) Investigation of a TiO2 photocatalyst immobilized with poly(vinyl alcohol). Catal Today 284:179–186. https://doi.org/10.1016/j.cattod.2016.11.050

Hoeng F, Denneulin A, Bras J (2016) Use of nanocellulose in printed electronics: a review. Nanoscale 8:13131–13154. https://doi.org/10.1039/C6NR03054H

Humayun M, Raziq F, Khan A, Luo W (2018) Modification strategies of TiO$_2$ for potential applications in photocatalysis: a critical review. Green Chem Lett Rev 11(2):86–102. https://doi.org/10.1080/17518253.2018.1440324

Ibanez JG, Rincoń ME, Gutierrez-Granados S, Chahma M, Jaramillo-Quintero OA, Frontana-Uribe BA (2018) Conducting polymers in the fields of energy, environmental remediation, and chemical-chiral sensors. Chem Rev 118(9):4731–4816. https://doi.org/10.1021/acs.chemrev.7b00482

Inamuddin NM, Rangreez TA, ALOthman ZA (2015) Ion-selective potentiometric determination of Pb(II) ions using PVC-based carboxymethyl cellulose Sn(IV) phosphate composite membrane electrode. Desalin Water Treat 56:806–813. https://doi.org/10.1080/19443994.2014.941307

Jahurul Islam M, Kim HK, Amaranatha Reddy D, Kim Y, Ma R, Baek H, Kim J, Kim TK (2017) Hierarchical BiOI nanostructures supported on a metal organic framework as efficient photocatalysts for degradation of organic pollutants in water. Dalton Trans 46 (18):6013–6023. https://doi.org/10.1039/c7dt00459a

Julkapli NM, Bagheri S (2017) Nanocellulose as a green and sustainable emerging material in energy applications: a review. Polym Adv Technol 28(12):1583–1594. https://doi.org/10.1002/pat.4074

Kan WQ, Liu B, Yang J, Liu YY, Ma JF (2012) A series of highly connected metal-organic frameworks based on triangular ligands and d10 metals: syntheses, structures, photoluminescence, and photocatalysis. Cryst Growth Des 12(5):2288–2298. https://doi.org/10.1021/cg2015644

Katančić Z, Šuka S, Vrbat K, Tašić A, Hrnjak-Murgić Z (2017) Synthesis of PEDOT/ZnO Photocatalyst: validation of Photocatalytic activity by degradation of Azo RR45 dye under solar and UV-A irradiation. Chem Biochem Eng Q 31(4):385–394. https://doi.org/10.15255/CABEQ.2017.1124

Kou J, Lu C, Wang J, Chen Y, Xu Z, Varma RS (2017) Selectivity enhancement in heterogeneous Photocatalytic transformations. Chem Rev 117(3):1445–1514. https://doi.org/10.1021/acs.chemrev.6b00396

Kuila A, Surib NA, Mishra NS, Nawaz A, Leong KH, Sim LC, Saravanan P, Ibrahim S (2017) Metal organic frameworks: a new generation coordination polymers for visible light Photocatalysis. ChemistrySelect 2(21):6163–6177. https://doi.org/10.1002/slct.201700998

Lee M, Chen BY, Walter Den W (2015) Chitosan as a natural polymer for heterogeneous catalysts support: a short review on its applications. Appl Sci 5(4):1272–1283. https://doi.org/10.3390/app5041272

Li F, Jiang X, Zhao J, Zhang S (2015) Graphene oxide: a promising nanomaterial for energy and environmental applications. Nano Energy 16:488–515. https://doi.org/10.1016/j.nanoen.2015.07.014

Li Y, Xu H, Ouyang S, Ye J (2016) Metal–organic frameworks for photocatalysis. Phys Chem Chem Phys 18(11):7563–7572. https://doi.org/10.1039/c5cp05885f

Li S, Hao X, Dai X, Tao T (2018) Rapid Photocatalytic degradation of pollutant from water under UV and sunlight via cellulose Nanofiber aerogel wrapped by TiO$_2$. J Nanomater 2018 (8752015):1–12. https://doi.org/10.1155/2018/8752015

Lin N, Dufresne A (2014) Surface chemistry, morphological analysis and properties of cellulose nanocrystals with gradiented sulfation degrees. Nanoscale 6(10):5384–5393. https://doi.org/10.1039/C3NR06761K

Liu J, McCarthy DL, Tong L, Cowan MJ, Kinsley JM, Sonnenberg L, Skorenko KH, Boyer SM, DeCoste JB, Bernier WE, Jones WE Jr (2016) Poly(3,4-ethylenedioxythiophene) (PEDOT) infused TiO2 nanofibers: the role of hole transport layer in photocatalytic degradation of phenazopyridine as a pharmaceutical contaminant. RSC Adv 6(115):113884–113892. https://doi.org/10.1039/c6ra22797j

Luo L, Yang LC, Xiao M, Bian L, Yuan B, Liu Y, Jiang F, Pan X (2015) A novel biotemplated synthesis of TiO$_2$/wood charcoal composites for synergistic removal of bisphenol a by adsorption and photocatalytic degradation. Chem Eng J 262:1275–1283. https://doi.org/10.1016/j.cej.2014.10.087

Mahata P, Madras G, Natarajan S (2006) Novel Photocatalysts for the decomposition of organic dyes based on metal-organic framework compounds. J Phys Chem B 110(28):13759–13768. https://doi.org/10.1021/jp0622381

Meng W, Xu Z, Ding J, Wu D, Han X, Hou H, Fan Y (2014) A systematic research on the synthesis, structures, and application in Photocatalysis of cluster-based coordination complexes. Cryst Growth Des 14(2):730–738. https://doi.org/10.1021/cg401601d

Menon MP, Selvakumar R, Kumar PS, Ramakrishna S (2017) Extraction and modification of cellulose nanofibers derived from biomass for environmental applications. RSC Adv 7 (68):42750–42773. https://doi.org/10.1039/C7RA06713E

Mohammad A, Inamuddin AA et al (2012) Forward ion-exchange kinetics of heavy metal ions on the surface of carboxymethyl cellulose Sn(IV) phosphate composite nano-rod-like cation exchanger. J Therm Anal Calorim 110:715–723. https://doi.org/10.1007/s10973-011-1887-9

Mohammed N, Grishkewich N, Tam KC (2018) Cellulose nanomaterials: promising sustainable nanomaterials for application in water/wastewater treatment processes. Environ Sci Nano 5 (3):623–658. https://doi.org/10.1039/C7EN01029J

Mon M, Bruno R, Ferrando-Soria J, Armentano D, Pardo E (2018) Metal-organic framework technologies for water remediation: towards a sustainable ecosystem. J Mater Chem A 6 (12):4912–4947. https://doi.org/10.1039/c8ta00264a

Nechyporchuk O, Belgacem MN, Bras J (2016) Production of cellulose nanofibrils: a review of recent advances. Ind Crop Prod 93:2–25. https://doi.org/10.1016/j.indcrop.2016.02.016

Ohtani N, Tonoi M (2014) Improved photoluminescence lifetime of organic emissive materials embedded in organic-inorganic hybrid thin films fabricated by sol-gel method using Tetraethoxysilane. Mol Cryst Liq Cryst 599:132–138. https://doi.org/10.1080/15421406.2014.935975

Ong W, Tan L, Ng YH, Yong S, Chai S (2016) Graphitic Carbon Nitride (g-C$_3$N$_3$)-Based Photocatalysts for artificial photosynthesis and environmental remediation: are we a step closer to achieving sustainability? Chem Rev 116(12):7159–7329. https://doi.org/10.1021/acs.chemrev.6b00075

Peng S, Zhu P, Mhaisalkar SG, Ramakrishna S (2012) Self-supporting three-dimensional ZnIn$_2$S$_4$/PVDF–Poly (MMA-co-MAA) composite Mats with hierarchical nanostructures for high Photocatalytic activity. J Phys Chem C 116(26):13849–13857. https://doi.org/10.1021/jp302741c

Peng Y, Qian L, Ding J, Zheng T, Zhang Y, Li B, Li H (2018) Syntheses, structures and photocatalytic degradation of organic dyes for two isostructural copper coordination polymers involving in situ hydroxylation reaction. J Coord Chem 71(9). https://doi.org/10.1080/00958972.2018.1460664

Rajinipriya M, Nagalakshmaiah M, Robert M, Elkoun S (2018) Importance of agricultural and industrial waste in the field of Nanocellulose and recent industrial developments of wood based Nanocellulose: a review. ACS Sustain Chem Eng 6(3):2807–2828. https://doi.org/10.1021/acssuschemeng.7b03437

Rameshthangam P, Solairaj D, Arunachalam G, Ramasamy P (2017) Chitin and Chitinases: biomedical and environmental applications of chitin and its derivatives. J Enzym 1(1):20–43

Rathod M, Moradeeya PG, Haldar S, Basha S (2018) Nanocellulose/TiO$_2$ composites: preparation, characterization and application in the photo-catalytic degradation of a potential endocrine disruptor, mefenamic acid, in aqueous media. Photochem Photobiol. https://doi.org/10.1039/c8pp00156a

Romero-Sáez M, Jaramillo LY, Saravanan R, Benito N, Pabón E, Mosquera E, Gracia F (2017) Notable photocatalytic activity of TiO$_2$-polyethylene nanocomposites for visible light degradation of organic pollutants. Express Polym Lett 11(11):899–909. https://doi.org/10.3144/expresspolymlett.2017.86

Saravanan R, Gracia F, Stephen A (2017) Basic principles, mechanism, and challenges of Photocatalysis. In: Khan M, Pradhan D, Sohn Y (eds) Nanocomposites for visible light-induced Photocatalysis, Springer series on polymer and composite materials. Springer, Cham. https://doi.org/10.1007/978-3-319-62446-4-2

Sayed M, Khan JA, Shah LA, Shah NS, Shah F, Khan HM, Zhang P, Arandiya H (2018) Solar light responsive poly(vinyl alcohol)-assisted hydrothermal synthesis of immobilized TiO2/Ti film with the addition of Peroxymonosulfate for Photocatalytic degradation of ciprofloxacin in aqueous media: a mechanistic approach. J Phys Chem C 122(1):406–421. https://doi.org/10.1021/acs.jpcc.7b09169

Sha Z, Sun J, Chan HSO, Jaenicke S, Wu J (2015) Enhanced Photocatalytic activity of the AgI/UiO-66(Zr) composite for Rhodamine B degradation under visible-light irradiation. ChemPlusChem 80(8):1321–1328. https://doi.org/10.1002/cplu.201402430

Shaghaleh H, Xu X, Wang S (2018) Current progress in production of biopolymeric materials based on cellulose, cellulose nanofibers, and cellulose derivatives. RSC Adv 8(2):825–842. https://doi.org/10.1039/C7RA11157F

Shahabuddin S, Sarih NM, Ismail FH, Shahid MM, Huang NM (2015) Synthesis of chitosan grafted-polyaniline/Co$_3$O$_4$ nanocube nanocomposites and their photocatalytic activity toward methylene blue dye degradation. RSC Adv 5:83857–83867. https://doi.org/10.1039/c5ra11237k

Shahat A, Awual MR, Khaleque MA et al (2015) Large-pore diameter nano-adsorbent and its application for rapid lead (II) detection and removal from aqueous media. Chem Eng J 273:286–295. https://doi.org/10.1016/j.cej.2015.03.073

Singh S, Mahalingam H, Singh PK (2013) Polymer-supported titanium dioxide photocatalysts for environmental remediation: a review. Appl Catal A Gen 462-463:178–195. https://doi.org/10.1016/j.apcata.2013.04.039

Singh S, Singh PK, Mahalingam H (2014) Novel floating Ag$^+$-Doped TiO$_2$/polystyrene Photocatalysts for the treatment of dye wastewater. Ind Eng Chem Res 53(42):16332–16340. https://doi.org/10.1021/ie502911a

Sivakumar K, Senthil-Kumar V, Shim JJ, Haldorai Y (2014) Conducting copolymer/ZnO Nanocomposite: synthesis, characterization, and its Photocatalytic activity for the removal of pollutants. Synth React Inorg Met Org Nano Met Chem 44(10):1414–1420. https://doi.org/10.1080/15533174.2013.809743

Su X, Liao Q, Liu L, Meng R, Qian Z, Gao H, Yao J (2017) Cu$_2$O nanoparticle-functionalized cellulose-based aerogel as high-performance visible-light photocatalyst. Cellulose 24(2):1017–1029. https://doi.org/10.1007/s10570-016-1154-0

Swager T (2017) 50th anniversary perspective: conducting/semi-conducting conjugated polymers. A personal perspective on the past and the future. Macromolecules 50(13):4867–4886. https://doi.org/10.1021/acs.macromol.7b00582

Tennakone K, Tilakaratne CTK, Kottegoda IRM (1995) Photocatalytic degradation of organic contaminants in water with TiO$_2$ supported on polythene films. J Photochem Photobiol A Chem 87:177–179. https://doi.org/10.1016/1010-6030(94)03980-9

Wang H, Wu Y, Wu P, Chen S, Guo X, Meng G, Peng B, Wu J, Liu Z (2017a) Environmentally benign chitosan as reductant and supporter for synthesis of Ag/AgCl/chitosan composites by one-step and their photocatalytic degradation performance under visible-light irradiation. Front Mater Sci 11(2):130–138. https://doi.org/10.1007/s11706-017-0383-y

Wang X, Yao C, Wang F, Li Z (2017b) Cellulose-based nanomaterials for energy applications. Small 13(42):1–19. https://doi.org/10.1002/smll.201702240

Xamena F, Corma A, Garcia H (2007) Applications for metal – organic frameworks (MOFs) as quantum dot semiconductors. J Phys Chem C 111(1):80–85. https://doi.org/10.1021/jp063600e

Xia Q, Yu X, Zhao H, Wang S, Wang H, Guo Z, Xing H (2017) Syntheses of novel lanthanide metal−organic frameworks for highly efficient visible-light-driven dye degradation. Cryst Growth Des 17(8):4189–4195. https://doi.org/10.1021/acs.cgd.7b00504

Xu J, Hu Y, Zeng C, Zhang Y, Huang H (2017) Polypyrrole decorated BiOI nanosheets: efficient photocatalytic activity for treating diverse contaminants and the critical role of bifunctional polypyrrole. J Colloid Interface Sci 505:719–727. https://doi.org/10.1016/j.jcis.2017.06.054

Yu C, Wu R, Fu Y, Dong X, Ma H (2012) Preparation of Polyaniline supported TiO_2 Photocatalyst and its Photocatalytic property. Adv Mater Res 356–360:524–528. https://doi.org/10.4028/www.scientific.net/AMR.356-360.524

Yuan S, Feng L, Wang K, Pang J, Bosch M, Lollar C, Sun Y, Qin J, Yang X, Zhang P, Wang Q, Zou L, Zhang Y, Zhang L, Fang Y, Li J, Zhou H (2018) Stable metal-organic frameworks: design, synthesis, and applications. Adv Mater 1704303. https://doi.org/10.1002/adma.201704303

Zanrosso CD, Piazza D, Azario Lansarin M (2018) Polymeric hybrid films with photocatalytic activity under visible light. J Appl Polym Sci 135(23):46367. https://doi.org/10.1002/app.46367

Zargar V, Asghari M, Dashti A (2015) A review on chitin and chitosan polymers: structure, chemistry, solubility, derivatives, and applications. ChemBioEng Rev 2(3):1–24. https://doi.org/10.1002/cben.201400025

Zhang C, Liu J (2017) Stable chitin whisker/Ag_3PO_4 composite Photocatalyst with enhanced visible light induced Photocatalytic activity. Mater Lett 196:91–94. https://doi.org/10.1016/j.matlet.2017.02.120

Zhang Z, Wang W, Gao E (2014) Polypyrrole/Bi_2WO_6 composite with high charge separation efficiency and enhanced Photocatalytic activity. J Mater Sci 49(20):7325–7332. https://doi.org/10.1007/s10853-014-8445-3

Zhang H, Liu G, Shi L, Liu H, Wang T, Ye J (2016) Engineering coordination polymers for photocatalysis. Nano Energy 22:149–168. doi.org/10.1016/j.nanoen.2016.01.029

Zhang G, Chen L, Fu X, Wang H (2018a) Cellulose microfiber-supported TiO_2@Ag Nanocomposites: a dual-functional platform for Photocatalysis and in situ reaction monitoring. Ind Eng Chem Res 57(12):4277–4286. https://doi.org/10.1021/acs.iecr.8b00006

Zhang M, Wang L, Zeng T, Shang Q, Zhou H, Pan Z, Cheng Q (2018b) Two pure MOF-photocatalysts readily prepared for the degradation of methylene blue dye under visible light. Dalton Trans 47(12):4251–4258. https://doi.org/10.1039/c8dt00156a

Zhao H, Xia Q, Xing H, Chen D, Wang H (2017) Construction of pillared-layer MOF as efficient visible-light Photocatalysts for aqueous Cr(VI) reduction and dye degradation. ACS Sustain Chem Eng 5(5):4449–4456. https://doi.org/10.1021/acssuschemeng.7b00641

Zhu L, Liu X, Jiang H, Sun L (2017) Metal−organic frameworks for heterogeneous basic catalysis. Chem Rev 117(12):8129–8176. https://doi.org/10.1021/acs.chemrev.7b00091

Chapter 8
Role of Conducting Polymer Nanostructures in Advanced Photocatalytic Applications

D. Duraibabu and Y. Sasikumar

Contents

Abstract Conducting polymers (CPs) have been widely used as electronic materials, electromagnetic devices, electrocatalysts, and photocatalysts in energy-related systems, sensors, and environmental protection. Generally, their high conductivity, promising catalytic activity, and electrochemical mechanical and optical properties are considered to be unique. In addition, CPs are cheap and convenient to prepare on a large scale via chemical or electrochemical techniques. Recently, CPs as photosensitizers have been proven to immensely enhance photodegradation by their excellent photocatalytic activity under both ultraviolet light and natural sunlight irradiation, which is not possible using semiconductors alone. Current advanced

D. Duraibabu (✉)
The Key Laboratory of Low-Carbon Chemistry & Energy Conservation of Guangdong Province/State Key Laboratory of Optoelectronic Materials and Technologies, School of Materials Science and Engineering, Sun Yat-Sen University, Guangzhou, People's Republic of China

Y. Sasikumar
Laboratory of Experimental and Applied Physics (LAFEA), Centro Federal de Educação Tecnológica (CEFET/RJ), Celso Suckow da Fonseca, Maracanã Campus, Rio de Janeiro, Brazil

© Springer Nature Switzerland AG 2020   189
M. Naushad et al. (eds.), *Green Photocatalysts*, Environmental Chemistry
for a Sustainable World 34, https://doi.org/10.1007/978-3-030-15608-4_8

techniques consist of synthesis in a new method for CPs, such as high-performance applications. These applications are based on CP-based inherent nanocomposite catalysts derived from heteroatom-doped carbon catalysts. This chapter introduces the mechanisms of catalysis with practical importance.

Keywords Conducting polymers · Conjugated polymers · Nanostructure · Photocatalysis · Mechanism · Photosensitizer · Applications

8.1 Introduction

Conducting polymers (CPs) have delocalized conjugated structures with unique one-dimensional structures. CPs possess outstanding electrical, optical, and electrochemical properties, which are widely used for photocatalytic applications (Kumar et al. 2017a; Li et al. 2009; Heinze et al. 2010). In addition, CPs are catalysts with excellent biocompatibility and can be readily fabricated with flexible polymer films and nanostructures (Henry et al. 2009). In particular, the high electroactive conductivity properties of CPs make them suitable to catalyze redox reactions involved in dye-sensitized solar cells (DSSCs), biosensors, and fuel cell applications. (Yun et al. 2014; Winther Jensen and Mac Farlane 2011; Hao et al. 2013; Yuan et al. 2013; Gerard et al. 2002; Cosnier and Holzinger 2011; Wen et al. 2000). Chemical structures with conjugated linear chains consisting of five- and six-member rings have been widely used as polymers. These CPs are also insulators in the neutral state, which exhibit strong ultraviolet (UV)-visible absorption from the range of 10^{-10}–10^{-5} S cm^{-1}.

To achieve high conductivity and improve the properties of CPs, conjugation in the form of chain structures can be achieved through doping methods by redox reactions and protonation (Hatchett et al 2008). Nevertheless, if an electron is added or withdrawn from the conducting polymers, a deformation will occur around the charge, affecting the elastic energy and establishing a lower electronic charge state. In addition, synthesized CPs with naturally conjugated monomers can be maintained during the synthesis process through electrochemical, chemical oxidation, or interfacial polymerization processes (Huang et al. 2003; Huang and Kaner 2004). For example, a conducting polymer that contained polythiophene derivatives showed semiconducting, electronic, optical, and mechanical characteristics.

As emerging nanostructures, CPs with multiple functionality of energy field conversion and storage have been investigated for the development of novel material applications (Zhang et al. 2013). Semiconductor nanoparticles are widely used for clean energy and to solve environmental pollution issues through solar energy methods (Goswami et al. 2004; Agarwala et al. 2010; Tacca et al. 2012; Agarwala and Ho 2012). To overcome drawbacks, metal oxide-based (TiO$_2$) semiconductors are being developed by surface modification with co-catalysts (Grabowska et al. 2013; Borges et al. 2016; Zhang et al. 2016a, b, c; Mamba et al. 2015; Gnanasekaran et al. 2015, 2016). Based upon these methods, a valuable elevation of TiO$_2$ by photocatalytic activity has been displayed. However, for large-scale applications, the photocatalytic activity of the modified materials was not found to be satisfactory. It has been reported that the hybrid materials can meet the requirements for optimal solar energy converters, including being well matched with the photo absorption of the solar spectrum; their suitable band gap energy and efficient charge separation may substantially improve photocatalytic activity (Weng et al. 2014).

In noble metal nanoparticles (NPs), surface plasmon resonance (SPR) causes concerted oscillations of conduction band electrons, with strong SPR lying in or near the visible range of the spectrum (Zhang et al. 2015). Noble metal NPs have longer wavelengths or harvest light to develop "hot electrons." Ultrafast hot electron injection into a semiconductor improves the lifetime and encourages hot electrons in photocatalytic reactions (Wang et al. 2016; Feizpoor et al. 2017). In metal oxides, the semiconductors are organic conducting polymer nanostructures (CPNs) materials (with multimodal capabilities) that have excellent thermal stability, lower energy photons, high carrier mobility, and chemical stability (Ghosh et al. 2015a, b). In general, CP materials have received much attention because of their low cost and easy availability for renewable energy applications, such as multifunctional coatings, electrolytes, and photovoltaic systems (Bella et al. 2016, 2017; Gerosa et al. 2016; Scalia et al. 2017).

The visible-light photocatalytic activity of the CPs poly (3,4-ethylenedioxythiophene) (PEDOT) and poly(diphenylbutadiyne) (PDPB) has been investigated for the elimination of organic contaminants (Ghosh et al. 2015a, b). PEDOT has outstanding properties such as environmental stability, biocompatibility, high optical transparency, and high conductivity (Ghosh et al. 2015a, b). According to density functional theory calculations, CPNs (organic semiconductors) could serve as low-bandgap options with wide absorption spectra under visible light, making them active photocatalysts (Liu et al. 2015; Ghasimi et al. 2016; Liu et al. 2016).

The introduction of nanomaterials in close contact with CPNs can enhance the interfacial interactions between both polymers and nanomaterials. However, CPs extended with metal plasmatic nanoparticles have not been used for photocatalytic applications. Similarly, the multifunctional nanostructures allow the simultaneous sensing/elimination of organic contaminants. Electrochemical sensing applications with metal/polymer nanocomposites that are a synthesis of PEDOT/gold-NP thin films use a multistep electrochemical method (Xue et al. 2013; Radhakrishnan et al. 2013; Yusoff et al. 2015). Herein, the synthesis of CPs is reported using various methods of chemical and electrochemical polymerization. In addition, we highlight some specific issues pertaining to the new development of CP nanostructures for advanced photocatalytic applications.

8.2 Conducting Polymers

Conducting polymers are largely used because of their unique electrical properties, electrochemical properties, conducting mechanisms, and doping/dedoping processes. Some common CPs are polypyrrole (PPy), poly(p-phenylenevinylene)

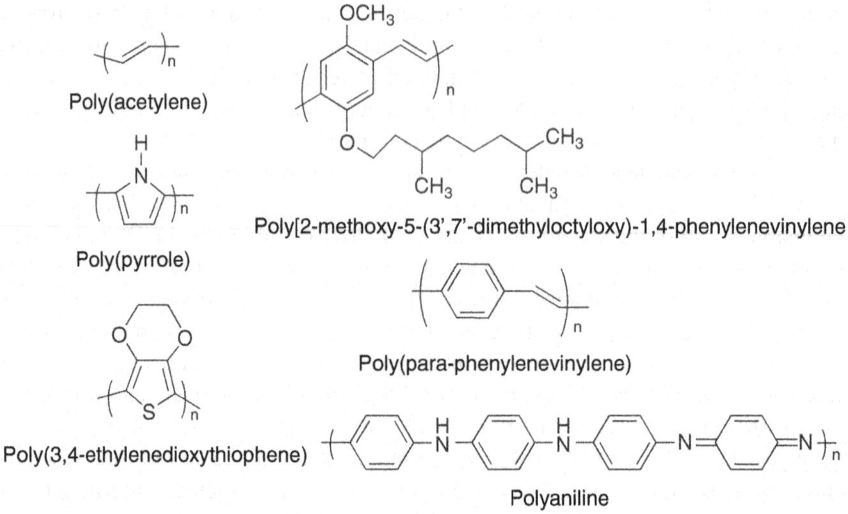

OCH₃

Poly(acetylene)

Poly(pyrrole)

Poly[2-methoxy-5-(3',7'-dimethyloctyloxy)-1,4-phenylenevinylene]

Poly(3,4-ethylenedioxythiophene)

Poly(para-phenylenevinylene)

Polyaniline

Fig. 8.1 Molecular structures of various conducting polymers

(PPV), poly(3,4-ethylenedioxythiophene (PEDOT), polyaniline (PANI), polyfuran (PF), and polythiohohene (PTh) derivatives, as shown in Fig. 8.1. (Lu et al. 2010; Li et al. 2009; Wan 2009; Stejskal et al. 2010; Laslau et al. 2010; MacDiarmid 2001; Heeger et al. 2001, 2010).

In many cases, CPs can absorb visible light because of a delocalized π-system, which has applications in organic photonics and electronics (Pathania et al. 2015; Xu et al. 2015; Zhang et al. 2016a, b, c; Heitzer et al. 2014; Wang et al. 2015). Linear poly(p-phenylene) conjugated photocatalysts for H_2 evolution were reported by Yanagida and co-workers in 1985. Since then, organic conjugated semiconductors have been extensively studied for their photocatalytic H_2 evolution owing to low activity and unfavorable strength.

8.2.1 Conjugated Polymers

Conjugated polymers are commonly controlled through kinetic reactions, which can be irreversibly formed by covalent bonds, as illustrated in Fig. 8.2. Conjugated polymers are linear, planar, and three-dimensional architectures that have been synthesized by a variety of monomers and synthetic procedures. The as-synthesized methods contain π-conjugated systems, novel linkages, and promising porosity, which could allow the balancing of π-conjugated skeletons and nanopores for functional exploration (Ding and Han 2015).

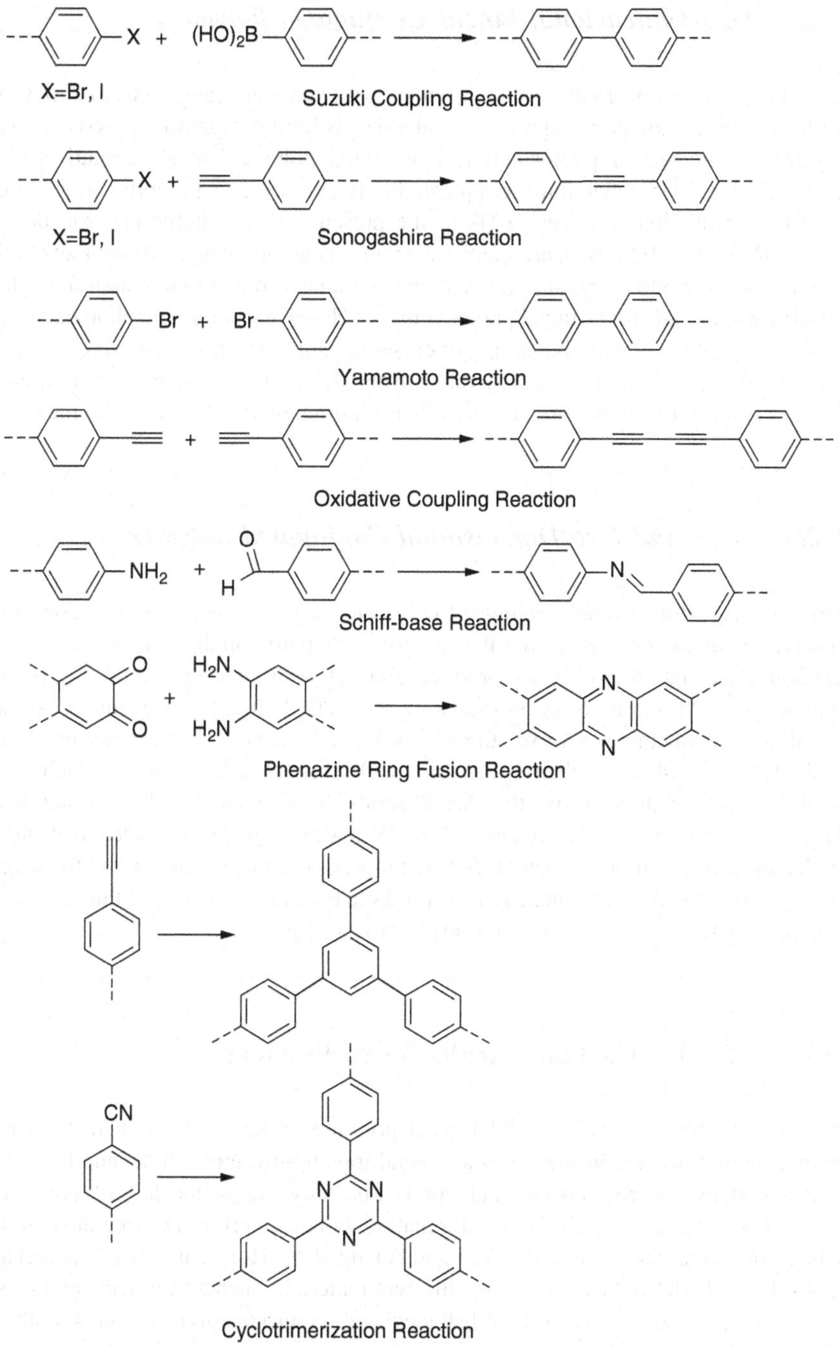
X
(HO)2B
X=Br, I
Suzuki Coupling Reaction
X
X=Br, I
Sonogashira Reaction
Br
Br
Yamamoto Reaction
Oxidative Coupling Reaction
NH2
O
H
N
Schiff-base Reaction
O
O
H2N
H2N
N
N
Phenazine Ring Fusion Reaction
CN
N
N
N
Cyclotrimerization Reaction

Fig. 8.2 Schematic representation of conjugated synthesis polymers

8.2.2 One-Dimensional Linear Conjugated Polymers

One-dimensional linear conjugated polymers with bandgaps ranging from 2 to 5 eV with various conjugation degrees are suitable visible-light photo catalysts for H_2 evolution. The use of photocatalytic conjugated polymers in H_2 evolution was established in 1985, when linear conjugated polymers were first reported to catalyze the photo reduction of water to H_2 in the presence of a suitable electron donor (Yanagida et al. 1985). Because there are a wide range of conjugated polymers with different compositions and structures, they offer a new route in the search for light-absorbing materials for artificial photosynthesis. However, polymeric photocatalysts endure a challenge with linear polymer semiconductors which showed a large binding energy of excitons in the range from 0.5 to 1 eV, other than inorganic (ZnO) crystal semiconductors of which their binding energy was up to 60 meV.

8.2.3 One- and Two-Dimensional Conjugated Polymers

One- and two-dimensional conjugated polymers show a reduction in the Coulomb binding energy, which dissociates the electron–hole pairs and therefore improves the exciton dissociation yields to produce free charges that expedite the redox photocatalytic reaction processes (Schwarz et al. 2012, 2013). For example, conjugated poly(azomethine) (ANW) three-dimensional networks have been prepared with 1,3,5-tris (4-aminophenyl)-benzene with difunctional aromatic aldehydes. With organic photocatalysts, the ANW series showed enhanced time stability. However, the catalytic performance of ANW materials were associated with their molecular composition and optoelectronic properties. Thus, it is important to design new polymers with long dimensions to promote exciton splitting and thus increase photocatalytic activities (H_2 generation) (Schwab et al. 2010).

8.3 Graphitic Carbon Nitride–Based Polymers

Graphitic carbon nitride (g-C3N4)–based polymers were discovered in 2009 for photocatalytic water splitting. They are metal free, lightweight, stable, and low cost with excellent properties (Wang et al. 2009). Similarly, the photocatalytic activity of pure g-C3N4 has serious charge recombination and much effort has been devoted to addressing issues (Sui et al. 2013; Yan and Huang 2011; Hou et al. 2013; Wang et al. 2014; Li et al. 2013). Generally, two different materials, such as s-triazine and tri-s-triazine (heptazine), coexist to establish ideal carbon nitride polymeric units; both of these active polymers have been used for photocatlaytic H_2 evolution (Fig. 8.3).

Figure 8.3a, c shows that graphitic carbon nitride (g-CN) polymers are a stable allotrope of carbon nitride binary materials under room-temperature conditions.

Fig. 8.3e reveals the imperfect denomination of precursors due to the formation of a tri-s-trazine–based melon structure consisting of in-plane infinite one-dimensional chains with Nitrogen and Hydrogen-bridged melem oligomer chains to create a zigzag motif, as well as variable chains that are well connected with hydrogen bonds (Lotsch et al. 2007). From the existence of hydrogen bonds with a carbon nitride framework, the block of electron transfer can be seen over the plane, which has an advantage of low conductivity. This may be due to the completely condensed, crystalline g-CN. The crystallinity of the photocatalysts and their activity owing to their weaknesses (performing recombination inside for electron-hole pairs) can be improved when the structure is reduced. (Diebold 2011; Maeda 2013).

A new technique has been established to synthesize crystalline g-CN through the self-condensation of dicyandiamide using salt to melt high-temperature solvents. Researchers have developed high crystallinity with the use of a salt melt to alter the structure of a carbon nitride-based trazine polymer structure (poly(triazine imide) (PTI), as shown in Fig. 8.3g. Consequently, tri-s-triazine–based crystalline g-CN could be synthesized using an ionothermal method. In contrast, triazine-based PTI and tris-s-triazine–based g-CN (Fig. 8.3c)are expected to be photocatalytic because of their fully condensed conjugation structures, which can enhance the sunlight-harvesting capability and stabilize π-electrons for rapid charge mobility due to a decrease of the bandgap. As a result, the s-triazine unit has a more energy than the tri-s-triazine unit. Furthermore, it was thermodynamically challenging to synthesize carbon nitride based upon s-triazine units. PTI and melon, which have related characteristics to s-triazine and tri-s-triazine structures, respectively, can be synthesized experimentally to achieve carbon nitride binary polymers with two-dimensional electronic structures.

8.4   Nanostructure Conducting Polymers

A new method has been developed for nanostructured conducting polymers using nanofiller-like nanowires and nanotubes. The preparation of nanostructure conducting polymers can use electrospinning, physical methods, and chemicals methods, such as a template free method (Table 8.1), interfacial polymerization, reverse emulsion polymerization, and dilute polymerization. Nanostructure CPs have been used in a wide range of applications, including biosensors, field emissions, field-effect transistors, electrochromic display devices, supercapacitors, and biosensors (Strenger Smith 1998; Kang et al. 1998; Gospodinova and Terlemezyan 1998; Palaniappan and John 2008; Bhadra et al. 2009; Pud et al. 2003).

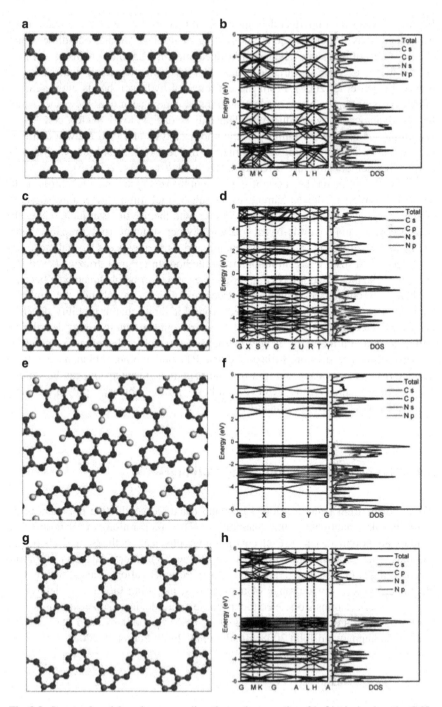

Fig. 8.3 Structural models and corresponding electronic properties of (**a**, **b**) triazine-based g-C$_3$N$_4$, (**c**, **d**) tri-s-triazine-based g-C$_3$N$_4$, (**e**, **f**) melon, and (**g**, **h**) PTI. The gray, blue, and white spheres denote C, N, and H atoms, respectively. (Lin et al. 2016)

Table 8.1 Hard, soft and free template fabrication approaches to synthesize nanostructure conducting polymers

Hard template	Soft template	Free template
Nanocomposites Existing nanostructures Inside pores or channels Voids of colloidal arrangements	Over self-assembled (monomer, surfactants and dopant), air bubbles	Interfacial polymerization Higher current densities

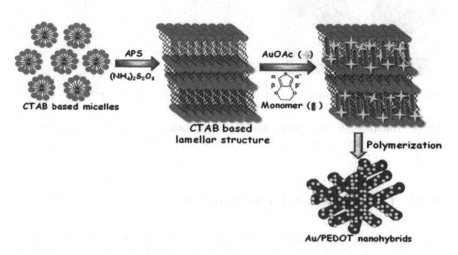

Fig. 8.4 Schematic diagram of the synthesis of Au/PEDOT nanohybrids. (Ghosh et al. 2018)

8.4.1   Nanofiber-Based Polymers

Nano-hybrid conducting polymers have been developed with a colloidal route method using gold nanoparticles (Au-NPs) with PEDOT (Ghosh et al. 2018), as presented in Fig. 8.4. The Au/PEDOT nano-hybrid polymer shows improved photocatalytic activity (~5.6 times greater compared with basic polymers) for organic pollutant degradation under visible light irradiation. Au-NPs are better electron reservoirs, with enriched separation of both electrons and holes via photo-induced interfacial charge transfer. At the same time, the photo-generated excitons are decayed at the interface of the metal and hybrid polymer system to produce charge species (Wang et al. 2012; Clavero 2014). Figure 8.5 depicts the occurrence of the electron transfer process from photo-excited PEDOT nanofibers to Au-NPs, in addition to improved charge separation. The charge separation efficiency of the photocatalytic degradation of methyl orange for the AU/PEDOT has been compared with PEDOT polymer nanofibers under visible light. Experimental results were correlated with the role of O_2^- photo-induced h^+ and $HO^{\cdot}$ radicals, which can be generated using photo-induced degradation of organic pollutants with Au/PEDOT NHs (Fig. 8.5). From the experimental data, Au/PEDOT nano-hybrid polymers seem

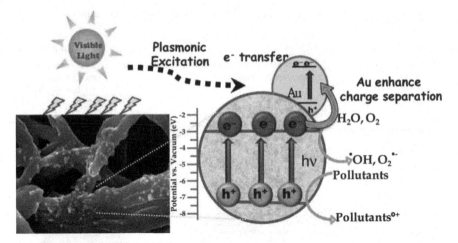

Fig. 8.5 The photocatalytic process of Au/PEDOT nanohybrids. (Ghosh et al. 2018)

to be suitable for the photocatalytic degradation of organic pollutants, as photocatalysts, and in other energy conversion applications.

8.4.2 One-Dimensional Nanostructures

One-dimensional nanostructures were developed by photopolymerization using 1,4-diphenyl butadiyne through a soft-template method (Ghosh et al. 2015a, b). PDPB nanofibers are composed of hexagonal mesophases of oil, arranged with swollen surfactant tubes on a triangular lattice in water under UV irradiation (Mackiewicz et al. 2011; Pena dos Santos et al. 2005). Higher concentrations (up to 20%) of diphenyl butadiyne monomer have a hydrophobic domain of the mesophases, which enhances polymerization through photo-irradiation in the presence of a free-radical initiator, benzoin methyl ether (1%). After polymerization, the doped mesophases were yellow in color (Fig. 8.6)

8.4.3 Photocatalysis Mechanisms

Figure 8.7 shows the photocatalytic mechanism and energy levels of polymer structures through functional density theory. Under UV light excitation, PDPB polymers generated charge transporters in the reactive oxygen species during the degradation method. $O_2^{\cdot-}$ plays a role in the parallel holes (h^+) and also diffuses on the surface. The energy from the band valance is calculated to drive the oxygen-evolution of a half reaction at pH 7 (Young et al. 2012). The holes can be oxidized directly from the pollutant molecules through a catalytic degradation reaction (Kumar et al. 2017b).

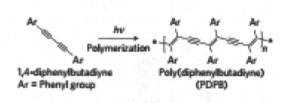

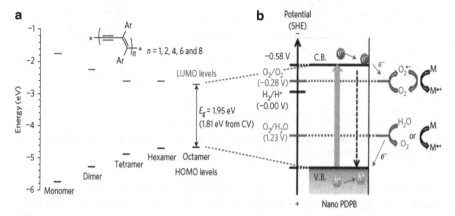

Fig. 8.6 Synthesis and characterization of PDPB nanofibers. (a) Photographs from before polymerization (transparent gel) and after polymerization (yellow gel) by ultraviolet irradiation. (b) Schematic of the polymerization of diphenyl butadiyne (DPB) by ultraviolet irradiation. (Ghosh et al. 2015a, b)

Fig. 8.7 Schematic diagram of the photocatalytic mechanisms. (a) Energy diagram representing the evaluated highest occupied molecular orbital and lowest unoccupied molecular orbital of the PDPB polymer. (b) Possible photocatalysis mechanism with charge separation in nano-PDPB, with electron-reducing oxygen and hole-oxidizing water; the holes and generated oxidative radicals can oxidize organic pollutants (noted as M). V.B. and C.B. represent the valence band and the conduction band of the PDPB polymer, respectively. (Ghosh et al. 2015a, b)

When examining the recombination of the electron-hole method in photocatalyst degradation, the addition of a scavenger with an electron hole recombination was found to decrease in the presence of isopropanol (a hole scavenger) (Schaming et al. 2010). The decomposition kinetics of PDPB nanofibers considerably increased in the presence of isopropanol (0 .1 M) (Fig. 8.8). This is expected to be an aggressive reaction from additional holes with isopropanol, which will reduce the recombination rate from the outstanding availability of oxidative radical $O_2{}^{\cdot-}$ extra electrons for moderate degradation. Based on the bandgap structures of polymers with scavengers, the experiments showed a feasible mechanism for organic pollutant degradation.

Figure 8.7b depicts the generation of charge carriers under irradiation and the formation of the oxidative radicals that are accountable for the oxidation and mineralization of pollutants. The energy levels of the band valence, conduction

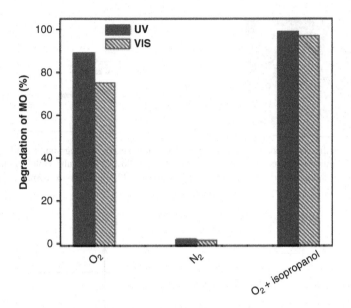

Fig. 8.8 Effect of oxygen, argon, and 0.1 M isopropanol on the photocatalytic activity of PDPB nanofibers. PDPB nanofibers degrade methyl orange under UV-visible and visible light (λ > 450 nm) irradiation in the presence of oxygen, argon, and oxygen and 0.1 M isopropanol after 270 min of exposure to light. (Ghosh et al. 2015a, b)

band, and PDPB nanostructures may be calculated to thermodynamically encourage a water-splitting reaction. The PDPB nanofibers were obtained from the π-staking of oliogomers in the presence of high photocatalytic mechanisms under UV and visible light, which are highly stable with cycling. As a result, the development of a new generation of conducting polymer nanofibers is possible, with efficient and cheap visible light-driven photocatalysts for eco-friendly protection.

8.4.4 Cerium Oxide Nanoparticles

Cerium oxide nanoparticles (CeONPs) have a wide range of applications in the field of catalysis, as well as the biomedical field. However, they have some drawbacks as photocatalysts due to a large bandgap. Hence, we developed an active catalyst support for the enhancement of photocatalytic activity with CeONPs. Polyaniline (PANI)/CeONP nanocomposites were prepared using surface spherical-shaped CeONPs via in-situ polymerization of an aniline monomer using a hydrothermal method (Samai and Bhattacharya et al. 2018). From the experimental results, it can be seen that PANI/CeNOP composites showed significantly improved photocatalytic efficiency compared to that of CeNOPs or polyaniline. This may be due to the synergistic effect of both PANI and CeONP from the energy level match between the highest occupied molecular orbital (HOMO), lowest unoccupied molecular orbital (LUMO; PANI), valance band (VB), and the conduction band (CB) of CeO_2. In addition, under UV light illumination, the electrons present in the HOMO (PANI) are excited to the LUMO (PANI) by the π to $\pi*$ transition, thus forming a hole in the HOMO. Simultaneously, the electrons present in the VB (CeO_2) are excited to the CB (CeO_2), thus forming holes in the VB.

During the photogeneration process, the hole in VB (CeO_2) is transferred to the HOMO (PANI). In the LUMO (PANI), it migrates to the CB (CeO_2) (Guo et al. 2014). Therefore, the photogenerated electrons and holes accumulate in the CB (CeO_2) and HOMO (PANI). Hence, the photocatalytic activity of CeO_2 is delayed due to its large bandgap at 3.36 eV, which causes rapid recombination of the photogenerated charge carriers (i.e., the electron and hole). This may be due to the synergistic effect between PANI and CeO_2, which avoids the rapid recombination of photogenerated charge carriers in CeO_2. In addition, the gathered electrons in the CB (CeO_2) can reduce the adsorbed O_2 to $O_2\cdot^-$ from the subsequent CB potential of CeO_2 (−0.61 eV), which is more negative than the standard redox potential of $O_2/O_2\cdot^-$ (−0.046 eV vs. a normal hydrogen electrode) (Zhang et al. 2014). The generated $O_2\cdot^-$ reacts with H_2O to form $HO\cdot$, which degrades a rhodamine B (RhB) molecule. Hence, additional probability accumulates in the hole of the HOMO (PANI) and also degrades RhB molecules. The enriched photocatalytic activity is depicted in Fig. 8.9. In addition, the PANI/CeONP nanocomposites were set to one step forward as a photocatalyst than CeONP.

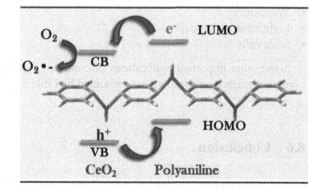

Fig. 8.9 Mechanisms of electron-hole transfer between CeO_2/PANI. (Samai and Bhattacharya et al. 2018)

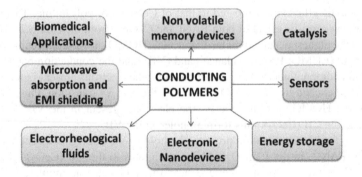

Fig. 8.10 Applications of conducting polymers

8.5 Applications of Conducting Polymers

Conducting polymers exhibit unique properties for electrocatalysis, nanoelectronic devices, and chemical/biological sensor, among others, as presented in Fig. 8.10. Potential applications of conducting polymers include the following:

- Rechargeable batteries
- Sensors for low concentrations of phenol and O_2 (among others)
- Transistors
- Telecommunication systems
- Photovoltaic cells
- Antistatic coatings
- Corrosion inhibitors
- Compact capacitors
- Anti-static substances for photographic film
- Electromagnetic shields for computer "smart windows"
- Transistors
- Light-emitting diodes
- Solar cells

Some other important applications of conducting polymers and their advanced high-performance applications are presented in Table 8.2.

8.6 Conclusion

Conducting polymers are widely used as electrocatalysts and photocatalysts. CPs have promising applications in DSSCs, fuel cells, electrochemical sensors, water splitting, and the degradation of organic pollutants. CPs can also be obtained as nanofibers with high photocatalytic activity under ultraviolet and visible light, which can be derived from the π-staking of oligomers. In the near future, basic knowledge

Table 8.2 Properties of the conducting polymers

Conducting polymer	Structure	Properties	Applications
Polythiophenes (PT)		Good electrical conductivity and optical properties	Biosensors, food industry
Polypyrrole (PPy)		High conductivity, opaque, brittle, amorphous material	Biosensors, antioxidants, drug delivery, bioactuators, neural prosthetics, cardiovascular applications
Polyaniline (PANI)		Semi-flexible, high conductivity at 100 S cm^{-1}	Biosensors, antioxidants, food industry, drug delivery, cardiovascular applications, bioactuators
Poly(3,4-ethylene dioxy thiophene) (PEDOT)		Chemical synthesis and electrochemical properties	Biosensors, antioxidants, drug delivery, neural prosthetics

of the mechanisms of conducting polymers should lead to the achievement of practical applications. However, CP-based catalysts still face several challenges because of their catalytic mechanisms, which have not yet been extensively explored. CPs were found to be very stable with cycling and are widely used as nanostructures in the field of photocatalysis. In addition, they are important for the understanding and further development of photocatalysts that are active under visible-light irradiation. Hence, CP nanofibers play a significant role in the development of a new generation of inexpensive, visible-light-driven photocatalysts and associated photocatalytic applications.

References

Agarwala S, Ho GW (2012) Self-ordering anodized nanotubes: enhancing the performance by surface plasmon for dye-sensitized solar cell. J Solid State Chem 189:101–107. https://doi.org/10.1016/j.jssc.2011.11.047

Agarwala S, Kevin M, Wong ASW, Peh CKN, Thavasi V, Ho GW (2010) Mesophase ordering of TiO_2 film with high surface area and strong light harvesting for dye-sensitized solar cell. ACS Appl Mater Interfaces 2:1844–1850. https://doi.org/10.1021/am100421e

Bella F, Lamberti A, Bianco S, Tresso E, Gerbaldi C, Pirri CF (2016) Floating photovoltaics: floating, flexible polymeric dye-sensitized solar-cell architecture: the way of near-future photovoltaics. Adv Mater Technol 1:1600002. https://doi.org/10.1002/admt.201600002

Bella F, Galliano S, Falco M, Viscardi G, Barolo C, Grätzel M, Gerbaldi C (2017) Approaching truly sustainable solar cells by the use of water and cellulose derivatives. Green Chem 19:1043–1051. https://doi.org/10.1039/C6GC02625G

Bhadra S, Khastgir D, Singha NK, Lee JH (2009) Progress in preparation, processing and applications of polyaniline. Prog Polym Sci 34(8):783–810. https://doi.org/10.1016/j.progpolymsci.2009.04.003

Borges ME, Sierra M, Cuevas E, García RD, Esparza P (2016) Photocatalysis with solar energy: sunlight-responsive photocatalyst based on TiO_2 loaded on a natural material for wastewater treatment. Sol Energy 135:527–535. https://doi.org/10.1016/j.solener.2016.06.022

Clavero C (2014) Plasmon-induced hot-electron generation at nanoparticle/metal-oxide interfaces for photovoltaic and photocatalytic devices. Nat Photonics 8:95–103. https://doi.org/10.1038/nphoton.2013.238

Cosnier S, Holzinger M (2011) Electrosynthesized polymers for biosensing. Chem Soc Rev 40:2146–2156. https://doi.org/10.1039/c0cs00090f

Diebold U (2011) Photocatalysts: closing the gap. Nat Chem 3(4):271–272. https://doi.org/10.1038/nchem.1019

Ding X, Han BH (2015) Metallophthalocyanine-based conjugated microporous polymers as highly efficient photosensitizers for singlet oxygen generation. Angew Chem Int Ed Engl 54 (22):6536–6539. https://doi.org/10.1002/anie.201501732

Feizpoor S, Yangjeh AH, Vadivel S (2017) Novel TiO_2/Ag_2CrO_4 nanocomposites: efficient visible-light-driven photocatalysts with n–n heterojunctions. J Photochem Photobiol A Chem 341:57–68. https://doi.org/10.1016/j.jphotochem.2017.03.028

Gerard M, Chaubey A, Malhotra BD (2002) Application of conducting polymers to biosensors. Biosens Bioelectron 17(5):345–359. https://doi.org/10.1016/S0956-5663(01)00312-8

Gerosa M, Sacco A, Scalia A, Bella F, Angelica C, Quaglio M, Tresso E, Bianco S (2016) Toward totally flexible dye-sensitized solar cells based on titanium grids and polymeric electrolyte. IEEE J Photovolt 6:498–505. https://doi.org/10.1109/JPHOTOV.2016.2514702

Ghasimi S, Lfester K, Zhang KAI (2016) Water compatible conjugated microporous polyazulene networks as visible-light photocatalysts in aqueous medium. Chem Cat Chem 8:694–698. https://doi.org/10.1002/cctc.201501102

Ghosh S, Kouamé NA, Ramos L, Remita S, Dazzi A, Deniset-Besseau A, Beaunier P, Goubard F, Aubert PH, Remita H (2015a) Visible-light active conducting polymer nano structures with superior photocatalytic activity. Nat Mater 14:505–511. https://doi.org/10.1038/nmat4220

Ghosh S, Kouamé NA, Ramos L, Remita S, Dazzi A, Deniset-Besseau A, Beaunier P, Goubard F, Aubert PH, Remita H (2015b) Conducting polymer nanostructures for photocatalysis under visible light. Nat Mater 14:505–511. https://doi.org/10.1038/NMAT4220

Ghosh S, Mallik AK, Basu Rajendra N (2018) Enhanced photocatalytic activity and photoresponse of poly(3,4-ethylenedioxythiophene) nanofibers decorated with gold nanoparticle under visible light. Sol Energy 159:548–560. https://doi.org/10.1016/j.solener.2017.11.036

Gnanasekaran L, Hemamalini R, Ravichandran K (2015) Synthesis and characterization of TiO_2 quantum dots for photocatalytic application. J Saudi Chem Soc 19:589–594. https://doi.org/10.1016/j.jscs.2015.05.002

Gnanasekaran L, Hemamalini R, Saravanan R, Ravichandran K, Gracia F, Gupta VK (2016) Intermediate state created by dopant ions (Mn, co and Zr) into TiO_2 nanoparticles for degradation of dyes under visible light. J Mol Liq 223:652–659. https://doi.org/10.1016/j.molliq.2016.08.105

Gospodinova N, Terlemezyan L (1998) Conducting polymers prepared by oxidative polymerization: polyaniline. Prog Polym Sci 23(8):1443–1484. https://doi.org/10.1016/S0079-6700(98)00008-2

Goswami DY, Vijayaraghavan S, Lu S, Tamm G (2004) New and emerging developments in solar energy. Sol Energy 76:33–43. https://doi.org/10.1016/S0038-092X(03)00103-8

Grabowska E, Zaleska A, Sorgues S, Kunst M, Etcheberry A, Colbeau-Justin C, Remita H (2013) Modification of titanium (IV) dioxide with small silver nanoparticles:application in photocatalysis. J Phys Chem C 117:1955–1962. https://doi.org/10.1021/jp3112183

Guo N, Liang Y, Lan S, Liu L, Zhang J, Ji GGan S (2014) Microscale hierarchical three-dimensional flowerlike TiO_2/PANI composite: synthesis, characterization, and its remarkable photocatalytic activity on organic dyes under UV-light and sunlight irradiation. J Phys Chem C 118(32):18343–18355. https://doi.org/10.1021/jp5044927

Hao F, Dong P, Luo Q, Li JB, Lou J, Lin H (2013) Recent advances in alternative cathode materials for iodine-free dye-sensitized solar cells. Energy Environ Sci 6:2003–2019. https://doi.org/10.1039/C3EE40296G

Hatchett DW, Josowicz M (2008) Composites of intrinsically conducting polymers as sensing nanomaterials. Chem Rev 108(2):746–769. https://doi.org/10.1021/cr068112h

Heeger AJ (2001) Semiconducting and metallic polymers: the fourth generation of polymeric materials. J Phys Chem B 105(36):8475–8491. https://doi.org/10.1021/jp011611w

Heeger AJ (2010) Semiconducting polymers: the third generation. Chem Soc Rev 39(7):2354–2371. https://doi.org/10.1039/B914956M

Heinze J, Frontana-Uribe BA, Ludwigs S (2010) Electrochemistry of conducting polymers--persistent models and new concepts. Chem Rev 110(8):4724–4471. https://doi.org/10.1021/cr900226k

Heitzer HM, Savoie BM, Marks TJ, Ratner MA (2014) Direct evidence of a surface quenching effect on size-dependent luminescence of upconversion nanoparticles. Angew Chem Int Ed 53:7456–7460. https://doi.org/10.1002/anie.201003959

Henry D, Li TD, Kane RB (2009) One-dimensional conducting polymer nanostructures: bulk synthesis and applications. Adv Mater 21:1487–1499. https://doi.org/10.1002/adma.200802289

Hou Y, Wen ZH, Cui SM, Guo XR, Chen JH (2013) Constructing 2D porous graphitic C_3N_4nanosheets/nitrogen-doped graphene/layered MoS_2 ternary nanojunction with enhanced photoelectrochemical activity. Adv Mater 25(43):6291–6297. https://doi.org/10.1002/adma.201303116

Huang J, Kaner RB (2004) A general chemical route to polyaniline nanofibers. J Am Chem Soc 126(3):851–855. https://doi.org/10.1021/ja0371754

Huang J, Virji S, Weiller BH, Kaner RB (2003) Polyaniline nanofibers: facile synthesis and chemical sensors. J Am Chem Soc 125(2):314–315. https://doi.org/10.1021/ja028371y

Kang ET, Neoh KG, Tan KL (1998) Polyaniline: a polymer with many interesting redox states. Prog Polym Sci 23(2):277–324. https://doi.org/10.1016/S0079-6700(97)00030-0

Kumar A, Kumar A, Sharma G et al (2017a) Solar-driven photodegradation of 17-β-estradiol and ciprofloxacin from waste water and CO2 conversion using sustainable coal-char/polymeric-g-C3N4/RGO metal-free nano-hybrids. New J Chem 41:10208–10224. https://doi.org/10.1039/c7nj01580a

Kumar A, Shalini SG et al (2017b) Facile hetero-assembly of superparamagnetic Fe3O4/BiVO4 stacked on biochar for solar photo-degradation of methyl paraben and pesticide removal from soil. J Photochem Photobiol A Chem 337:118–131. https://doi.org/10.1016/j.jphotochem.2017.01.010

Laslau C, Zujovic Z, Travas-Sejdic J (2010) Theories of polyaniline nanostructure self-assembly: towards an expanded, comprehensive multi-layer theory (MLT). Prog Polym Sci 35(12):1403–1419. https://doi.org/10.1016/j.progpolymsci.2010.08.002

Li C, Bai H, Shi GQ (2009) Conducting polymer nanomaterials: electrosynthesis and applications. Chem Soc Rev 38:2397–2409. https://doi.org/10.1039/B816681C

Li YB, Zhang HM, Liu PR, Wang D, Li Y, Zhao HJ (2013) Cross-linked g-C3N4/rGO nanocomposites with tunable band structure and enhanced visible light photocatalytic activity. Small 9(19):3336–3344. https://doi.org/10.1002/smll.201203135

Lin LL, Ou HH, Zhang YF, Wang XC (2016) Tri-s-triazine-based crystalline graphitic carbon nitrides for highly efficient hydrogen evolution Photocatalysis. ACS Catal 6(6):3921–3931. https://doi.org/10.1021/acscatal.6b00922

Liu J, Liu Y, Liu N, Han Y, Zhang X, Huang H, Lifshitz Y, Lee ST, Zhong J, Kang Z (2015) Metal-free efficient photocatalyst for stable visible water splitting via a two-electron pathway. Science 347:970–974. https://doi.org/10.1126/science.aaa3145

Liu D, Wang J, Bai X, Zong R, Zhu Y (2016) Self-assembled PDINH supramolecular system for photocatalysis under visible light. Adv Mater 28:7284–7290. https://doi.org/10.1002/adma.201601168

Lotsch BV, Döblinger M, Sehnert J, Seyfarth L, Senker J, Oeckler O, Schnick W (2007) Unmasking melon by a complementary approach employing electron diffraction, solid- state NMR spectroscopy, and theoretical calculations-structural characterization of a carbon nitride polymer. Chemistry 13(17):4969–4980. https://doi.org/10.1002/chem.200601759

Lu XF, Zhang WJ, Wang C, Wen TC, Wei Y (2010) One-dimensional conducting polymer nanocomposites: synthesis, properties and applications. Prog Polym Sci 36:671–712. https://doi.org/10.1016/j.progpolymsci.2010.07.010

MacDiarmid AG (2001) Synthetic metals: a novel role for organicpolymers. Angew Chem Int Ed 40(14):2581–2590. https://doi.org/10.1002/1521-3773(20010716)40

Mackiewicz N, Gravel E, Garofalakis A, Ogier J, John J, Dupont DM, Gombert K, Tavitian B, Doris E, Ducongé F (2011) Tumor-targeted polydiacetylene micelles for in vivo imaging and drug delivery. Small 7(19):2786–2792. https://doi.org/10.1002/smll.201100212

Maeda K (2013) Z-scheme water splitting using two different semiconductor Photocatalysts. ACS Catal 3(7):1486–1503. https://doi.org/10.1021/cs4002089

Mamba G, Mbianda XY, Mishra AK (2015) Photocatalytic degradation of diazo dye naphthol blue black in water using MWCNT/Gd, N, S-TiO2 nanocomposite under simulated solar light. J Environ Sci 33:219–228. https://doi.org/10.1016/j.jes.2014.06.052

Palaniappan S, John A (2008) Polyaniline materials by emulsion polymerization pathway. Prog Polym Sci 33(7):732–758. https://doi.org/10.1016/j.progpolymsci.2008.02.002

Pathania D, Sharma G, Kumar A et al (2015) Combined sorptional–photocatalytic remediation of dyes by polyaniline Zr(IV) selenotungstophosphate nanocomposite. Toxicol Environ Chem 97:526–537. https://doi.org/10.1080/02772248.2015.1050024

Pena dos Santos E, Tokumoto MS, Surendran G, Remita H, Bourgaux C, Dieudonné P, Prouzet E, Ramos L (2005) Existence and stability of new Nanoreactors: highly swollen hexagonal liquid crystals. Langmuir 21(10):4362–4369. https://doi.org/10.1021/la047092g

Pud A, Ogurtsov N, Korzhenko A, Shapoval G (2003) Some aspects of preparation methods and properties of polyaniline blends and composites with organic polymers. Prog Polym Sci 28 (12):1701–1753. https://doi.org/10.1016/j.progpolymsci.2003.08.001

Radhakrishnan S, Sumathi C, Dharuman V, Wilson J (2013) Gold nanoparticles functionalized poly (3,4-ethylenedioxythiophene) thin film for highly sensitive label free DNA detection. Anal Methods 5:684–689. https://doi.org/10.1039/C2AY26143J

Samai B, Bhattacharya SC (2018) Conducting polymer supported cerium oxide nanoparticle: enhanced photocatalytic activity for waste water treatment. Mater Chem Phys 220:171–181. https://doi.org/10.1016/j.matchemphys.2018.08.050

Scalia A, Bella F, Lamberti A, Bianco S, Gerbaldi C, Tresso E, Pirri CF (2017) A flexible and portable power pack by solid-state supercapacitor and dye-sensitized solar cell integration. J Power Sources 359:311–321. https://doi.org/10.1016/j.jpowsour.2017.05.072

Schaming D, Costa-Coquelard C, Sorgues S, Ruhlmann L, Lampre I (2010) Photocatalytic reduction of Ag_2SO_4 by electrostatic complexes formed by tetracationic zinc porphyrins and tetracobalt Dawson-derived sandwich polyanion. Appl Catal A 373(1–2):160–167. https://doi.org/10.1016/j.apcata.2009.11.010

Schwab MG, Hamburger M, Feng XL, Shu J, Spiess HW, Wang XC, Antonietti M, Mgllen K (2010) Photocatalytic hydrogen evolution through fully conjugated poly(azomethine) networks. Chem Commun 46:8932–8934. https://doi.org/10.1039/c0cc04057f

Schwarz C, Bässler H, Bauer I, Koenen JM, Preis E, Scherf U, Köhler A (2012) Does conjugation help exciton dissociation? A study on poly(p-phenylene)s in planar heterojunctions with C_{60} or TNF. Adv Mater 24(7):922–925. https://doi.org/10.1002/adma.201104063

Schwarz C, Tscheuschner S, Frisch J, Winkler S, Koch N, Bässler H, Köhler A (2013) Role of the effective mass and interfacial dipoles on exciton dissociation in organic donor-acceptor solar cells. Phys Rev B 87(15):155205–155213. https://doi.org/10.1103/PhysRevB.87.155205

Stejskal J, Sapurina I, Trchova M (2010) Polyaniline nanostructures and the role of aniline oligomers in their formation. Prog Polym Sci 5(12):1420–1481. https://doi.org/10.1016/j.progpolymsci.2010.07.006

Strenger-Smith JD (1998) Intrinsically electrically conducting polymers. Synthesis, characterization and their applications. Prog Polym Sci 23(1):57–79. https://doi.org/10.1016/S0079-6700 (97)00024-5

Sui Y, Liu J, ZhangY TX, Chen W (2013) Dispersed conductive polymer nanoparticles on graphitic carbon nitride for enhanced solar-driven hydrogen evolution from pure water. Nanoscale 5:9150–9155. https://doi.org/10.1039/C3NR02413J

Tacca A, Meda L, Marra G, Savoini A, Caramori S, Cristino V, Bignozzi CA, Pedro VG, Boix PP, Gimenez S, Bisquert J (2012) Photoanodes based on nanostructured WO_3 for water splitting. Chem Phys Chem 13:3025–3034. https://doi.org/10.1002/cphc.201200069

Wan MX (2009) Some issues related to polyaniline micro−/nanostructures. Macromol Rapid Commun 30:963–975. https://doi.org/10.1002/marc.200800817

Wang X, Maeda K, Thomas A, Takanabe K, Xin G, Carlsson JM, Domen K, Antonietti M (2009) A metal-free polymeric photocatalyst for hydrogen production from water under visible light. Nat Mater 8:76–80. https://doi.org/10.1038/nmat2317

Wang S, Huang Z, Wang J, Li Y, Tan Z (2012) Thermal stability of several polyaniline/rare earth oxide composites (I): polyaniline/CeO_2 composite. J Therm Anal Calorim 107(3):1199–1201. https://doi.org/10.1007/s10973-011-1777

Wang D, Zhang Y, Chen W (2014) A novel nickel–thiourea–triethylamine complex adsorbed on graphitic C_3N_4 for low-cost solar hydrogen production. Chem Commun 50:1754–1756. https://doi.org/10.1039/C3CC48141G

Wang ZJ, Ghasimi S, Landfester K, Zhang KAI (2015) Molecular structural Design of Conjugated Microporous Poly(Benzooxadiazole) networks for enhanced photocatalytic activity with visible light. Adv Mater 27:6265–6270. https://doi.org/10.1002/adma.201502735

Wang M, Ye M, Locozzia J, Lin C, Lin Z (2016) Plasmon-mediated solar energy conversion via photocatalysis in noble metal/semiconductor composites. Adv Sci 3(6):1600024 1–14. https://doi.org/10.1002/advs.201600024

Wen C, Hasegawa K, Kanbara T, Kagaya S, Yamamoto TJ (2000) Visible light-induced catalytic degradation of iprobenfos fungicide by poly(3-octylthiophene-2,5-diyl) film. J Photochem Photobiol A: Chem 133(1–2):59–66. https://doi.org/10.1016/S1010-6030(00)00206-9

Weng L, Zhang H, Govorov AO, Ouyang M (2014) Hierarchical synthesis of noncentrosymmetric hybrid nanostructures, enabled plasmon-driven photocatalysis. Nat Commun 2(5):4602–4792. https://doi.org/10.1038/ncomms5792

Winther-Jensen B, Mac Farlane DR (2011) New generation, metal-free electrocatalysts for fuelcells, solar cells and water splitting. Energy Environ Sci 4:2790–2798. https://doi.org/10.1039/c0ee00652a

Xu H, Gao J, Jiang DL (2015) Stable, crystalline, porous, covalent organic frameworks as a platform for chiral organocatalysts. Nat Chem 7:905–912. https://doi.org/10.1038/nchem.2352

Xue C, Hans Q, Wang Y, Wu J, Wen TT, Wang RY, Hong JL, Zhou XM, Jiang HJ (2013) Amperometric detection of dopamine in human serum by electrochemical sensor based on gold nanoparticles doped molecularly imprinted polymers. Biosens Bioelectron 49:199–203. https://doi.org/10.1016/j.bios.2013.04.022

Yan H, Huang Y (2011) Polymer composites of carbon nitride and poly(3-hexylthiophene) to achieve enhanced hydrogen production from water under visible light. Chem Commun 47:4168–4170. https://doi.org/10.1039/C1CC10250H

Yanagida S, Kabumoto A, Mizumoto K, Pac C, Yoshino K (1985) Poly(p-phenylene)-catalysed photoreduction of water to hydrogen. J Chem Soc Chem Commun 0:474–475. https://doi.org/10.1039/C39850000474

Young KJ, Martini LA, Milot RL, SnoebergerIII RC, Batista VS, Schmuttenmaer CA, CrabtreeGary RH, Brudvig W (2012) Light-driven water oxidation for solar fuels. Coord Chem Rev 256(21–22):2503–2520. https://doi.org/10.1016/j.ccr.2012.03.031

Yuan XX, Ding XL, Wang CY, Ma ZF (2013) Use of polypyrrole in catalysts for low temperature fuel cells. Energy Environ Sci 6:1105–1124. https://doi.org/10.1039/c3ee23520c

Yun SN, Hagfeldt A, Ma TL (2014) Pt-free counter electrode for dye-sensitized solar cells with high efficiency. Adv Mater 26:6210–6237. https://doi.org/10.1002/adma.201402056

Yusoff N, Pikumar A, Ramaraj R, Lim HN, Huang NM (2015) Gold nanoparticle based optical, electrochemical sensing of dopamine. Microchim Acta 182:2091–2114. https://doi.org/10.1007/s00604-015-1609-2

Zhang Q, Uchaker E, Celaria SL, Cao G (2013) Nanomaterials for energy conversion, storage. Chem Soc Rev 42:3127–3171. https://doi.org/10.1039/C3CS00009E

Zhang S, Li J, Wang X, Huang Y, Zeng M, Xu J (2014) In situ ion exchange synthesis of strongly coupled Ag@AgCl/g-C3N4 porous nanosheets as lasmonic photocatalystfor highly efficient visible-light photocatalysis. ACS Appl Mater Interfaces 6(24):22116–22125. https://doi.org/10.1021/am505528c

Zhang P, Wang T, Gong J (2015) Mechanistic understanding of the plasmonic enhancement for solar water splitting. Adv Mater 27:5328–5342. https://doi.org/10.1002/adma.201500888

Zhang G, Lami V, Rominger F, Vaynzof Y, Mastalerz M (2016a) Rigid conjugated twisted Truxene dimers and trimers as Electron acceptors. Angew Chem Int Ed 55:3977–3981. https://doi.org/10.1002/anie.201511532

Zhang G, Lan ZA, Wang X (2016b) Conjugated polymers: catalysts for photocatalytic hydrogen evolution. Angew Chem Int Ed Engl 55(51):15712–15727. https://doi.org/10.1002/anie.201607375

Zhang JY, Xiao FX, Xiao GC, Liu B (2016c) Linker-assisted assembly of 1D TiO_2 nanobelts/3D CdS nanospheres hybrid heterostructure as efficient visible light photocatalyst. Appl Catal A-Gen 521:50–56. https://doi.org/10.1016/j.apcata.2015.10.046

Chapter 9
Heterogeneous Type-I and Type-II Catalysts for the Degradation of Pollutants

J. Nimita Jebaranjitham and Baskaran Ganesh Kumar

Contents

Abstract Water is a major source for living systems including terrestrial, aquatic, and aerial flora and fauna. But the growing industrialization releases a huge amount of toxic and obnoxious chemicals into the water stream. Thus, the inadequate access

J. N. Jebaranjitham
P.G. Department of Chemistry, Women's Christian College, Chennai, TN, India

B. G. Kumar (✉)
Department of Electrical and Electronics Engineering, Koc University, Istanbul, Turkey

Department of Chemistry, PSR Arts and Science College, Sivakasi, TN, India

© Springer Nature Switzerland AG 2020
M. Naushad et al. (eds.), *Green Photocatalysts*, Environmental Chemistry
for a Sustainable World 34, https://doi.org/10.1007/978-3-030-15608-4_9

to clean water is an extremely important problem throughout the world. Solving this water problem requires sincere research to identify robust new methods. The new method is expected to be a technology for all, that is, a technology for developed and developing countries and resource-limited settings. To meet the demand, the method should be effective, simple, cheap, and environmentally friendly, and one such method is the photocatalytic degradation of water. Photocatalytic degradation of water provides the unprecedented opportunities for the purification of water by providing clean technology. Considerable effort has been made to design, fabricate, and manipulate materials for photocatalytic degradation. The present work summarizes the established and recent progress in heterogeneous catalysis and its types. Based on the band alignment of semiconductors, the heterogeneous catalyst is divided into type-I and type-II. The type-I photocatalyst having a straddle band gap between two semiconductors, and the type-II photocatalyst having a staggered band gap between two semiconductors. We described the prominent and effective catalysts in each type and explained with advanced examples. The mechanism of degradation of organic impurities is also discussed. The factors influencing the photocatalytic performance are also summarized in detail. To guide the user, design considerations for the photocatalyst are also suggested. We suggested a range of future directions necessary to promote the photocatalytic degradation as an effective method among water purification technologies and to promote the breakthrough photocatalytic degradation phenomenon.

Keywords Photocatalysts · Heterogeneous photocatalysts · Water purification · Water degradation · Environment · Toxicity · Eco-friendliness · Organic impurities · Photocatalytic degradation · Type-I band alignment · Type-II band alignment

9.1 Introduction

9.1.1 Water Purification: A Universal Problem

Establishment of industries is an integral part of technology and has a profound impact on human society. Inherently, industries are generating large amounts of industrial waste during the manufacturing processes, and they contaminate the ecosystem, e.g., land and water (Kumar et al. 2018; Naushad et al. 2015). The impurities must be removed or destroyed before being discharged to the environment. Otherwise, the water contamination can harm human, animal, plants, and all other biological processes because of their inherent toxicity. The quality of water is associated with the development index of society, and much attention is given to technologies which can purify water. Hence, extensive investments made in water purification and stringent regulations are adopted. We need a practical and efficient solution for water purification (Awual et al. 2015; Pugazhendhi et al. 2017). The chemical impurities can be removed by physical, chemical, and biological processes. Among all, the chemical process of degradation has been proven as a highly cost-effective way of purifying water by degradation. One such chemical technology is heterogeneous photocatalytic degradation which is based on degrading organic impurities using light and catalyst. The use of photocatalysts for the degradation of water impurities is very promising and an intuitively efficient approach toward water purification. Herein, we described the available heterogeneous photocatalysis technologies for purification of water and highlighted the novel photocatalysts and robust new materials for water purification.

9.1.2   Impurities and Purification of Water

The chemical pollutants present in the industrial waste stream are inorganic micropollutants such as metals and metalloids and a wide variety of synthetic organic compounds (Gupta et al. 2017; Pandiyarajan et al. 2017; Saravanan et al. 2015). The most common organic pollutants of water are usually organic materials such as dyes (methylene blue, rhodamine B, methyl orange, fluorescein), polymers (PVC, polyethylene), haloalkanes (trichloroethylene and perchloroethylene), haloaromatics (p-dichlorobenzene), and many more (Mills et al. 1993). The common impurities of water and typical photocatalysts for purification are presented in Table 9.1. Drinking water standards only allow 1–50 ppb of toxic substances (Ollis 1985). An ideal water purification process should completely mineralize all toxic organic substances without releasing any toxic side-product residues. Hence, the removal of organic toxic substances and purification of the contaminants in surface water and ground water are intriguing problems to both industry and academia. In this regard, water quality control standards and regulation against toxic materials are established throughout the world including the United States and the European Union.

The purification process of water is the degradation of organic toxic substances into nontoxic or less toxic substances. The purification processes of industrial wastewater include noncatalytic, catalytic (Saravanan et al. 2013a, b, c), and incineration (thermal destruction). Among all, the catalytic process is a less energy-intensive and economically feasible method for water purification. The catalyst can be of a homogeneous (same phase with water) or heterogeneous phase (different phase than water). If the homogeneous catalyst is used for water purification, it should be separated after the water purification. Separating the catalyst of the same phase in industrial scale is scientifically tedious and economically unachievable. Whereas heterogeneous catalysts that are in a different phase (solid) with that of water can be separated by simple filtration. Thus, using heterogeneous photocatalysts for the water purification is energetically efficient. The catalyst enhances the degradation rate by providing the alternative route involving different transition states of lower energy. Hence, the heterogeneous catalyst provides clean technology for water purification and will have a great impact on human life.

Table 9.1 Types of organic impurities and popular degrading catalysts

Impurities family	Typical examples	Known catalysts for degradation
Dyes	Rhodamine B	CuS nanostructures, Srinivas et al. (2015)
	Congo red	TiO_2-nonwoven paper, Guillard et al. (2003)
	Methylene blue	TiO_2-activated carbon fiber, Fu et al. (2004)
	Methyl Orange	Degussa P-25-coated sand, Matthews (1991)
	Textile dye Reactive Black 5	Degussa P-25 TiO_2, Poulios and Tsachpinis (1999)
	Commercial azo dyes	Degussa P-25 TiO_2, Tanaka et al. (2000)
Aliphatic compounds	Perchloroethylene	Anatase TiO_2, Gupta and Tanaka (1995)
	Glycolic acid	TiO_2 Degussa, Mazzarino and Piccinini (1999)
	Dichloromethane	TiO_2 and titanium, Tanguay et al. (1989)
	Trichloroethylene	Anatase TiO_2, Ahmed and Ollis (1984)
	CCl_4	Degussa P-25 TiO_2, Choi and Hoffmann (1996)
	Chloroform	Degussa P-25 TiO_2, Choi and Hoffmann (1996),
	$CHBr_3$	Ag-TiO_2, Kondo and Jardim (1991)
Aromatic compounds	Aniline	TiO_2/Ni plate, Wenhua et al. (2000)
	4-Nitrophenol	TiO_2-Eurotitania, Andreozzi et al. (2000)
	Toluene	Anatase TiO_2, Butler and Davis (1993)
	Benzoic acid	TiO_2 Degussa P-25, Ajmera et al. (2002)
	2,4-Dichlorophenol	Anatase TiO_2, Ku and Hsieh (1992)
	Chlorobenzene	TiO_2, Pelizzetti et al. (1993)
	Nitrobenzene	TiO_2 Degussa P-25/Glass, Matthews (1987) or CdS, Davis and Huang (1990)
	Phenol	TiO_2 Degussa P-25/Glass, Herrmann et al. (1997),
	Malic acid	Ag-TiO_2, Herrmann et al. (1997)
Pesticides/ insecticides	Parathion	TiO_2-SiO_2, Shifu and Gengyu (2005)
	Dicarzol	TiO_2-nonwoven paper, Guillard et al. (2003)
	Diuron	TiO_2-β-SiC foam, Hajiesmaili et al. (2010)
	Alachlor	TiO_2/glass tube, Ryu et al. (2003)
	Methamodiphos	Re-TiO_2, Umar and Aziz (2013)
	Carbendazim	Umar and Aziz (2013)
	Trichlorfon	
	Turbophos	

9.1.3 Heterogeneous Catalysts and Types

Heterogeneous photocatalysis uses a continuously illuminated, photoexcitable solid catalyst to convert the reactants absorbed (organic impurities) on the photocatalyst surface. The photocatalysis uses a synergetic combination of irradiation and catalysis at ambient temperature and pressure (Hoffmann et al. 1995; Mills and Le Hunte 1997). Photocatalysis is more effective when compared to individual catalysis and

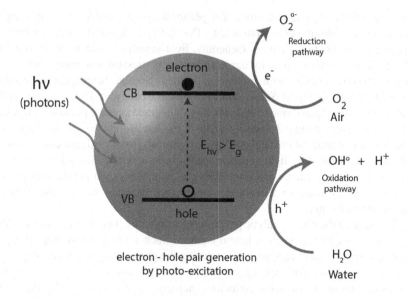

Fig. 9.1 Basic process of degradation mechanism. Initially, a hole-electron pair is generated by high-energy photons. Then, the hole and electron react with water and air to form hydroxyl and superoxide radicals, respectively. Then, the radicals attack the organic pollutants multiple times and degrade impurities to water, carbon dioxide, and more

photolysis. Simply, photocatalysis involves photoreaction which occurs at the surface of the photocatalyst. The photocatalyst absorbs the photon when the energy of the incident photon is equal to or higher than the band gap of the catalyst. As a result, the electrons from the valence band of the photocatalyst excite to the conduction band. The process generates electron-hole pairs that have high oxidation and reduction potential. Hence, the photoexcited electron-hole pairs can assist the degradation process through various photochemical reactions and destruct almost all organic pollutants (Fig. 9.1). The photogenerated holes react with water to generate hydroxyl radicals and consecutively mineralize the organic impurities. Similarly, the photogenerated electrons react with oxygen (air) and generate the superoxide radical and degrade the organic impurities. The photocatalyst is capable of accelerating the degradation without being consumed. The catalyst degrades wide varieties of organic pollutants by oxidation. Generally, light-sensitive d-block elements exhibit photocatalytic activity due to the coordinatively unsaturated nature which can promote redox reaction with organic impurities. Notable heterogeneous catalysts are TiO_2, ZnO_2, Fe_2O_3, CdS, and ZnS. Among these catalysts, TiO_2 is first described and widely investigated due to its capability to degrade organic pollutants, low cost, and robust photochemical stability (Fujishima and Honda 1972). In the present context, we define photocatalysis as a chemical (toxins) degradation reaction which is accelerated by light-sensitive catalytic compounds. Since photons are the most stable energy source and photocatalysis is nature's own purification process, future water purification technology completely depends on the photocatalytic purification of water.

To improve the photocatalytic performance, normally, two different semiconductors are coupled to generate a heterojunction. Such a combination improves both light absorption and charge separation during the photocatalysis. Hence, depending upon band positions, the heterogeneous photocatalyst is further classified based on the band alignment of semiconductors, namely, type-I and type-II. Figure 9.2 explains the types of heterogeneous catalyst with band diagram. Type-I photocatalysts possess a straddle band alignment between two semiconductor materials. In type-I heterojunction, the valence band of semiconductor 2 is lower than the valence band of semiconductor 1. Similarly, the conduction band of semiconductor 2 is higher than the conduction band of semiconductor 1. Hence, photoexcited electrons and holes are confined in semiconductor 1. Hence, due to the band alignment of both electrons and holes is confined, and efficient charge separation is not possible. Type-II photocatalysts possess staggered band alignment between two semiconductors. In type-II heterogeneous catalysts, the valence and conduction bands of semiconductor 2 are higher than that of semiconductor 1. The most common type-I and type-II catalysts are presented in Tables 9.2 and 9.3. Type-II band structure provides an optimum band alignment for charge separation and hence photocatalysis. In type-II alignment, photoexcited electrons can transfer from CB (2) to CB (1) due to favorable energetics; similarly, holes can transfer from VB (1) and VB (2). Hence, the holes and electrons are separated from each other and make the recombination process a less probable event (Marschall 2014). Overall, based on band alignment, heterogeneous catalysts are classified into two types, and type-II is ideal for the water degradation process.

9.2 Operational Parameters Affecting Photocatalytic Activity

To increase the effectiveness of the catalyst, studying the factors which influence the photocatalyst is necessary. Photocatalytic activity depends on several key factors such as concentration of catalysts, nature of impurities, nature of photons, surface

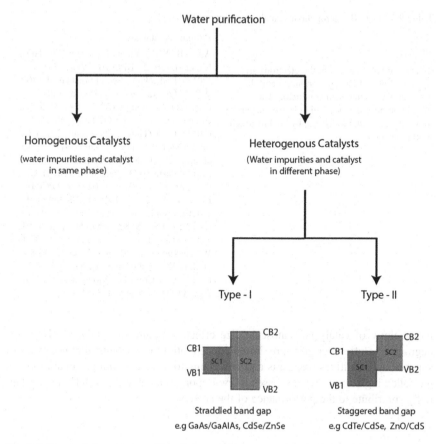

Fig. 9.2 Types of photocatalysts. Type-I photocatalysts possess a straddle band alignment between two semiconductors (*CB* conduction band, *VB* valence band, *SC* semiconductor). Type-II photocatalysts possess staggered band alignment between two semiconductors

Table 9.2 Type-I heterogeneous catalysts and typical examples

Types	Prominent catalysts
Type-I Straddle heterojunction: electrons and holes confined in semiconductor I	Bi_2S_3/CdS, Fang et al. (2011); Fe_2O_3/$SrTiO_3$, Schultz et al. (2012); CdSe/CdS, Thibert et al. (2011); $NaTaO_3$/$SrTiO_3$, Nazir and Schwingenschlögl (2011); ZnS/Fe_2S_2, Wen et al. (2012); I-BiOCl/I-BiOBr, Jia et al. (2017); V_2O_5/$BiVO_4$, Su et al. (2011); CdS/ZnS, Huang et al. (2013a); In_2O_3/In_2S_3, Yang et al. (2013); CdS/CdSe, Tongying et al. (2012); TiO_2/WO_3, Ostermann et al. (2009); TiO_2/MoS_2, King et al. (2013); TiO_2/Nb_2O_5, Furukawa et al. (2011); TiO_2/ZrO_2, Siedl et al. (2009); TiO_2/Fe_2O_3, Jeon et al. (2011); TiO_2/$BiVO_4$, Hu et al. (2011); TiO_2/$K_2Ti_4O_9$, Xiong and Zhao (2012); TiO_2/$ZnMn_2O_4$, Bessekhouad et al. (2005)

Table 9.3 Type-II heterogeneous catalysts and typical examples

Types	Prominent catalysts
Type-II Staggered heterojunction: the electrons are confined in the first semiconductor and the holes in the second semiconductor. The configuration provides best hole-electron separation efficiency due to the built-in field formed in the junction	Ag_2O/Bi_2WO_6, Yu et al. (2012); CdS/TiO_2, Liu et al. (2012); Bi_2O_3/Bi_2WO_6, Li et al. (2012); Bi_2S_3/TiO_2, Bessekhouad et al. (2004); WO_3/TiO_2, Ramos-Delgado et al. (2013); CdS/NiO, Khan et al. (2012); $CdS/CdSnO_3$, Chen et al. (2012); Fe_2O_3/TiO_2, Mu et al. (2011); In_2O_3/TiO_2[18], Chen et al. (2012); CdS/SnO_2, Nasr et al. (1997); $ZnFe_2O_4/TiO_2$, Li et al. (2011); CdS/ZnO, Barpuzary et al. (2011); Fe_2O_3/WO_3, Bi and Xu (2013); Cu_2O/CuO, Zhang and Wang (2012); $ZnO/CdTe$, Liu et al. (2013); SnO_2/V_2O_5, Shahid et al. (2010); ZnO/In_2S_3, Khanchandani et al. (2013); ZnO/CdS, Khanchandani et al. (2012); $SrTiO_3/TiO_2$, Cao et al. (2011); $BiVO_4/CeO_2$, Wetchakun et al. (2012); $\alpha\text{-}Fe_2O_3/CdS$, Shi et al. (2012); $CdS/ZnFe_2O_4$, Xiong et al. (2013); $CdS/CoFe_2O_4$, Xiong et al. (2013); Ag_3VO_4/TiO_2, Wang et al. (2012b)

morphology of catalysts, nanoscale size effect, structure of catalysts, pH of the degrading solution, temperature, initial concentration of reactants, doping, effect of support, and additives (oxidants and reductants for effective charge separation). In the following section, we discussed a few important parameters which can substantially contribute to the performance of the catalyst.

9.2.1 Concentration of Catalyst

The loading or concentration of the catalyst directly influences the overall performance of the catalyst. The relationship is linear and increases the overall performance of the degradation. It is understandable that more amount of catalyst provides more surface area for degradation. The more amount of catalyst also provides more absorption centers for photons. Typically, an optimized concentration of the catalyst is required for the efficient degradation process and effective use of a precious metal-based catalyst (Srinivas et al. 2015). At low concentration, the organic impurities adsorb few places at the surface, and a large active absorption center in the catalyst is unoccupied. Hence, the effective catalyst usage has not occurred. If a very high concentration of the catalyst was used, undesirable light scattering or photon-blocking (turbidity impedes the light path) occurs in UV wavelengths and decreases the degradation efficiency (Chun et al. 2000). Additionally, at higher concentration, agglomeration is very much probable, and the catalyst may be agglomerated and consecutively precipitated (Sohrabnezhad et al. 2009). In summary, the best performance of catalyst occurs during photocatalysis when all surfaces are exposed to adsorbate molecules and the total surface area is illuminated.

9.2.2   Nature and Concentration of the Organic Impurities

Organic impurities (substrates) degrade on the surface of the catalyst. Hence, the molecules attached directly to the surface will undergo faster reaction. Hence, the molecules which have the functional groups can anchor on the surface and promote complete degradation. Pangarkar et al. (Bhatkhande et al. 2004) reported that nitrophenol undergoes faster degradation than phenol due to better anchoring. Since the surface of the catalyst is electron-rich in character, the functional groups which have electron-withdrawing nature (e.g., $-NO_2$) can anchor effectively on the photocatalyst surface and promote the catalysis. The initial concentration of the substrate is also an important parameter in the photocatalytic process. If more amounts of organic impurities are present, they will occupy the active centers of the photocatalyst and act as barriers between the incident photons and the catalytic centers (Shaban et al. 2018). Hence, the nature and concentration of organic impurities influence the catalytic performance.

9.2.3   Nature and Intensity of Photons

The intensity of photons used for the photocatalysis will have a considerable impact on the degradation properties. The intensity of photons will have a direct relationship with the rate of the reaction ($\sqrt{I}$) (Srinivas et al. 2015). Few photons are sufficient to initiate the photodegradation of impurities in water. But to realize the full potential of the catalyst, more photons should be used. The reaction rate increases along with the intensity of the photons absorbed. This is due to the fact that light intensity determines the photogenerated electron-hole pair concentration which leads to more radical formation and more photodegradation successively. It should be noted that the reaction rate increases without a change in the degradation mechanism or pathway. Sometimes, a very high intensity of light can promote the recombination process of electron and hole and hence decrease the reaction rate (Malato et al. 2009). Hence, the intensity must be optimized against the degradation rate.

Generally, the wavelength between 380–750 nm is desirable for photocatalytic applications for the real world (effective sunlight region). The wavelength should be within the threshold of band gap energy of the catalyst. However, the absorption of the photocatalyst can be tuned by mere doping, and hence the importance of wavelength over reaction rate is normally ignored. Normally, low-energy light is most desirable for photocatalysis because it reduces the photolysis side products (Füldner et al. 2010). Hence, the energy of radiation can be used for product selectivity rather than the efficiency of the catalyst. Another important optical factor to consider is quantum yield, which is the ratio of the reaction rate in molecules per second to the photonic flux in photons per second. The performance of the catalyst is directly proportional to the quantum yield. In summary, when compared to the wavelength of photocatalytic absorption, the quantum yield has greater importance for photodegradation.

9.2.4 pH of the Reaction Solution

Photocatalysis is a surface phenomenon and happens only on the surface which has different charge densities due to the surface dangling bonds. Hence, the pH of the solution will have a greater impact on the catalytic performance. If the solution is acidic (pH < 6.9), the surface of the catalyst will be positively charged. If the solution is basic (pH > 6.9), the surface of the catalyst will be negatively charged. Depending on the electrostatic attraction or repulsion between the photocatalytic surface and organic impurities, the reaction rate is enhanced or inhibited by the pH of the solution. Sun et al. (2006) reported that at acidic pH, surfaces of titanium are positively charged, and the oxidation performance of TiO_2 is considerably increased when compared to basic pH values. pH should be optimized for every degradation reaction. In general, through the electrostatic interaction, pH can control the adsorption and dissociation of organic molecules.

The solution pH also affects the aggregate formation of photocatalysts. The catalyst has neutral or zero surface charge; hence, the catalyst with a nanoparticle size aggregates to a big size to avoid more surface area. But if the surface has the positive or negative charge by pH, it provides an electrostatic repulsion between adjacent particles and prevents the aggregation. The charge created by pH also avoids precipitation due to aggregation because larger particles sediment quickly from the suspension. A similar behavior was observed in the TiO_2 photocatalytic system where the particles of size 300 nm agglomerate to 2–4 μm at the zero-point charge. The better photocatalytic performance was observed at acidic or basic pH values (Malato et al. 2009). Adjustment of pH is not practical in real-world applications due to a large amount of industrial waste. Hence, the optimization of pH values should be within the range of industrial waste stream pH (4–10). Overall, the pH of the solution can be maintained between acidic and basic for better catalytic performance, and zero-point charge must be avoided to prevent the absorption.

9.2.5 Temperature

Temperature is an old and a known parameter for increasing the rate of chemical degradation. The band gap of semiconductors used in photocatalysis is very high when compared to the supplied thermal energy. But the temperature can increase the degradation rate due to change in the adsorption of organic molecules/substrates, because an increase in temperature will increase the collision frequency between the organic impurities on the surface of the photocatalyst. At the same time, a very high temperature has a negative effect because it promotes the recombination of charge carriers and decreases the catalytic performance. Additionally, adsorption is an exothermic process, and an increase in the surface temperature promotes the desorption of the substrates from the catalytic centers and decreases the catalytic performance (Fu et al. 1996). In the organic mechanistic point of view, the high temperature is not desired due to the poor selectivity of the reactants/catalysts at high temperatures. Hence, the formation of many side products is possible at high temperatures. At low temperatures, there was an increase in the activation energy of the adsorbed impurities, and the reaction rate decreased, and the desorption of the final products is also not possible (Malato et al. 2009). Overall, the degradation rate is dependent on temperature, but it should be optimized between 20 and 80 °C.

9.2.6 Surface Properties (Size, Shape, Structure, and Morphology)

Catalysis is a surface phenomenon and can be influenced by the particle size of the catalyst. The size of the nanoparticles can influence the surface area, surface imperfection, and considerably change in the electronic state of the semiconductor. Normally, the smaller the size of the photocatalyst, the better the catalytic performance due to the large surface area. If the surface area is larger, more catalytic centers are available for degradation. Normally, the size effects will be observed with the photocatalysts with the particle size well below 10 nm. The properties of nanoparticles are not only size dependent but are also linked to the actual shape. Advances in the synthesis of nanoparticles help to prepare nanoparticles with different asymmetric shapes such as tube, prism, hexagon, octahedron, disk, wire, rod, tube, etc. Kundu et al. (2017) reported that CdS nanoneedles have higher photocatalytic activity than nanowires due to the reduced charge recombination. Similarly, Qingshan et al. (Ren et al. 2014) reported that rose-shaped Cu_2ZnSnS_4 nanoparticles had higher photocatalytic activity than sphere-shaped Cu_2ZnSnS_4 nanoparticles because of efficient charge separation.

It is well known that the structure of the catalyst plays an important role in photocatalytic performance. Typically, TiO_2 has three phase structures, namely, anatase, rutile, and brookite. Among the three, the anatase phase has high catalytic activity due to its structural stability. Additionally, the photoexcited electron can be trapped in the available vacancies of TiO_2, and hence the hole and electron are separated (Chen et al. 2010). As a result, the possibility of recombination decreased, and the overall efficiency will increase. Within the structures, energetically stable planes show better photocatalytic activity than the weak unstable planes. The (1 0 1) planes of TiO_2 are energetically high and show better photocatalytic performance of reduction of O_2 to $O_2{}^{\circ-}$ than the (0 0 1) planes (Wu et al. 2010). In the same way, Yurong et al. (Su et al. 2014) reported that Bi_5O_7I rods have a different photocatalytic activity than Bi_5O_7I sheets. They found that while exposing the (0 0 1) and (1 0 0) facets of rods to light, synergetic and internal electric field effects occurred. The effects promote a better photocatalytic activity of rods than the sheets. In the crystal structure, different facets have different kinetics of the degradation process. Gang et al. (Wang et al. 2012a) synthesized Ag/Ag_2O nanostructures and analyzed the reactivity of different facets. They found that the rate of degradation is of the order of (1 0 0) > (1 1 0) > (1 1 1). They found that the larger value of the weighted average of the effective mass of holes and electrons in the (1 0 0) facets of cubic nanoparticles promoted better catalytic activity than the other facets.

The surface morphology of catalytic particles can also influence the rate of the reaction. Saravanan et al. reported that ZnO nanoparticles with a spherical shape show superior catalytic performance when compared to the spindle- and rod-shaped ZnO (Saravanan et al. 2013a). At the surface, the sudden disruption of lattice generates the dangling bonds. Hence, the properties of bulk (e.g., photon absorption, electronic states) are different from that of the surface atoms. Hence, the semi-conductors having completely passivated surfaces could significantly promote the unified properties for both bulk and surface. Moreover, photocatalysis occurs at the solid/electrolyte interface which is at the surface of the catalyst. Charge transfer and recombination occur at the surface of the catalyst. Hence, the catalyst surface should be sufficiently passivated for better catalytic performance. Recently, hierarchical and porous nanostructures gained much attention due to the largely accessible surface morphology (Xia et al. 2016). The unique surface morphology of hierarchical nanostructures provides better permeability for the reactant molecule from the surface to the core. The hierarchical nanostructures not only have more active adsorption sites and photocatalytic reaction sites but also have active sites not only on the surface but everywhere in the nanostructure. Surface defects positively influence the photocatalytic performance due to high surface energy. Xinyu et al. (Zhang et al. 2014) tuned the surface defects of ZnO nanorods by modifying the aspect ratio. They found high levels of the surface ratio having high level of surface defects which promote high photocatalytic performance.

9.3 Photocatalysts

For water purification (degradation), type-II band structure is ideal for its better optical and electronic properties than type-I. The reason is that in type-II photocatalyst, the chemical potential difference between the two semiconductors provides a built-in field and separates the photogenerated hole-electron pair. There are numerous type-I and type-II photocatalysts to be considered. Herein, we describe the typical and prominent examples of photocatalysts based on the composition of the catalyst.

9.3.1 Common Metal Oxide Photocatalysts

Various metal oxides are very common photocatalysts long been used for photocatalytic degradation of organic impurities in water (TiO_2, ZnO, Fe_2O_3, SnO_2, ZrO_2, CeO_2, and many more). Initially, much attention has been given to TiO_2-based photocatalysts due to their photostability, absorption, and low cost. But

Table 9.4 Summary of parameters influencing the photocatalytic properties

S. No	Key parameter	Description
1.	The concentration of photocatalyst	Degradation rate $\propto$ concentration
		Optimization is required because:
		Low concentration – not all the absorption centers are occupied
		High concentration – scattering dominates
2.	Nature and concentration of the organic impurities	Impurities have functional groups which can anchor on the surface of the photocatalyst and promote complete degradation
3.	Nature and intensity of photons	Degradation rate $\propto$ I (at low intensity)
		Degradation rate $\propto \sqrt{I}$ (at high intensity)
		Degradation rate $=$ constant (at a very high intensity)
		Degradation rate $\propto$ quantum yield of the photocatalyst
4.	pH of the reaction solution	pH can influence the surface of the catalyst and influence absorption of organic impurities by its charge; pH also improves the stability of the photocatalyst
5.	Temperature	Degradation rate $\propto$ T (at high T, due to increased collision)
		Degradation rate $= 1/T$ (very high T enhances recombination of photogenerated electron-hole and poor selectivity of the reactants)
6.	Surface properties (size, shape, structure, and morphology)	1. Degradation rate $\propto$ 1/photocatalyst particle size
		2. Shape of the nanocatalyst can control the charge recombination
		3. Catalyst crystal structure can show performance due to the difference in the reactivity of the surface facets
		4. Surface morphology can influence the performance of the catalyst due to reactant reactivity

the charge-separation kinetics of TiO_2 are relatively weak and usually coupled with narrow band gap semiconductors to effectively overcome this issue which resulted in type-II heterostructures such as CdS/TiO_2 (Liu et al. 2012), Bi_2S_3/TiO_2 (Bessekhouad et al. 2004), and many more. A typical example for the TiO_2-based type-II heterostructure is CdS/TiO_2 (Liu et al. 2012). The association of CdS with TiO_2 not only improves the charge-separation kinetics but also extends the spectral response of the photocatalyst. Liu et al. (Liu et al. 2012) reported CdS/TiO_2 photocatalyst based on the solvothermal method and analyzed the selective oxidation of various functional group-containing organic impurities such as alcohols to aldehydes. When comparing to bare TiO_2, the band alignment of TiO_2 and CdS provided effective charge separation by the prolonged lifetime of photogenerated charge carriers. Recently, MoS_2/TiO_2 and WS_2/TiO_2 heterostructures (Wang et al. 2013) exhibited the high photocatalytic activity of photodegradation of 4-chlorophenol than bare TiO_2. When compared to bare TiO_2 or MoS_2, WS_2

provided a suitable band alignment of charge separation due to the quantum confinement and extended absorption spectrum. There are numerous metal oxide catalysts being developed with the same idea, such as Ag_2O and V_2O_5, and exhibited excellent photocatalytic performance (Wang et al. 2013). Additionally, metal ions in catalysts are doped with transition metals such as Fe^{3+}, Mo^{5+}, Re^{5+}, V^{4+}, Al^{3+}, Ru^{3+}, and OS^{3+}. The doped photocatalyst enhances the performance of catalysts by influencing the charge carrier recombination dynamics and interfacial electron-transfer rates.

Other than TiO_2, the widely investigated catalyst is ZnO-based photocatalysts because of their similar degradation mechanism and charge kinetics. In a few cases, higher photocatalytic performance than TiO_2 is reported (Wang et al. 2013). Few notable combinations are ZnO/C_3N_4, ZnO/SnO_2, ZnO/ZnSe, ZnO/In_2O_3, etc. Xing et al. (Huang et al. 2013b) reported the preparation of ZnO/SnO_2 photocatalysts by thermal evaporation approach. The combination of ZnO and SnO_2 offered remarkable photocatalytic degradation of rhodamine B dye than pure ZnO. The hetero-interface of ZnO and SnO_2 provides efficient charge separation and hence effectively reduces the recombination. Similarly, Wu et al. reported ZnO/ZnSe photocatalyst which showed excellent performance against photodegradation of rhodamine B. The coupling of ZnO with ZnSe provides extended absorption of the catalyst and effective separation of the photoinduced electron-hole pairs. In summary, metal oxide-based photocatalysts are well-studied photocatalytic materials and used as models for all other theoretical/empirical understanding.

9.3.2   Multicomponent Oxides

Recently other than titanium, metal oxides were developed with the multicomponent materials such as Bi_2O_3/Bi_2WO_6 (Li et al. 2012). Bismuth-based multicomponent oxides were widely investigated because of their responsivity toward visible light and have a similar performance that of titanium dioxide photocatalysts. The narrow band gap in the visible-light region of bismuth is due to its valence band consisting of Bi-6 s orbitals and O-2p orbitals. The configuration is much lower in the band gap diagram than the O-2p orbital of TiO_2, and hence narrow band gap is possible. Similarly, the following are multicomponent type-II catalysts widely investigated for the photodegradation of organic impurities, namely, Ag_2O/Bi_2WO_6 (Yu et al. 2012), $CdS/CdSnO_3$ (Chen et al. 2012), $ZnFe_2O_4/TiO_2$ (Li et al. 2011), $SrTiO_3/TiO_2$ (Cao et al. 2011), $BiVO_4/CeO_2$ (Wetchakun et al. 2012), $CdS/ZnFe_2O_4$ (Xiong et al. 2013), $CdS/CoFe_2O_4$ (Xiong et al. 2013), Ag_3VO_4/TiO_2 (Wang et al. 2012b). Yu et al. (2012) reported that the Ag_2O/Bi_2WO_6 is the photocatalyst for the photodegradation of methyl orange. Ag_2O/Bi_2WO_6 showed excellent photocatalytic performance than individual Ag_2O and Bi_2WO_6. The enhanced photocatalytic activity is due to the improved absorption of the catalyst by the plasmonic absorption centers. At the same time, Xiong et al. (Xiong et al. 2013) claimed that the $CdS/ZnFe_2O_4$ photocatalyst showed a high photocatalytic degradation ability of 4-chlorophenol and also high photostability than the individual components. The high photodegradation ability of catalyst is attributed to intense absorption of light and efficient charge separation. In summary, the multicomponent catalyst provides the effective photodegradation ability due to multifunctionality like magnetic separation and band gap tuning.

9.3.3 Metals

Metal-based photocatalysts are evolving because of the new phenomenon called the plasmonic effect of metals. The plasmonic effect is generated when the frequency of incident photons matches the natural frequency of surface electrons of metals such as Au, Ag, and Cu. The plasmonic materials cannot involve directly in photocatalytic degradation, but with the supportive materials such as metal oxides like CeO_2. Hence, the photocatalytic activity is collectively decided by both plasmonic metal and supportive materials. The materials like ZrO_2, Al_2O_3, or SiO_2 are normally used along with plasmonic Au, Ag, and Cu to enable the effective photocatalytic activity. The mechanism of photocatalytic degradation is slightly different from that of conventional photocatalysis. Initially, incident light excites the electrons in the plasmonic states. Then the electrons are transferred to the supported metal oxide conduction band and finally to the electron acceptors such as O_2. The hole would be quenched during the entire photocatalytic cycle. Yogi et al. reported the Au/TiO_2-based photocatalytic effective degradation of methylene blue. They attributed the higher photocatalytic activity to the excellent band alignment with Au and TiO_2 and effective charge separation (Yogi et al. 2008). Similarly, Quanjun et al. (Xiang et al. 2010) reported the Ag/TiO_2-based catalyst for the photocatalytic degradation of rhodamine B. They reported that the combination of Ag and TiO_2 has a higher photocatalytic performance than bare TiO_2. The high photocatalytic activity is due to the surface plasmon absorption of Ag particles under visible light and effective charge separation of photogenerated charge carrier. In summary, plasmonic photocatalysts offer effective charge separation by plasmon generation and charge injection for enhanced photocatalytic activity.

9.3.4 Nanocomposites

Recently nanocomposites gained much attention due to tunable gaps. Normally nanocomposites are multicomponent oxides intercalated with C_{60}, polyaniline, g-C_3N_4, and various types of carbon nanotubes (Dong et al. 2015). The tunable band gap provides an opportunity for extending the absorption spectrum to the entire visible region and more. Since composite materials contain two individual components and the band alignment, complete charge separation and transport are possible. Recently Zhao et al. reported that the C_{60} composite along with Bi_2MoO_6 exhibited a dramatic photocatalytic performance for the degradation of bromate ion under visible-light irradiation (Zhao et al. 2011). The good performance is due to the interaction between C_{60} and Bi_2MoO_6 which increased the mobility of the photogenerated charge carrier and transferred the valence electrons from the valence band to the conduction band. Additionally, the organic materials like carbon nanotubes have excellent adsorption toward bulky organic molecules due to the pores and very high adsorption sites. Hence, organic impurities with anchoring functional groups such as –COOH, -OH, and -NH_2 can adsorb on the surface of the carbon nanotubes or C_{60}. Currently, many nanocomposite photocatalysts are being developed. Few notable catalysts are $C_3N_4/SmVO_4$, $Fe_{0.01}Ni_{0.01}Zn_{0.98}O$/ PANI, and $ZnFe_2O_4$/carbon nanotubes (Dong et al. 2015). Therefore, diverse nanocomposites can be easily prepared by the different combinations of multicomponent oxides and organic functional materials, having better photocatalytic activity.

9.3.5   Core-Shell Nanoparticles

As the name implies, the nanoparticles contain two semiconductors, namely, the core (inner material) and shell (out layer material). Both the semiconductor at the core and the semiconductor at the shell collectively decide the photocatalytic properties. The core-shell nanoparticles have the property of nanoparticles as well as the property of the core and shell materials. Few notable core-shell nanoparticle photocatalysts are ZnO/SiO_2, CuS/ZnS, ZnO/ZnS, Cu_2O/Si nanowires, Fe_2O_3/TiO_2, Fe_2O_3/ZnO, and many more (Thuy et al. 2014; Zhai et al. 2010). Generally, magnetic core-shell particles are preferred due to the ease of magnetic separation of the catalyst after degradation. Hence, materials having excellent physical and chemical properties for photodegradation are combined with magnetic nanoparticles. Zing et al. used ZnO/SiO_2 core-shell particles for the photocatalytic degradation of rhodamine B. The core-shell nanoparticles showed better catalytic performance than ZnO bare nanoparticles due to the hydrophobic nature of SiO_2 and complete surface passivation (Zhai et al. 2010). Similarly, Ung et al. (Thuy et al. 2014) demonstrated the CuS/ZnS nanocrystal-based photocatalytic degradation of the rhodamine B dye solution. They reported that ZnS shell increased the photocatalytic performance by suppressing the dangling bonds and increased CuS core stability. The main issue with the core-shell nanoparticle is that the properties (surface, crystal quality) of core-shell particles entirely depend on the synthetic process, and the synthesis has a complex process to scale up.

9.3.6   Photocatalyst with Supports

During the photocatalysis, the catalytic molecules tend to aggregate and precipitate out. Hence, supports are used as anchors for catalytic molecules to prevent the aggregation. The most common supports are activated carbon, polymers, glass, metals, and silica materials. The photocatalyst support is expected to have the following characteristics: (i) irreversible adhesion between the catalyst and support; (ii) retention of the catalytic properties after the adhesion; (iii) sufficient adhesion centers for the adsorption of organic impurities; and (iii) preservation of the surface area of the photocatalyst.

Anderson et al. reported that silica-supported TiO_2/Al_2O_3 photocatalysts showed better catalytic performance of degradation of Rhodamine-6G than the unsupported TiO_2/Al_2O_3 photocatalysts alone. The silica support not only prevents the agglomeration but also provides stability to the photocatalyst. Similarly, Zhou et al. (2016) reported the N-TiO_2/AC (nitrogen-doped TiO_2 and active carbon-supported) photocatalyst for the degradation of methyl orange solution. They have claimed that the activated support converted the catalyst into a very robust one having excellent recyclability. The main disadvantage is the preparation of the support photocatalyst because of the physical methods and the catalyst uniformity of the surface of the support. Hence, the deposition widely influences the quality and performance of the photocatalyst.

9.4 Design of Active Photocatalysts

Heterogeneous catalysts gained the broad attention from materials science, photophysics, photochemistry, electrochemistry, and many more. To prepare an effective photocatalyst, general guidelines are required. Hence, many factors should be considered at various stages of photocatalytic technology (Fig. 9.3). We have classified guidelines based on before using, during, and after using the catalyst. Apart from high activity, high selectivity, long lifetime, and durability, the following criteria should be considered while designing photocatalysts.

First, the very important thing to consider is the photon-harvesting centers and catalytic sites. Currently, sunlight is an effective and free photon source for photocatalysis. Hence, while designing harvesting centers, it should be tuned to such a way that it should absorb light in the entire sunlight spectrum. Sunlight consists of 5% UV, 43% visible, and 52% IR light. Another very important factor that should be considered for photocatalytic design is the electronic structure of the catalyst which entirely decides the optical and band structure of catalysts. To initiate the degradation of organic impurities through oxidation and reduction reaction, the catalyst requires sufficient redox potential for the active sites at the surface. Hence, the surface potential can be tuned by introducing electron-withdrawing or electron-donating peripheral ligands. In general, the degradation processes are a series of continuous reactions. Thus, the energy levels of the catalytic active sites and the redox potential of the adsorbed molecules should be matched to initiate the degradation process. The energy levels of the catalytically active sites and the redox potentials of the adsorbed reactant molecules should be matched to initiate a reduction or oxidation catalytic reaction. Additionally, the catalyst surface should provide a wide scope for a large number of organic pollutants. In recent times,

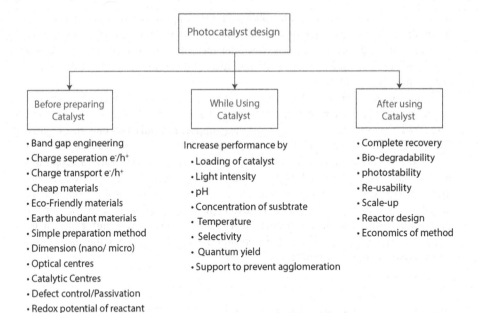

Fig. 9.3 Design consideration of a photocatalyst

crystallinity is also considered as the key feature for photocatalysis. Perfect crystalline materials mean fewer crystal defects which reduce the recombination of electron-hole after photoexcitation. Last but not least, the photocatalyst should operate efficiently in ambient conditions or mild conditions.

While using a photocatalyst, improvising the performance should be considered as the design parameter. Photocatalyst has many handles for the enhancement of photocatalytic activity such as loading of the catalyst, light intensity, pH of the solution, the concentration of substrates, temperature, selectivity, quantum yield, and support to prevent agglomeration. All these parameters and their influences are summarized in Table 9.3. While using a catalyst, all the efforts should capture the opportunities to enhance the performance of the photocatalyst.

The main advantage of a heterogeneous catalyst is recovery; hence, proper recovery mechanism should be identified after water degradation. Usually, it is achieved by the association of magnetic materials along with the photocatalyst. The recovered catalyst should have the capability of immediate reuse after recovery without any activation. To achieve proper industrial-level identification scale-up procedures, a complete automated reactor should be designed. Another important parameter is biodegradability and the environmental impact of the catalyst. After the preparation, photocatalyst reactivity, the health hazard, and fire hazard should be assessed based on the Hazardous Materials Identification System (HMIS). Very importantly, the economics of preparation has been a long ignored factor. The catalyst price should be evaluated based on the precursor materials, the cost of preparation, co-catalysts, and donors/acceptors. Even though the area of photocatalytic design is considerably developed, effective photocatalyst design provides a unique opportunity for the advances of new and exciting photocatalyst technologies.

9.5   Conclusions

9.5.1   Summary

Heterogeneous catalysts are fascinating; each year, a wide variety of heterogeneous photocatalysts are being developed, and numerous works increasingly appear in publications. Through the present work, we acknowledged conventional catalysts and suggested the novel materials and briefly reported the representative heterogeneous catalysts. We attempted to depict the current important areas of type-I and type-II heterogeneous photocatalysts for water degradation and identified the recent efforts. The heterogeneous photocatalyst-based water purification is the proven technology for the degradation of toxic organic substances in water. Hence, the conventional photocatalyst-based research attained a certain level of maturity and thus will find increasing applications in industry. At the same time, progress is anticipated in the areas of organometallic catalysts, nanocatalysts, biofriendly catalysts, crystalline catalysts, and highly efficient catalysts. The water purification problem is universal, and heterogeneous catalysts may outperform all other water purification methods through their simplicity and effectiveness.

Despite many advances made in photocatalytic water degradation, preparation of photocatalysts has so far been a trial-and-error method based on the accumulated chemical knowledge, experience, and common sense. Until now, there are no specific guidelines for the selection of heterogeneous catalysts for water degradation. But for the next generation of photocatalysts, we need a rational and systematic approach. Hence, we suggested guidelines for designing catalysts for water degradation (Fig. 9.3). The directive guideline was created based on every part of the catalytic process such as before preparation, while using, and after using. We also suggested the operational parameters such as handles to effectively control the photocatalytic degradation such as pH and the concentration of the catalyst/organic impurities. The guideline helps the user/researcher to select the required catalyst or narrow down the available material choices for the degradation of water.

9.5.2   Future Direction: Where to Next?

A large number of photocatalysts have been studied for the purification of water at ambient temperature and pressure. The future direction of photocatalyst is not only to focus on new materials but also to address the remaining key challenges. The following suggestions deserve even more research interest and will most likely become a very important area of research: (1) Besides precious metals, the catalyst made out of earth-abundant materials should meet the industrial-scale demand and global need. (2) The catalyst should be made up of less toxic materials. The catalyst itself should not generate any type of toxicity, and hence it should be made out of environment-friendly materials such as zinc and copper. Additionally, the catalyst made out of eco-friendly materials should reduce the disposal risk. (3) The catalyst life should be increased by discovering a new leaching and regeneration process. It is also possible by designing catalysts with chemical stability and with the resistance to poisoning. (4) Photocatalysts should be designed for resource-limited settings (Sia et al. 2004). The water purification problem is universal, and the photocatalyst-based technology should reach poor people in poor countries. Hence, it should be cheap, simple to prepare, and safe to handle. (5) Tailoring a catalyst for every organic toxic material is practically not possible. Hence, multifunctional catalysts with degradation capability of various organic impurities must be developed to achieve the broad generality of the catalyst. Since organocatalysts are being developed recently, multifunctionality is possible and hence the degradation. (6) There should be efficiency enhancements from the understanding of photocatalytic pathways and catalytic interfaces. It is possible by mechanistic understanding at the depth of molecular and atomic levels with the support of computational and spectroscopic analysis. (7) There should be more efficient photocatalysts based on metamaterials (nanosized, porous, and hollow) and new structures such as perovskites (Royer et al. 2014) and metal-organic frameworks (MOFs) (Lee et al. 2009). (8) The photocatalyst's spectral sensitivity should not be limited to a few nanometers of the UV, visible, or NIR region. Hence, the spectral sensitivity of the photocatalyst must be extended to cover the entire UV-visible-NIR region. (9) Surface science analysis must be developed to identify the surface phenomena such as association/dissociation/activation of molecules on the surface and to identify surface-active sites. Such analysis must lead to an empirical approach to surface engineering and molecular design. (10) There should be improvising of the kinetics of photocatalytic degradation. Nowadays, identifying the degradation pathways is easier due to modern instrumentation such as ultrahigh-resolution NMR spectroscopy. Hence, intermediates and transition states can be easily identified, and the reaction rate could be enhanced. (11) Developing large-scale high-throughput synthesis and its automated reactor would be a meaningful step for industrialization of photocatalysts. (12) Last but not least, the standard economic analysis should be established. Many projects are devoted to technical analysis and technical comparison, but there is no detailed study on comparing photocatalysis with the existing water purification technologies. Hopefully, new developments and achievements in these areas would substantially contribute to water degradation. In summary, heterogeneous photocatalyst technologies at the pre-industrial level and the above challenges limit the widespread use of heterogeneous photocatalysts, and advances in this direction propel the research on water purification.

Acknowledgments B.G.K wishes to dedicate the present work to his PhD advisor Prof. K. Muralidharan, School of Chemistry, University of Hyderabad, India. We thank the editor Dr. Saravanan Rajendran for valuable suggestions, feedbacks, and discussions.

References

Ahmed S, Ollis DF (1984) Solar photoassisted catalytic decomposition of the chlorinated hydrocarbons trichloroethylene and trichloromethane. Sol Energy 32:597–601. https://doi.org/10.1016/0038-092X(84)90135-X

Ajmera AA, Sawant SB, Pangarkar VG, Beenackers AA (2002) Solar-assisted photocatalytic degradation of Benzoic acid using Titanium dioxide as a photocatalyst. Chem Eng Technol 25:173–180. https://doi.org/10.1002/1521-4125(200202)25:2<173::AID-CEAT173>3.0.CO;2-C

Andreozzi R, Caprio V, Insola A, Longo G, Tufano V (2000) Photocatalytic oxidation of 4-nitrophenol in aqueous TiO2 slurries: an experimental validation of literature kinetic models. J Chem Technol Biotechnol 75:131–136. https://doi.org/10.1002/(SICI)1097-4660(200002)75:2<131::AID-JCTB191>3.0.CO;2-F

Awual MR, Hasan MM, Shahat A et al (2015) Investigation of ligand immobilized nano-composite adsorbent for efficient cerium(III) detection and recovery. Chem Eng J 265:210–218. https://doi.org/10.1016/j.cej.2014.12.052

Barpuzary D, Khan Z, Vinothkumar N, De M, Qureshi M (2011) Hierarchically grown urchinlike CdS@ ZnO and CdS@ Al2O3 heteroarrays for efficient visible-light-driven photocatalytic hydrogen generation. J Phys Chem C 116:150–156. https://doi.org/10.1021/jp207452c

Bessekhouad Y, Robert D, Weber J (2004) Bi2S3/TiO2 and CdS/TiO2 heterojunctions as an available configuration for photocatalytic degradation of organic pollutant. J Photochem Photobiol A Chem 163:569–580. https://doi.org/10.1016/j.jphotochem.2004.02.006

Bessekhouad Y, Robert D, Weber J-V (2005) Photocatalytic activity of Cu2O/TiO2, Bi2O3/TiO2 and ZnMn2O4/TiO2 heterojunctions. Catal Today 101:315–321. https://doi.org/10.1016/j.cattod.2005.03.038

Bhatkhande DS, Kamble SP, Sawant SB, Pangarkar VG (2004) Photocatalytic and photochemical degradation of nitrobenzene using artificial ultraviolet light. Chem Eng J 102:283–290. https://doi.org/10.1016/j.cej.2004.05.009

Bi D, Xu Y (2013) Synergism between Fe2O3 and WO3 particles: photocatalytic activity enhancement and reaction mechanism. J Mol Catal A Chem 367:103–107. https://doi.org/10.1016/j.molcata.2012.09.031

Butler EC, Davis AP (1993) Photocatalytic oxidation in aqueous Titanium dioxide suspensions: the influence of dissolved transition metals. J Photochem Photobiol A Chem 70:273–283. https://doi.org/10.1016/1010-6030(93)85053-B

Cao T, Li Y, Wang C, Shao C, Liu Y (2011) A facile in situ hydrothermal method to SrTiO3/TiO2 nanofiber heterostructures with high photocatalytic activity. Langmuir 27:2946–2952. https://doi.org/10.1021/la104195v

Chen X, Shen S, Guo L, Mao SS (2010) Semiconductor-based photocatalytic hydrogen generation. Chem Rev 110:6503–6570. https://doi.org/10.1021/cr1001645

Chen Y-C, Pu Y-C, Hsu Y-J (2012) Interfacial charge carrier dynamics of the three-component In2O3–TiO2–Pt heterojunction system. J Phys Chem C 116:2967–2975. https://doi.org/10.1021/jp210033y

Choi W, Hoffmann MR (1996) Novel photocatalytic mechanisms for CHCl3, CHBr3, and CCl3CO2-degradation and the fate of photogenerated trihalomethyl radicals on TiO2. Environ Sci Technol 31:89–95. https://doi.org/10.1021/es960157k

Chun H, Yizhong W, Hongxiao T (2000) Destruction of phenol aqueous solution by photocatalysis or direct photolysis. Chemosphere 41:1205–1209. https://doi.org/10.1016/S0045-6535(99)00539-1

Davis AP, Huang C (1990) The removal of substituted phenols by a photocatalytic oxidation process with cadmium sulfide. Water Res 24:543–550. https://doi.org/10.1016/0043-1354(90)90185-9

Dong S et al (2015) Recent developments in heterogeneous photocatalytic water treatment using visible light-responsive photocatalysts: a review. RSC Adv 5:14610–14630. https://doi.org/10.1039/C4RA13734E

Fang Z et al (2011) Epitaxial growth of CdS nanoparticle on Bi2S3 nanowire and photocatalytic application of the heterostructure. J Phys Chem C 115:13968–13976. https://doi.org/10.1021/jp112259p

Fu X, Clark LA, Zeltner WA, Anderson MA (1996) Effects of reaction temperature and water vapor content on the heterogeneous photocatalytic oxidation of ethylene. J Photochem Photobiol A Chem 97:181–186. https://doi.org/10.1016/1010-6030(95)04269-5

Fu P, Luan Y, Dai X (2004) Preparation of activated carbon fibers supported TiO2 photocatalyst and evaluation of its photocatalytic reactivity. J Mol Catal A Chem 221:81–88. https://doi.org/10.1016/j.molcata.2004.06.018

Fujishima A, Honda K (1972) Electrochemical photolysis of water at a semiconductor electrode. Nature 238:37. https://doi.org/10.1038/238037a0

Füldner S, Mild R, Siegmund HI, Schroeder JA, Gruber M, König B (2010) Green-light photocatalytic reduction using dye-sensitized TiO 2 and transition metal nanoparticles. Green Chem 12:400–406. https://doi.org/10.1039/B918140G

Furukawa S, Shishido T, Teramura K, Tanaka T (2011) Photocatalytic oxidation of alcohols over TiO2 covered with Nb2O5. ACS Catal 2:175–179. https://doi.org/10.1021/cs2005554

Guillard C et al (2003) Solar efficiency of a new deposited titania photocatalyst: chlorophenol, pesticide and dye removal applications. Appl Catal B Environ 46:319–332. https://doi.org/10.1016/S0926-3373(03)00264-9

Gupta H, Tanaka S (1995) Photocatalytic mineralisation of perchloroethylene using titanium dioxide. Water Sci Technol 31:47–54. https://doi.org/10.1016/0273-1223(95)00405-C

Gupta VK, Saravanan R, Agarwal S, Gracia F, Khan MM, Qin J, Mangalaraja R (2017) Degradation of azo dyes under different wavelengths of UV light with chitosan-SnO2 nanocomposites. J Mol Liq 232:423–430. https://doi.org/10.1016/j.molliq.2017.02.095

Hajiesmaili S, Josset S, Bégin D, Pham-Huu C, Keller N, Keller V (2010) 3D solid carbon foam-based photocatalytic materials for vapor phase flow-through structured photoreactors. Appl Catal A Gen 382:122–130. https://doi.org/10.1016/j.apcata.2010.04.044

Herrmann J-M, Tahiri H, Ait-Ichou Y, Lassaletta G, Gonzalez-Elipe A, Fernandez A (1997) Characterization and photocatalytic activity in aqueous medium of TiO2 and Ag-TiO2 coatings on quartz. Appl Catal B Environ 13:219–228. https://doi.org/10.1016/S0926-3373(96)00107-5

Hoffmann MR, Martin ST, Choi W, Bahnemann DW (1995) Environmental applications of semiconductor photocatalysis. Chem Rev 95:69–96. https://doi.org/10.1021/cr00033a004

Hu Y et al (2011) BiVO4/TiO2 nanocrystalline heterostructure: a wide spectrum responsive photocatalyst towards the highly efficient decomposition of gaseous benzene. Appl Catal B Environ 104:30–36. https://doi.org/10.1016/j.apcatb.2011.02.031

Huang L, Wang X, Yang J, Liu G, Han J, Li C (2013a) Dual cocatalysts loaded type I CdS/ZnS core/shell nanocrystals as effective and stable photocatalysts for H2 evolution. J Phys Chem C 117:11584–11591. https://doi.org/10.1021/jp400010z

Huang X et al (2013b) Type-II ZnO nanorod–SnO 2 nanoparticle heterostructures: characterization of structural, optical and photocatalytic properties. Nanoscale 5:3828–3833. https://doi.org/10.1039/C3NR34327H

Jeon TH, Choi W, Park H (2011) Photoelectrochemical and photocatalytic behaviors of hematite-decorated titania nanotube arrays: energy level mismatch versus surface specific reactivity. J Phys Chem C 115:7134–7142. https://doi.org/10.1021/jp201215t

Jia X, Cao J, Lin H, Zhang M, Guo X, Chen S (2017) Transforming type-I to type-II heterostructure photocatalyst via energy band engineering: a case study of I-BiOCl/I-BiOBr. Appl Catal B Environ 204:505–514. https://doi.org/10.1016/j.apcatb.2016.11.061

Khan Z, Khannam M, Vinothkumar N, De M, Qureshi M (2012) Hierarchical 3D NiO–CdS heteroarchitecture for efficient visible light photocatalytic hydrogen generation. J Mater Chem 22:12090–12095. https://doi.org/10.1039/C2JM31148H

Khanchandani S, Kundu S, Patra A, Ganguli AK (2012) Shell thickness dependent photocatalytic properties of ZnO/CdS core–shell nanorods. J Phys Chem C 116:23653–23662. https://doi.org/10.1021/jp3083419

Khanchandani S, Kundu S, Patra A, Ganguli AK (2013) Band gap tuning of ZnO/In2S3 core/shell nanorod arrays for enhanced visible-light-driven photocatalysis. J Phys Chem C 117:5558–5567. https://doi.org/10.1021/jp310495j

King LA, Zhao W, Chhowalla M, Riley DJ, Eda G (2013) Photoelectrochemical properties of chemically exfoliated MoS 2. J Mater Chem A 1:8935–8941. https://doi.org/10.1039/C3TA11633F

Kondo MM, Jardim WF (1991) Photodegradation of chloroform and urea using Ag-loaded titanium dioxide as catalyst. Water Res 25:823–827. https://doi.org/10.1016/0043-1354(91)90162-J

Ku Y, Hsieh C-B (1992) Photocatalytic decomposition of 2, 4-dichlorophenol in aqueous TiO2 suspensions. Water Res 26:1451–1456. https://doi.org/10.1016/0043-1354(92)90064-B

Kumar A, Kumar A, Sharma G et al (2018) Quaternary magnetic BiOCl/g-C3N4/Cu2O/Fe3O4 nano-junction for visible light and solar powered degradation of sulfamethoxazole from aqueous environment. Chem Eng J 334:462–478. https://doi.org/10.1016/j.cej.2017.10.049

Kundu J, Khilari S, Pradhan D (2017) Shape-dependent photocatalytic activity of hydrothermally synthesized cadmium sulfide nanostructures. ACS Appl Mater Interfaces 9:9669–9680. https://doi.org/10.1021/acsami.6b16456

Lee J, Farha OK, Roberts J, Scheidt KA, Nguyen ST, Hupp JT (2009) Metal–organic framework materials as catalysts. Chem Soc Rev 38:1450–1459. https://doi.org/10.1039/B807080F

Li X, Hou Y, Zhao Q, Chen G (2011) Synthesis and photoinduced charge-transfer properties of a ZnFe2O4-sensitized TiO2 nanotube array electrode. Langmuir 27:3113–3120. https://doi.org/10.1021/la2000975

Li X, Huang R, Hu Y, Chen Y, Liu W, Yuan R, Li Z (2012) A templated method to Bi2WO6 hollow microspheres and their conversion to double-shell Bi2O3/Bi2WO6 hollow microspheres with improved photocatalytic performance. Inorg Chem 51:6245–6250. https://doi.org/10.1021/ic300454q

Liu S, Zhang N, Tang Z-R, Xu Y-J (2012) Synthesis of one-dimensional CdS@ TiO2 core–shell nanocomposites photocatalyst for selective redox: the dual role of TiO2 shell. ACS Appl Mater Interfaces 4:6378–6385

Liu D, Zheng Z, Wang C, Yin Y, Liu S, Yang B, Jiang Z (2013) CdTe quantum dots encapsulated ZnO nanorods for highly efficient photoelectrochemical degradation of phenols. J Phys Chem C 117:26529–26537. https://doi.org/10.1021/jp410692y

Malato S, Fernández-Ibáñez P, Maldonado MI, Blanco J, Gernjak W (2009) Decontamination and disinfection of water by solar photocatalysis: recent overview and trends. Catal Today 147:1–59. https://doi.org/10.1016/j.cattod.2009.06.018

Marschall R (2014) Semiconductor composites: strategies for enhancing charge carrier separation to improve photocatalytic activity. Adv Funct Mater 24:2421–2440. https://doi.org/10.1002/adfm.201303214

Matthews RW (1987) Solar-electric water purification using photocatalytic oxidation with TiO2 as a stationary phase. Sol Energy 38:405–413. https://doi.org/10.1016/0038-092X(87)90021-1

Matthews RW (1991) Photooxidative degradation of coloured organics in water using supported catalysts. TiO2 on sand. Water Res 25:1169–1176. https://doi.org/10.1016/0043-1354(91)90054-T

Mazzarino I, Piccinini P (1999) Photocatalytic oxidation of organic acids in aqueous media by a supported catalyst. Chem Eng Sci 54:3107–3111. https://doi.org/10.1016/S0009-2509(98)00430-8

Mills A, Le Hunte S (1997) An overview of semiconductor photocatalysis. J Photochem Photobiol A Chem 108:1–35. https://doi.org/10.1016/S1010-6030(97)00118-4

Mills A, Davies RH, Worsley D (1993) Water purification by semiconductor photocatalysis. Chem Soc Rev 22:417–425. https://doi.org/10.1039/CS9932200417

Mu J et al (2011) Enhancement of the visible-light photocatalytic activity of In2O3–TiO2 nanofiber heteroarchitectures. ACS Appl Mater Interfaces 4:424–430. https://doi.org/10.1021/am201499r

Nasr C, Hotchandani S, Kim WY, Schmehl RH, Kamat PV (1997) Photoelectrochemistry of composite semiconductor thin films. Photosensitization of SnO2/CdS coupled nanocrystallites with a ruthenium polypyridyl complex. J Phys Chem B 101:7480–7487. https://doi.org/10.1021/jp970833k

Naushad M, ALOthman ZA, Awual MR et al (2015) Adsorption kinetics, isotherms, and thermodynamic studies for the adsorption of Pb2+ and Hg2+ metal ions from aqueous medium using Ti
(IV) iodovanadate cation exchanger. Ionics (Kiel) 21:2237–2245. https://doi.org/10.1007/
s11581-015-1401-7

Nazir S, Schwingenschlögl U (2011) High charge carrier density at the NaTaO3/SrTiO3 hetero-
interface. Appl Phys Lett 99:073102. https://doi.org/10.1063/1.3625951

Ollis DF (1985) Contaminant degradation in water. Environ Sci Technol 19:480–484. https://doi.
org/10.1021/es00136a002

Ostermann R, Sallard S, Smarsly BM (2009) Mesoporous sandwiches: towards mesoporous
multilayer films of crystalline metal oxides. Phys Chem Chem Phys 11:3648–3652. https://
doi.org/10.1039/b820651c

Pandiyarajan T, Saravanan R, Karthikeyan B, Gracia F, Mansilla HD, Gracia-Pinilla M,
Mangalaraja RV (2017) Sonochemical synthesis of CuO nanostructures and their morphology
dependent optical and visible light driven photocatalytic properties. J Mater Sci Mater Electron
28:2448–2457. https://doi.org/10.1007/s10854-016-5817-2

Pelizzetti E, Minero C, Borgarello E, Tinucci L, Serpone N (1993) Photocatalytic activity and
selectivity of titania colloids and particles prepared by the sol-gel technique: photooxidation of
phenol and atrazine. Langmuir 9:2995–3001. https://doi.org/10.1021/la00035a043

Poulios I, Tsachpinis I (1999) Photodegradation of the textile dye Reactive Black 5 in the presence
of semiconducting oxides. J Chem Technol Biotechnol 74:349–357. https://doi.org/10.1002/
(SICI)1097-4660(199904)74:4<349::AID-JCTB5>3.0.CO;2-7

Pugazhendhi A, Boovaragamoorthy GM, Ranganathan K et al (2017) New insight into effective
biosorption of lead from aqueous solution using Ralstonia solanacearum: characterization and
mechanism studies. J Clean Prod 174:1234. https://doi.org/10.1016/j.jclepro.2017.11.061

Ramos-Delgado N, Hinojosa-Reyes L, Guzman-Mar I, Gracia-Pinilla M, Hernández-Ramírez A
(2013) Synthesis by sol–gel of WO3/TiO2 for solar photocatalytic degradation of malathion
pesticide. Catal Today 209:35–40. https://doi.org/10.1016/j.cattod.2012.11.011

Ren Q, Wang W, Shi H, Liang Y (2014) Synthesis and shape-dependent visible-light-driven
photocatalytic activities of Cu2ZnSnS4 nanostructures. Micro Nano Lett 9:505–508. https://
doi.org/10.1049/mnl.2014.0142

Royer S, Duprez D, Can F, Courtois X, Batiot-Dupeyrat C, Laassiri S, Alamdari H (2014)
Perovskites as substitutes of noble metals for heterogeneous catalysis: dream or reality. Chem
Rev 114:10292–10368. https://doi.org/10.1021/cr500032a

Ryu CS, Kim M-S, Kim B-W (2003) Photodegradation of alachlor with the TiO2 film immobilised
on the glass tube in aqueous solution. Chemosphere 53:765–771. https://doi.org/10.1016/
S0045-6535(03)00506-X

Saravanan R, Gupta VK, Narayanan V, Stephen A (2013a) Comparative study on photocatalytic
activity of ZnO prepared by different methods. J Mol Liq 181:133–141. https://doi.org/10.1016/
j.molliq.2013.02.023

Saravanan R, Karthikeyan N, Gupta V, Thirumal E, Thangadurai P, Narayanan V, Stephen A
(2013b) ZnO/Ag nanocomposite: an efficient catalyst for degradation studies of textile effluents
under visible light. Mater Sci Eng C 33:2235–2244. https://doi.org/10.1016/j.msec.2013.01.046

Saravanan R, Thirumal E, Gupta V, Narayanan V, Stephen A (2013c) The photocatalytic activity of
ZnO prepared by simple thermal decomposition method at various temperatures. J Mol Liq
177:394–401. https://doi.org/10.1016/j.molliq.2012.10.018

Saravanan R, Gupta VK, Mosquera E, Gracia F, Narayanan V, Stephen A (2015) Visible light
induced degradation of methyl orange using β-Ag0. 333V2O5 nanorod catalysts by facile
thermal decomposition method. J Saudi Chem Soc 19:521–527. https://doi.org/10.1016/j.jscs.
2015.06.001

Schultz AM, Salvador PA, Rohrer GS (2012) Enhanced photochemical activity of α-Fe 2 O 3 films
supported on SrTiO 3 substrates under visible light illumination. Chem Commun 48
(14):2012–2014. https://doi.org/10.1039/c2cc16715h

Shaban M, Ashraf AM, Abukhadra MR (2018) TiO 2 Nanoribbons/Carbon nanotubes composite with enhanced photocatalytic activity; fabrication, characterization, and application. Sci Rep 8:781. https://doi.org/10.1038/s41598-018-19172-w

Shahid M, Shakir I, Yang S-J, Kang DJ (2010) Facile synthesis of core–shell SnO2/V2O5 nanowires and their efficient photocatalytic property. Mater Chem Phys 124:619–622. https://doi.org/10.1016/j.matchemphys.2010.07.023

Shi Y, Li H, Wang L, Shen W, Chen H (2012) Novel α-Fe2O3/CdS cornlike nanorods with enhanced photocatalytic performance. ACS Appl Mater Interfaces 4:4800–4806. https://doi.org/10.1021/am3011516

Shifu C, Gengyu C (2005) Photocatalytic degradation of organophosphorus pesticides using floating photocatalyst TiO2· SiO2/beads by sunlight. Sol Energy 79:1–9. https://doi.org/10.1016/j.solener.2004.10.006

Sia SK, Linder V, Parviz BA, Siegel A, Whitesides GM (2004) An integrated approach to a portable and low-cost immunoassay for resource-poor settings. Angew Chem Int Ed 43:498–502. https://doi.org/10.1002/anie.200353010

Siedl N, Elser MJ, Bernardi J, Diwald O (2009) Functional interfaces in pure and blended oxide nanoparticle networks: recombination versus separation of photogenerated charges. J Phys Chem C 113:15792–15795. https://doi.org/10.1021/jp906368f

Sohrabnezhad S, Pourahmad A, Radaee E (2009) Photocatalytic degradation of basic blue 9 by CoS nanoparticles supported on AlMCM-41 material as a catalyst. J Hazard Mater 170:184–190. https://doi.org/10.1016/j.jhazmat.2009.04.108

Srinivas B, Kumar BG, Muralidharan K (2015) Stabilizer free copper sulphide nanostructures for rapid photocatalytic decomposition of rhodamine B. J Mol Catal A Chem 410:8–18. https://doi.org/10.1016/j.molcata.2015.08.028

Su J et al (2011) Macroporous V2O5− BiVO4 composites: effect of heterojunction on the behavior of photogenerated charges. J Phys Chem C 115:8064–8071. https://doi.org/10.1021/jp200274k

Su Y, Wang H, Ye L, Jin X, Xie H, He C, Bao K (2014) Shape-dependent photocatalytic activity of Bi 5 O 7 I caused by facets synergetic and internal electric field effects. RSC Adv 4:65056–65064. https://doi.org/10.1039/C4RA08431D

Sun J, Wang X, Sun J, Sun R, Sun S, Qiao L (2006) Photocatalytic degradation and kinetics of Orange G using nano-sized Sn (IV)/TiO2/AC photocatalyst. J Mol Catal A Chem 260:241–246. https://doi.org/10.1016/j.molcata.2006.07.033

Tanaka K, Padermpole K, Hisanaga T (2000) Photocatalytic degradation of commercial azo dyes. Water Res 34:327–333. https://doi.org/10.1016/S0043-1354(99)00093-7

Tanguay JF, Suib SL, Coughlin RW (1989) Dichloromethane photodegradation using titanium catalysts. J Catal 117:335–347. https://doi.org/10.1016/0021-9517(89)90344-8

Thibert A, Frame FA, Busby E, Holmes MA, Osterloh FE, Larsen DS (2011) Sequestering high-energy electrons to facilitate photocatalytic hydrogen generation in CdSe/CdS nanocrystals. J Phys Chem Lett 2:2688–2694. https://doi.org/10.1021/jz2013193

Thuy UTD, Liem NQ, Parlett CM, Lalev GM, Wilson K (2014) Synthesis of CuS and CuS/ZnS core/shell nanocrystals for photocatalytic degradation of dyes under visible light. Catal Commun 44:62–67. https://doi.org/10.1016/j.catcom.2013.07.030

Tongying P, Plashnitsa VV, Petchsang N, Vietmeyer F, Ferraudi GJ, Krylova G, Kuno M (2012) Photocatalytic hydrogen generation efficiencies in one-dimensional CdSe heterostructures. Journal Phys Chem Lett 3:3234–3240. https://doi.org/10.1021/jz301628b

Umar M, Aziz HA (2013) Photocatalytic degradation of organic pollutants in water. In: Organic pollutants-monitoring, risk and treatment. InTech, New York

Wang G et al (2012a) Controlled synthesis of Ag 2 O microcrystals with facet-dependent photocatalytic activities. J Mater Chem 22:21189–21194. https://doi.org/10.1039/C2JM35010F

Wang J et al (2012b) Highly efficient oxidation of gaseous benzene on novel Ag3VO4/TiO2 nanocomposite photocatalysts under visible and simulated solar light irradiation. J Phys Chem C 116:13935–13943. https://doi.org/10.1021/jp301355q

Wang Y, Wang Q, Zhan X, Wang F, Safdar M, He J (2013) Visible light driven type II heterostructures and their enhanced photocatalysis properties: a review. Nanoscale 5:8326–8339. https://doi.org/10.1039/C3NR01577G

Wen F, Wang X, Huang L, Ma G, Yang J, Li C (2012) A hybrid photocatalytic system comprising ZnS as light harvester and an [Fe2S2] hydrogenase mimic as hydrogen evolution catalyst. ChemSusChem 5:849–853. https://doi.org/10.1002/cssc.201200190

Wenhua L, Hong L, Sao'an C, Jianqing Z, Chunan C (2000) Kinetics of photocatalytic degradation of aniline in water over TiO2 supported on porous nickel. J Photochem Photobiol A Chem 131:125–132. https://doi.org/10.1016/S1010-6030(99)00232-4

Wetchakun N, Chaiwichain S, Inceesungvorn B, Pingmuang K, Phanichphant S, Minett AI, Chen J (2012) BiVO4/CeO2 nanocomposites with high visible-light-induced photocatalytic activity. ACS Appl Mater Interfaces 4:3718–3723. https://doi.org/10.1021/am300812n

Wu N et al (2010) Shape-enhanced photocatalytic activity of single-crystalline anatase TiO2 (101) nanobelts. J Am Chem Soc 132:6679–6685. https://doi.org/10.1021/ja909456f

Xia Y, Wang J, Chen R, Zhou D, Xiang L (2016) A review on the fabrication of hierarchical ZnO nanostructures for photocatalysis application. Crystals 6:148

Xiang Q, Yu J, Cheng B, Ong H (2010) Microwave-hydrothermal preparation and visible-light photoactivity of plasmonic photocatalyst Ag-TiO2 nanocomposite hollow spheres. Chem Asian J 5:1466–1474. https://doi.org/10.1002/asia.200900695

Xiong Z, Zhao XS (2012) Nitrogen-doped titanate-anatase core–shell nanobelts with exposed {101} anatase facets and enhanced visible light photocatalytic activity. J Am Chem Soc 134:5754–5757. https://doi.org/10.1021/ja300730c

Xiong P, Zhu J, Wang X (2013) Cadmium sulfide–ferrite nanocomposite as a magnetically recyclable photocatalyst with enhanced visible-light-driven photocatalytic activity and photostability. Ind Eng Chem Res 52:17126–17133. https://doi.org/10.1021/ie402437k

Yang X, Xu J, Wong T, Yang Q, Lee C-S (2013) Synthesis of In 2 O 3–In 2 S 3 core–shell nanorods with inverted type-I structure for photocatalytic H 2 generation. Phys Chem Chem Phys 15:12688–12693. https://doi.org/10.1039/C3CP51722E

Yogi C, Kojima K, Wada N, Tokumoto H, Takai T, Mizoguchi T, Tamiaki H (2008) Photocatalytic degradation of methylene blue by TiO2 film and Au particles-TiO2 composite film. Thin Solid Films 516:5881–5884. https://doi.org/10.1016/j.tsf.2007.10.050

Yu H, Liu R, Wang X, Wang P, Yu J (2012) Enhanced visible-light photocatalytic activity of Bi2WO6 nanoparticles by Ag2O cocatalyst. Appl Catal B Environ 111:326–333. https://doi.org/10.1016/j.apcatb.2011.10.015

Zhai J, Tao X, Pu Y, Zeng X-F, Chen J-F (2010) Core/shell structured ZnO/SiO2 nanoparticles: preparation, characterization and photocatalytic property. Appl Surf Sci 257:393–397. https://doi.org/10.1016/j.apsusc.2010.06.091

Zhang Z, Wang P (2012) Highly stable copper oxide composite as an effective photocathode for water splitting via a facile electrochemical synthesis strategy. J Mater Chem 22:2456–2464. https://doi.org/10.1039/C1JM14478B

Zhang X, Qin J, Xue Y, Yu P, Zhang B, Wang L, Liu R (2014) Effect of aspect ratio and surface defects on the photocatalytic activity of ZnO nanorods. Sci Rep 4:4596. https://doi.org/10.1038/srep04596

Zhao X, Liu H, Shen Y, Qu J (2011) Photocatalytic reduction of bromate at C60 modified Bi2MoO6 under visible light irradiation. Appl Catal B Environ 106:63–68. https://doi.org/10.1016/j.apcatb.2011.05.005

Zhou J et al (2016) Photodegradation performance and recyclability of a porous nitrogen and carbon co-doped TiO 2/activated carbon composite prepared by an extremely fast one-step microwave method. RSC Adv 6:84457–84463. https://doi.org/10.1039/C6RA19757D

Chapter 10
Advances and Challenges in BiOX (X: Cl, Br, I)-Based Materials for Harvesting Sunlight

David Contreras, Victoria Melin, Gabriel Pérez-González, Adolfo Henríquez, and Lisdelys González

Contents

D. Contreras (✉) · V. Melin (✉) · G. Pérez-González
Facultad de Ciencias Químicas, Centro de Biotecnología, Universidad de Concepción, Concepción, Chile
e-mail: dcontrer@udec.cl; victoriamelin@udec.cl; gperezg@udec.cl

A. Henríquez · L. González
Centro de Biotecnología, Universidad de Concepción, Concepción, Chile
e-mail: adohenriquez@udec.cl; lisdegonzalez@udec.cl

© Springer Nature Switzerland AG 2020
M. Naushad et al. (eds.), *Green Photocatalysts*, Environmental Chemistry
for a Sustainable World 34, https://doi.org/10.1007/978-3-030-15608-4_10

Abstract Photocatalysts can use sunlight to catalyze chemical reactions. These materials follow the global trend of performing chemical processes under the twelve principles of green chemistry and have emerged as a promising technology to demonstrate new ways of collecting sunlight for environmental applications and the removal of contaminants. Photocatalysts are being incorporated in green chemistry and used to improve chemical processes.

The use of photocatalytic processes with semiconductor photocatalysts has the advantage of allowing recovery and reuse of the photocatalyst for further applications. The most well-known photocatalyst material is titanium dioxide (TiO_2); however, in the past 10 years, bismuth oxyhalides have been widely studied. Among the advantages of bismuth oxyhalides is a wide range of band gap energies, reaching values in the visible spectrum. This is a great improvement as the solar spectrum has 50% of its radiation in the visible range, allowing more efficient collection of sunlight. Moreover, within the bismuth photocatalysts synthesized to date, oxyhalides have lower costs and use fewer polluting reagents; hence, they fit well into the "green photocatalysts" group.

In this chapter, the main characteristics of the synthesis routes and doping strategies of BiOX-based materials will be reviewed in detail. In addition, the main uses of BiOX in energy, environmental remediation, and green chemistry fields will be discussed.

Keywords Bismuth oxyhalides · Photocatalyst · Synthesis routes · Applications

10.1   Introduction

In 1912, the pioneer chemist Giacomo Ciamician, in his article "The photochemistry of the future", discussed how coal was a synonym of energy and wealth, but was also a limited resource. He emphasizes that coal will eventually be consumed or the price for extraction will rise. Thus, he asked the question, *Is solar energy the only one that may be used in modern life and civilization?*, and encouraged future generations to develop industrial chemical processes that could synthetize chemicals as plants do, using an inexpensive, safe, and unlimited energy source such as sunlight (Ciamician 1912). In this chapter, more than 100 years after Ciamician's famous article, we will discuss the best way to use sunlight and develop methods and processes using one of the cleanest energy sources: the Sun.

10.1.1   Photocatalysis

Photocatalysis is the use of light to catalyze chemical reactions; this term was introduced into the glossary of science in the early 1930s. However, it was not until the mid-1990s that the scientific community made an effort to properly define it as "a catalytic reaction involving light absorption by a catalyst or by a substrate" (IUPAC, 1996), prompted by significant development in this field in the 1980s and 1990s (Parmon 1997). Currently, the formal definition is very similar; for example, the International Union of Pure and Applied Chemistry (IUPAC), in their latest issue of "Glossary of terms used in photocatalysis and radiation catalysis" (2011), the term *photocatalysis* was defined as "either catalytic reactions proceeding under the action of light, or the overall phenomena connected both with photochemical and catalytic processes" (Braslavsky et al. 2011).

These photocatalytic processes have been used and studied in several fields, such as addressing energy issues (energy production and storage), environmental remediation, and green chemistry (which will be discussed in detail later). It is worth mentioning that wastewater treatment is one of the most widespread applications of photocatalytic processes. A good example is the generation of highly reactive species of oxygen (ROS) from hydrogen peroxide (H_2O_2) and UV light (Elmorsi et al. 2010; He, X. et al. 2014b). In addition, transition metal complexes, such as iron and copper (Andreozzi and Marotta 2004; Ciesla et al. 2004), can be mixed with H_2O_2 in a technique called the *photo-Fenton reaction* (Fe(II)/ H_2O_2/UV) (Pouran et al. 2015; Škodič et al. 2017). Despite the method used, the homogeneous photocatalysis mechanism is the same; i.e., in situ generation of ROS, such as hydroxyl radicals ($\cdot$OH) that further react with organic matter and recalcitrant compounds. For this reason, such methods are often used for decontamination of wastewater (Catalá et al. 2015; Miklos et al. 2018). Advantages of this method include complete mineralization of contaminants, no waste disposal issues, low cost, and mild conditions (Saravanan et al. 2017).

Heterogeneous photocatalytic processes involve semiconductor materials such as BIOX (X = Cl, Br, I), ZnO, and most commonly, TiO_2, which are suspended in a liquid and can be recovered for later use. The requisites for an ideal photocatalyst are low cost, high chemical stability, and highly oxidizing photogenerated holes to form active species, such as hydroxyl radicals, hydrogen peroxide, and super oxide anions (Fujishima et al. 2000; Rehman et al. 2009). The formation mechanism for these species depends on the ability of the photocatalyst to absorb a specific wavelength of light from the solar spectrum and form an excited electron, which can reduce dioxygen to superoxide, while the generated hole can produce hydroxyl radicals (Fig. 10.1) (Fujishima et al. 2000; Nakata and Fujishima 2012; Schneider et al. 2014). Similarly, in the case of homogeneous photocatalysis, these radicals can be used in wastewater treatment to remove hazardous substances. In addition, as the photocatalyst are solid particles, they can also be used in paints to form self-cleaning surfaces, or even air cleaning walls (e.g., for operating rooms and other medical applications), or removal of NO_x emitted by cars on roads and in tunnels. Applications also have been found in energy fields, such as solar cells for energy conversion, evolution of O_2 and H_2 from H_2O, and reduction of CO_2 with H_2O to form O_2 (Nakata and Fujishima 2012; Schneider et al. 2014). Such uses of photocatalysts are now contained within what is now called Green Chemistry, which is the current trend in the way we practice chemistry and will be discussed here in this chapter.

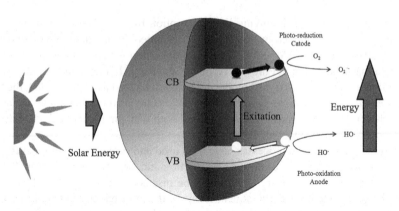

Fig. 10.1 Schematic diagram of a semiconductor photocatalyst, the yellow arrows show the path of an electron from the anode where photo-oxidation of hydroxyl anions forms hydroxyl radicals. The electron is excited by the solar energy from the valence band (VB) to the conduction band (CB). Finally, the hole (h+) generated can take other electron and the excited electron diffuses to the surface of the photocatalyst to photo-reduce oxygen, forming superoxide anion ($O_2^{\cdot -}$) at the cathode

10.1.2 Photocatalysis in Green Chemistry

In 1998, Paul Anastas and John Warner introduced the *twelve principles of green chemistry*, providing guidelines for sustainability at a molecular level, highlighting the importance of reducing waste generated by chemical processes worldwide (Anastas and Eghbali 2009; Sheldon 2016). Even though many efforts have been made in the past 20 years to perform scientific research following these principles, and various "green" chemical processes have been achieved, there is still a considerable amount of waste that we continue to dispose of in the environment. From soil contamination to air pollution, many types of industries, including textile, mining, pulp and paper mills, pharmaceutical, and food, contaminate the environment by releasing hazardous chemicals generated during processing (e.g., dyeing, bleaching, washing, and fertilizing) (Hajem et al. 2007; Husaini et al. 2008; Priya et al. 2009). Some of these materials are toxic, mutagenic, or carcinogenic (Lei et al. 2007; Kayan et al. 2010).

Environmental chemistry is a field parallel to green chemistry that includes decontamination of the environment to remove existing pollution (Ekar et al. 2016). To achieve this, there are several methods to treat wastewater, which can be separated into physical and chemical processes. Physical processes include heat treatment, distillation, reverse osmosis, microfiltration, slow sand filtering, and activated charcoal filtering. The chemical methods include the addition of chlorine, iodine, silver, potassium permanganate, and flocculation or coagulating agents. Both methods can be used independently or combined in order to remove solids, pathogens, organic matter, and nutrients (Cheremisinoff 2002; Prasse 2015). However, it is known that these methods are insufficient for removing anthropogenic contaminants present in the ng/L to µg/L range (Prasse 2015; Gupta and Thakur 2017; Matamoros et al. 2017); therefore, additional methods must be used to remove these contaminants. This is where photocatalytic processes combine with green chemistry as the most common method to remove recalcitrant contaminants is O_3/ UV irradiation or photo-Fenton systems (Cheremisinoff 2002; Davididou et al. 2018) and TiO_2 (Fujishima et al. 2000; Nakata and Fujishima 2012; Schneider et al. 2014).

Among heterogeneous photocatalysts, TiO_2 is the most widely used as it has nearly ideal photocatalytic behavior and can be recycled for further use. In TiO_2, the energy required to mobilize an electron from the conduction band (CB) to the valence band (VB) (band gap) is high, in the range of UV light; however, the amount of UV light reaching sea level is only 5–10% of the total solar spectrum, compared to almost 50% for visible light (Blesa et al. 2005; Bohórquez-Ballen and Pérez 2007). Hence, developing new materials with band gap energies in the range of visible light is promising for enhancing photocatalytic activity (Huizhong et al. 2008; Yoon et al. 2010; Etacheri et al. 2015). In the past 10 years, bismuth-based compounds have been reported as efficient photocatalysts for decontamination, where bismuth oxide (Bi_2O_3) can be used alone, or doped/modified with a wide range of transition metals, including gallium, thallium, tungsten, niobium, and vanadium (Kudo et al. 1999; Zou et al. 2001b; Luan et al. 2004; Tang et al. 2004; Yao et al. 2004; He and Gu 2006; Muktha et al. 2006; Zhang et al. 2006; Fu et al. 2007). Recently, metal-free bismuth oxyhalide compounds have been synthetized and characterized, showing several differences compared to metal-bismuth photocatalysts (Huizhong et al. 2008).

10.1.3 Bismuth Oxyhalides

The emergence of bismuth oxyhalides in the photocatalytic field is relatively recent, showing an increase of 1600% in publications regarding BiOX (X = Cl, Br, I) materials in the past decade (data extracted from Web of Science). However, these materials have been discussed since around 1900, and studies of the crystallography of BIOX and then their photoelectrical properties (Bannister and Hey 1935; Bletskan et al. 1972; Bastow and Whitfield 1977; Poznyak and Kulak 1990b) were performed before the scientific community realized their potential as photocatalysts.

Table 10.1 summarizes some of the common bismuth photocatalysts, such as Bi_2O_3, Bi_2WO_6, and Bi_2XYO_7 (X = Ga, Al, In; Y = Ta, Nb). BiOX-based catalysts have shown significantly better performance than these other materials, attributed to both their chemical-photocatalytic characteristics (such as the band gap, wider spectrum of photocatalytic activity allowing both water splitting and pollutant decontamination) and their synthesis methods (which use fewer toxic reagents and produce less waste). Finally, reagents used in BIOX synthesis are considerably less expensive than those required for other bismuth compounds (see Table 10.1). Thus, bismuth oxyhalides fit well within the concept of "ideal green photocatalysts." Some of the differences in band gap and surface area observed in the table are due to their

Table 10.1 Comparison of the properties of common bismuth photocatalysts synthetized in the last decade and those of TiO_2

Compound	Band gap (eV)[a]	Water splitting	Pollutant decontamination	Surface area (m²/g)[a]	Reagents used for synthesis	Total reagent price/kilogram (US$)[b]	References
TiO_2	3.0–3.2	✓	✓	2.82–108.0	Commercial	221	Wu et al. (1998), Khan et al. (2002), Serpone (2006) and Chen et al. (2009)
Bi_2O_3	2.8	N.R.	✓	0.01–72	Commercial	927	Zhang et al. (2006), Zhou et al. (2009) and Li et al. (2010)
Bi_2WO_6	2.69	✓	✓	21.5	Bi_2O_3, WO_3	4600	Tang et al. (2004) and Lv et al. (2016)
Bi_2GaTaO_7	2.19–3.0	✓	✓	0.71–1.7	Bi_2O_3, Ga_2O_3, Ta_2O_5	32,000	Luan et al. (2004) and Wang et al. (2005)
Bi_2AlNbO_7	2.86	✓	N.R.	0.51	$Bi_2(CO_3)_3$, Al_2O_3, Nb_2O_5	9474	Zou et al. (2001c) and Li et al. (2009)
Bi_2GaNbO_7	2.75	✓	N.R.	0.52	$Bi_2(CO_3)_3$, Ga_2O_3, Nb_2O_5	20,460	Zou et al. (2001b, c)
Bi_2InNbO_7	2.7	✓	N.R.	0.51	$Bi_2(CO_3)_3$, In_2O_3, Nb_2O_5	19,990	Zou et al. (2001b, c)
$Bi_{12}SiO_{20}$	2.6–2.8	N.R.	✓	4.87	$Bi(NO_3)_3$, $(C_2H_5O)_4Si$	1217	He and Gu (2006)
$Bi_{24}AlO_{39}$	2.46	N.R.	✓	N.R.	$Bi(NO_3)_3$, $Al_2(NO_3)_3$	8721	Yao et al. (2004)
$BiVO_4$	2.4–2.9	✓	N.R.	0.5–6.3	V_2O_5, $Bi(NO_3)_3$	34,121	Kudo et al. (1999)

BiNbO$_4$	2.6	N.R.	✓	< 1	Bi$_2$O$_3$, Nb$_2$O$_5$	7660	Zou et al. (2001a) and Muktha et al. (2006)
BiTaO$_4$	2.6	N.R.	✓	< 1	Bi$_2$O$_3$, Ta$_2$O$_5$	16,684	
BiOCl	3.22	✓	✓	0.19–36	Bi$_2$O$_3$, HCl	3637	Jiang et al. (2012), Ye et al. (2014) and Xiao et al. (2016)
BiOBr	2.7	✓	✓	4.36–22.36	Bi$_2$O$_3$, HBr	3763	Ye et al. (2014)
BiOI	1.79	✓	✓	5–15	Bi(NO$_3$)$_3$, KI	3062	Xiao and Zhang (2010), Hu et al. (2014) and Ye et al. (2014)

NR Not reported

[a]Differences can be attributed to the synthesis method or metal doping

[b]Prices for 1 kg of each reagent obtained from Sigma-Aldrich for synthesis grade reagents

synthesis methods, which will be discussed in detail in the next section. This chapter focuses on the synthesis routes, doping methods, and distinct characteristics of BiOX materials, as well as introducing present applications in energy, environmental remediation, and green chemistry.

10.2 Bismuth Oxyhalide Synthesis Routes

Bismuth oxyhalides, BiOX (X = F, Cl, Br, I), are semiconductor materials with the tetragonal matlockite (PbFCl) structure, where the central bismuth atom is surrounded by four oxygen atoms and four halogen atoms (Bannister and Hey 1935). Bismuth oxyhalides have a layered structure with $[Bi_2O_2]^{2+}$ slabs interleaved with negative halide slabs, resulting in an internal static electric field perpendicular to each layer. The tetragonal structures of bismuth oxyfluoride, bismuth oxychloride, bismuth oxybromide, and bismuth oxyiodide are shown in Fig. 10.2. This electronic structure allows these materials to effectively separate photogenerated electron–hole pairs (Bannister and Hey 1935).

The band gap of bismuth oxyhalides decreases as the atomic number of the halogen increases. Bismuth oxyfluoride has the widest band gap (3.6 eV) and does not show photocatalytic activity under visible light. The band gap of BiOCl is in the range of 3.2–3.5 eV, which is similar to that of TiO_2. Bismuth oxybromide and bismuth oxyiodide have band gaps of 1.8–1.9 and 2.6, respectively, and exhibit excellent photoactivity under visible light (He, R. et al. 2014a).This trend in optical behavior of bismuth oxyhalides has been explained by analyzing the density of states of the materials. In BiOF, the top of the valence band is dominated by O 2p states,

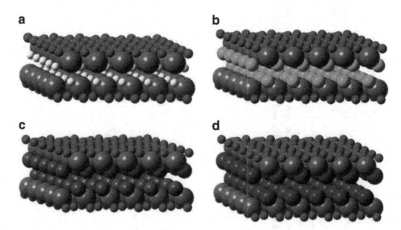

Fig. 10.2 Schematic diagram of the tetragonal structure of BiOX. (**a**). Bismuth oxyfluoride. (**b**). Bismuth oxychloride. (**c**). Bismuth oxybromide and (**d**). Bismuth oxyiodide. The Bi^{3+} ions are purple, O^{2-} are red, F^- are light blue, Cl^- are green, Br^- are dark red, and I^- are violet (Color figure online)

with the F 2p states found at higher binding energies. As the halide anion changes upon moving down through halogen group, the contribution of the halide p states to the valence band maximum increases. On the other hand, the conduction band minimum is dominated by Bi p states (Ganose et al. 2016).

Among the bismuth oxyhalides, bismuth oxyiodide attracts the most attention as it has a narrower band gap and the strongest absorption (and photocatalytic activity) under visible light irradiation (hence, sunlight) (Xu et al. 2017a, b, c). Qin et al. in 2013 evaluated the degradation of methyl orange dye to determine the photocatalytic performance of bismuth oxyhalides with hierarchical architecture under visible light irradiation ($\geq$ 420 nm). They found that such structures exhibited much higher photocatalytic efficiency than nitrogen-doped titanium dioxide, in the order BiOI > BiOBr > BiOCl (Qin et al. 2013).

It is well known that the synthesis method can affect the surface area, morphology, and size of photocatalytic materials; these properties determine the adsorption properties and photocatalytic activity of the materials (Ye et al. 2014). The main synthesis routes for bismuth oxyhalide materials include precipitation, solvothermal, and electrodeposition processes. Table 10.2 shows a summary of typical synthesis routes, morphologies, and photocatalytic activity for bismuth oxyiodide. Such

Table 10.2 Summary of typical synthesis methods, morphologies, and photocatalytic testing conditions of bismuth oxyiodide materials

Synthesis method	Morphology	Photocatalytic experimental conditions	References
Solvothermal process: Bi $(NO_3)_3\cdot5H_2O$, KI, ethylene glycol, 160 °C for 18 h.	Hierarchical microspheres	Substrate: gallic acid; 12 W Xe lamp (VIPHID 6000 K)	Mera et al. (2017)
Solvothermal method: Bi $(NO_3)_3\cdot5H_2O$, KI, ethanol, 180 °C for 2 h.	Nanosheet-assembled microspheres	Substrate: bisphenol A; 500 W Xe lamp	Pan et al. (2015)
Ionic liquid-assisted solvothermal process: Bi $(NO_3)_3\cdot5H_2O$, 1-hexyal-3-methylimidazoliumchloride, 160 °C for 6–12 h.	Hierarchical microspheres	Substrate: methyl orange; 300 W Xe arc lamp (PLS-SXE300)	Qin et al. (2013)
Solvothermal process: Bi $(NO_3)_3\cdot5H_2O$, KI, 150 °C for 15 h.	Hierarchical microspheres	Substrate: ethylene blue; 500 W Xe lamp and a UV cutoff filter of 420 nm	Luo et al. (2016)
Reactable ionic liquid-assisted process: Bi $(NO_3)_3\cdot5H_2O$, 1-butyl-3-methylimidazolium Iodine, room temperature, 1 h.	Nanosheet-assembled hollow microspheres	Substrate: rhodamine B, tetracycline and bisphenol A; 300 W Xe lamp and a UV cutoff filter of 400 nm	Di et al. (2014)
Electrochemical process: Bi $(NO_3)_3\cdot5H_2O$, KI, working electrode at −0.1 V vs. Ag/AgCl (4 M KCl)	Thin two-dimensional plate-like crystals		McDonald and Choi (2012)

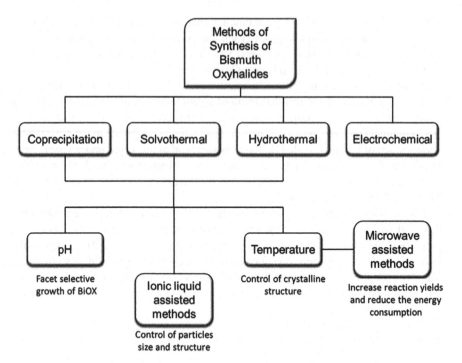

Fig. 10.3 Scheme of synthesis routes of bismuth oxyhalide-based materials

methods can be modified by using microwave irradiation and ionic liquids, which will be discussed in subsequent subsections. The scheme in Fig. 10.3 shows the different synthesis routes used to date to prepare bismuth oxyhalides.

10.2.1 Precipitation Process

The precipitation method used to synthesize bismuth oxyhalides involves the drop-wise addition of a halide solution to a bismuth salt solution, generally bismuth nitrate; the solutions are generally aqueous acidic solutions. This easy synthesis route can produce high-purity bismuth oxyhalides at room temperature and atmospheric pressure and allow preferential exposure of desired crystal facets by modifying the pH of the reaction media. This method of synthesis is considered green due to the use of water as a solvent, which is cheap, nontoxic, nonflammable, and environmentally benign; however, the produced bismuth oxyhalides have a relatively small specific surface area.

10.2.2 Hydrothermal and Solvothermal Processes

Solvothermal processes are characterized by reactions occurring in nonaqueous solvents inside of an autoclave reactor. This synthesis method controls the size, shape, and crystallinity of the synthesized materials. In the case of the hydrothermal processes, aqueous solvents are used.

10.2.3 Electrochemical Process

Anodic polarization of bismuth electrodes immersed in an electrolyte produce uniform, insulating oxide films which behave in a similar way to other oxide films deposited on tantalum, aluminum, titanium, and similar metals. When anodic oxidation of bismuth occurs in neutral aqueous solutions of halides, porous bismuth oxyhalides are formed on the electrode surface. In 1970, Poznyak and Kulak (Poznyak and Kulak 1990a) reported the synthesis of BiOCl using an electrochemical process. The obtained anodic films were highly porous and permeable to the electrolyte, with a BET surface area of 45 ± 2 m^2/g. The X-ray analysis revealed that bismuth oxychloride crystallized in its matlockite structure.

10.2.4 Effect of Solvent During the Synthesis of Bismuth Oxyhalides

The solvents chosen for use during chemical reactions and material synthesis are important as there is a wide range of diverse solvent with different properties. The most commonly used solvents are water and organic solvents. However, the narrow temperature range of aqueous media (0–100 °C) limits the application of water as a solvent for material synthesis.

Ionic-Liquid-Assisted Process

The solvent plays an important role in the determining the morphology of the synthesized material. This is important in the synthesis of photocatalytic materials as the photocatalytic activity depends on the morphology; however, many organic solvents have low boiling points and high vapor pressures, the solubility of inorganic reactants in organic solvents is usually low, and some organic solvents are highly toxic, flammable, and even explosive.

In recent years, growing environmental awareness has resulted in the use of ionic liquids as green media in the synthesis of inorganic materials. The main properties of ionic liquids are a low melting point, high polarity, solubilizes a wide range of compounds, chemical and thermal stability, zero vapor pressure, and wide electro-chemical window (Armand et al. 2009). The advantage of using ionic liquids in the synthesis of bismuth oxyhalide materials is that this acts as a halide source, template, and solvent. These properties allow ionic liquids to be used in solvothermal and electrochemical BiOX synthesis processes.

Materials with hierarchical structures have abundant spaces formed between adjacent nanounits, which facilitates the transport of reactants to the photocatalyst surface and offers a high specific surface and excellent adsorption capacity. In addition, such materials efficiently capture light due to multiple reflections of the light from the surfaces of neighboring nanounits. Hierarchical materials are easily separated and recycled from reaction media due to the microscale particles (Liu et al. 2013a, b).

10.2.5 Effects of pH and Temperature in BiOX Structure

Effect of pH

Tuning the pH value of the bismuth oxyhalide synthesis reaction enables selective growth of a layered structure with the desired crystal facet exposed, thereby improv-ing the photocatalytic performance of the material. The crystal facet is an important property of crystalline materials as it is associated with certain geometric and electronic structures (Jiang et al. 2012). Therefore, materials with different crystal facets exposed exhibit different intrinsic reactivity and physicochemical properties. Because of this, tailoring semiconductor photocatalysts with preferentially exposed facets has attracted attention as a method for enhancing the photocatalytic properties. A recent study by Wu et al. (2015) showed that it is possible to improve the photocatalytic activity of a BiOBr photocatalyst by adjusting the pH during the synthesis of the material. They showed that the addition of BiOBr with preferentially exposed 010 facets had higher photocatalytic activity than BiOBr with 001 facets preferentially exposed. Pan et al. (2015) demonstrated that bismuth oxyiodide photocatalysts with different exposed facets oxidized the bisphenol A molecule via different mechanisms. They examined the differences in radical formation mecha-nisms between BiOI with more (1 1 0) facets exposed and BiOI with more (0 0 1) facets exposed (BiOI-110 and BiOI-001, respectively). The radical formation mech-anisms were evaluated using EPR experiments with 5,5-dimethyl-1-pyrroline N-oxide (DMPO) as a spin-trap electron to follow the radical species generated in the photodegradation of bisphenol A. They observed that both BiOI-001 and BiOI-110 photocatalysts were able to generate superoxide radical anions, $O_2^{\bullet-}$.

Effect of Temperature

In addition to the pH value, the temperature and reaction time can affect the morphology, size, and exposed surface of bismuth oxyhalides materials. A recent report showed that increasing the hydrothermal temperature from 130 to 280 °C during the solvothermal synthesis of bismuth oxyiodide produced a gradual change in the composition of reaction products from BiOI to α-Bi$_2$O$_3$. Others authors reported that thermal treatment of bismuth oxyhalides materials changed the preferred exposed crystal facet (Jamett et al. 2017).

The microwave-assisted synthesis route is a green technique that improves the properties of synthesized materials by accelerating the chemical reactions under mild conditions. This is achieved via the microwaves efficiently heating target molecules without heating the entire reactor, increasing reaction yields and reducing the reaction times and energy consumption (Chen et al. 2017a, b, c).

The photocatalytic activity of a semiconductor photocatalyst depends on the charge separation between the valence and conduction band. In the photocatalytic process, the generated photoelectrons and photoholes need to be efficiently separated and transferred in opposite directions to improve the photocatalyst efficiency. In this context, there are different alternatives for increasing the photocatalytic efficiency of semiconductor materials, including doping, addition of co-catalysts, and facet engineering (Wu et al. 2017). Vacancies are point defects where an atom is missing from one of the lattice sites.

Oxygen vacancies (OV) play an important role in the photoreactivity of semiconductor photocatalysis materials as these defects could act as capture centers for excited electrons, improving the photo-induced electron and holes. Computational analysis of the density of states revealed that oxygen vacancies in bismuth oxyhalides resulted in the four bismuth atoms bonded to the lost oxygen atom becoming electropositive, acting as capture centers for induced photoelectrons, inhibiting their direct recombination with photoholes (Zhang et al. 2012).

10.3 Improvement of BiOX Performance by Structural Modification

The two most important factors for photocatalysis with semiconductor materials are photon absorption and efficient separation of photogenerated charge carriers. To enhance these properties, many modifications have been applied to BiOX, including doping, forming Bi-rich stoichiometry (Bi$_x$O$_y$X$_z$), addition of co-catalysts, and forming heterojunctions. These modification methods will be discussed briefly here.

10.3.1 Doping

Doping of BiOX materials has already been discussed in detail (Mieghem 1992; Pfeiffer et al. 2003; Walsh et al. 2008). In the synthesis of semiconductors, doping refers to the intentional addition of impurities to a pure semiconductor to change its electronic properties. Introducing impurities has been shown to be effective for modifying the band structure of semiconductors, where it is possible to achieve intermediate energy levels, allowing less energetic photons to induce photoexcitation. Semiconductors contain different energy bands, the valence band (VB) and conduction band (CB). Both bands arise from the overlap of atomic levels of the valence electrons and, according to their degree of occupation, they contain the highest occupied molecular orbital (HOMO) level, and lowest unoccupied molecular orbital (LUMO) level. The energy difference between the VB and CB bands is referred to as the band gap energy (E_g). The introduction of different dopants into the lattice can reduce the HOMO–LUMO gap (Shevlin and Woodley 2010). The energy band diagram illustrated in Fig. 10.4 shows the general mechanisms for the creation of valence band holes and conduction band electrons.

The band model shows that: (a) When impurity atoms with three valence electrons are included in the lattice, p-type semiconductors are formed, where the doping element introduces free holes in the VB. These atoms can be reduced by taking electrons from the VB and increasing the hole density (and decreasing the electron concentration) in the VB (Bent 2008); (b) When impurity atoms with five valence electrons are added (donor atoms), n-type semiconductors are produced, which enhance the conductivity due to the contribution of extra electrons. The excess electrons of donor atoms are located in states near the conduction band, these atoms can be oxidized, transferring electrons to the CB and increasing the electron density in the presence of the donor impurities. Therefore impurity doping introduces a donor band (when the dopant has more valence electrons) or an acceptor

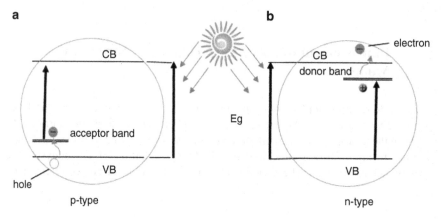

Fig. 10.4 Bands for doped semiconductors. The band model of n-type and p-type semiconductors shows that extra levels have been added by the impurities

Table 10.3 Changes in the band gap energy with increasing atomic ratio of the dopant in n-Type semiconductors

BiOX (X: Cl, Br, I)	Atomic ratio (Å)	E_g (eV)	References
Mn-BiOCl	1,72	2,75	Pare et al. (2011)
Fe-BiOCl	1,7	1,86	Yuan et al. (2017)
Zn-BiOCl	1,53	3,37	(Li, W. T. et al. (2015c)
Sn-BiOCl	1,72	2,91	Xie et al. (2014)
Y-BiOBr	2,27	2,53	He et al. (2015)
Al-BiOBr	1,82	2,75	Liu et al. (2014)
In-BiOI	2	1,71	Li, H. et al. (2017a)
Nd-BiOCl	2,08	3,51	Gnanaprakasam et al. (2017)
Er-BiOI	2,45	1,73	Peng et al. (2018)
Yb-BiOI	2,51	1,53	Zhang, L. et al. (2018a)

band (when it has fewer valence electrons) (Schubert 2005), resulting in higher efficiency of photogenerated charge carrier separation and a decrease in the band gap energy. In several studies, different dopants for bismuth oxyhalide semiconductors have been investigated in order to obtain n-type or p-type behavior (Maeda 2013; Chen et al. 2015; Yu et al. 2016).

Doping with Metals

The existing literature reports n-type semiconductors from the addition of, e.g., Al (Liu et al. 2014), In (Li et al. 2017a), Sn (Xie et al. 2014), Mn (Pare et al. 2011), Fe (Yuan et al. 2017), Y (He et al. 2015), Zn (Li et al. 2015c), Er (Peng et al. 2018), and Yb (Zhang et al. 2018a) to BiOX (Table 10.3).

When doping with d-transition elements, the states of d dopants insert a new band into the original band gap, which shifts to lower energy with increasing atomic number of the dopant (Umebayashi et al. 2002; Xiangchao Meng and Zhang 2016). The enhanced modulation of the energy band by doping BiOCl with transition metals (Fe ((Xia et al. 2013) and Mn (Pare et al. 2011)) has been reported. The band gap energies changed from 3.06 eV for pure BiOBr to 1.86 eV for Fe/BiOCl. The band gap of Fe/BiOCl porous microspheres was smaller than that of pure BiOCl; therefore, the doped Fe/BiOCl could be excited under visible light, resulting in an improvement in the photocatalytic activity for the degradation of RhB and MB under visible light irradiation. In addition, the solvothermal method was used in the presence of ionic liquids (ILs) and yttrium nitrate to produce a Y-BiOBr series of microspheres with different Y contents (0–10 at.%); the band gaps narrowed from 2.63 eV for pure BiOBr to 2.53 eV for Y (5 at.%)-BiOBr (He et al. 2015). The Y-doped BiOBr material showed enhanced photocatalytic behavior for the degradation of ciprofloxacin and Rhodamine B (RhB) compared with that of undoped samples under UV-vis irradiation.

When using metal dopants, localized d states can be inserted into the band gap, resulting in recombination centers for electrons and holes that reduce the photocatalytic efficiency (Xiangchao Meng and Zhang 2016). In addition, very small amounts of dopants (in the parts-per-million range) can dramatically affect the conductivity of semiconductors. Hence, it is necessary to use very pure semiconductor materials and careful doping (considering both the concentration and spatial distribution of impurity atoms) (Bent 2008). A series of Al-doped BiOBr microspheres (Al content of 1, 2, 4, and 6 at.%) was prepared using a solvothermal method (Liu et al. 2014). The enhanced photocatalytic activity of the Al-BiOBr microspheres was attributed to a large BET surface area and efficient separation of photogenerated electron–hole pairs. When the surface of the photocatalyst changes due to differences between the Bi^{3+} and Al^{3+} ions, oxygen vacancies with localized levels below the conduction edge are formed. However, in the case of excessive Al doping, an increase in the band gap energy was observed. This is the so-called Burstein-Moss (BM) effect, which results in a blue shift of the optical band gap (Liu et al. 2013a, b) due to an excess of charge carriers and Fermi level shifts inside the conduction band.

The rare earth metals have unique characteristics not shared by other metal elements; for example, they have f-orbitals that can act as shallow traps for photogenerated electrons and holes (Di et al. 2017). These ions of rare earth metals can extend the spectral response to the visible region of the electromagnetic spectrum (Peng et al. 2018; Zhang et al. 2018a, b), subsequently improving the photocatalytic activity. For example, Yb/Er co-doped BiOCl samples were prepared via a hydrothermal route and used to degrade RhB under visible light (Yu et al. 2016). The addition of the dopant clearly affected the crystal structure, interestingly in this case, due to the size and charge compatibility of Bi^{3+} with that of lanthanide ions (Er^{3+} and Yb^{3+}). After co-doping, x-ray diffraction peaks revealed that the Yb^{3+}/Er^{3+} was incorporated into the BiOCl lattice by substituting for Bi^{3+} with progressive decrease in cell volume. Then, band energy value decreases due to, the introduction of Yb^{3+}/Er^{3+} formed a sub-band to form below the conduction band (He et al. 2018), resulting localized energy levels, which trapped electrons and the recombination can efficiently prevented by Er^{3+} and/or Yb^{3+}-ions in BiOCl lattice. Subsequently, BiOCl sheets co-doped with 2.0% Yb^{3+}/0.5% Er^{3+} (BOC-2.5%) exhibited the highest degradation efficiency for RhB. Similar observations of additional absorption bands and decreased band gap were found for Er-BiOI-doped nanosheets (Peng et al. 2018) and Yb-BiOI-doped microspheres (Zhang et al. 2018a, b). The BM effect has been reported in rare earth metals (Gnanaprakasam et al. 2017); in this case, Nd-doped photocatalysts were prepared using the solvothermal method. In order to compare the photocatalytic behavior, photocatalysts were prepared with different dopant contents (Nd = 1% or 2%) and were compared to pure BiOCl. The 1% Nd-doped BiOCl clearly exhibited better degradation of Brilliant Green (BG) dye than the 2% Nd-doped BiOCl.

Doping with Nonmetals

In addition to doping with metals, doping with nonmetals has been used to modify the band structure of semiconductors. Several nonmetal elements have been successfully doped into BiOX to form semiconductors such as C-BiOCl (Yu et al. 2013a, b), N-BiOCl (Chen et al. 2015), and S-BiOCl (Jiang et al. 2015). When BiOX is doped with nonmetals, the occupied orbitals are introduced into the valence band, resulting in a narrower band gap. Yu et al. (Yu et al. 2013a, b) synthesized flower-like C-doped BiOCl nanostructures using a low-temperature wet-chemical method with polyacrylamide as a chelating agent and C source. The impurity energy level formed by carbon doping appeared above the valence band of BiOCl, inducing the wider light absorption range and effective usage of the light source. The band gap of C-doped BiOCl was 3.12 eV, lower than that of pure BiOCl (3.21 eV). Thus, the photodegradation rate of methyl orange organic dye and phenol was significantly enhanced by C-doping. In addition, a nitrogen-doped BiOCl (N-BiOCl) photocatalyst was synthesized using a hydrothermal method (Chen et al. 2015). N-doping enhanced absorption of visible light by elevating the valence band maximum to narrow the band gap with N 1 s states or form some localized N 1 s states within the band gap. The diffuse reflectance spectra of N-BiOCl samples exhibited an obvious red shift to a longer wavelength of 460 nm compared to the BiOCl sample. The samples that more effectively absorbed visible light showed enhanced degradation of RhB under visible light. Some important advantages of doping with nonmetals are reduced recombination of photogenerated charge carriers and a narrow band gap that induces high photocatalytic activity (Jiang et al. 2015).

10.3.2 $Bi_xO_yX_z$

In Bi-based compounds, increasing the Bi and O contents can affect the CB potential and regulate E_g (Di et al. 2017). Using this principle, a new class of materials known as Bi-rich $Bi_xO_yX_z$ (X: Cl, Br, I) compounds have been demonstrated. Many synthesis methods have been reported to create Bi-rich photocatalyst materials. Table 10.4 shows some of these methods and the most common strategies used to prepare different $Bi_xO_yX_z$ photocatalysts.

As shown in Table 10.4, a wide variety of methods have been used to synthesize Bi-rich structures, along with parallel strategies, such as alkalization, molecular precursors, and calcination (Jin et al. 2017). Solvents are important in material syntheses, where ethylene glycol (EG) and glycerol are the most commonly used solvents for synthesizing $Bi_xO_yX_z$. In addition, the reagents used for alkalization are NaOH and ammonia solutions. It has been clearly demonstrated that the fabrication of different $Bi_xO_yX_z$ microstructures is dependent on the solution pH (Wu et al.

Table 10.4 Synthesis method and strategies for preparing BixOyIz photocatalysts

$Bi_xO_yI_z$	Synthesis method	Strategies			References
		Alkalization	Molecular precursor	Calcination	
$Bi_4O_5I_2$ Bi_5O_7I	Hydrolysis			✓	Huang et al. (2017)
$Bi_4O_5Br_2$	Hydrolysis	✓			Ye et al. (2016a, b)
$Bi_{24}O_{31}Br_{10}$	Precipitation			✓	Shang et al. (2014)
$Bi_4O_5X_2$ (X: Br, I)	Hydrolysis				Bai et al. (2016a, b, c, d)
$Bi_{24}O_{31}Cl_{10}$	Precipitation			✓	Cui et al. (2016)
$Bi_{24}O_{31}Cl_{10}$	Solvothermal			✓	Yin et al. (2017)
Bi_3O_4Br, $Bi_{12}O_{17}Br_2$	Hydrothermal	✓			Li, K. L. et al. (2014c)
$Bi_{24}O_{31}Cl_{10}$	Hydrothermal	✓			Jin et al. (2015)
Bi_3O_4Br	Hydrothermal	✓	.		Wang et al. (2013)
$Bi_{12}O_{17}Cl_2$	Precipitation			✓	Zhao et al. (2016)
$Bi_4O_5Br_2$	Hydrothermal			✓	Zheng et al. (2017)
$Bi_4O_5I_2$	Solvothermal				He et al. (2016)

2018). Another important strategy is calcination, where the high-temperature treatment results in the replacement of the halogen atoms by oxygen. Different degrees of halogenation by calcination have been reported (Huang et al. 2017) when producing $Bi_4O_5I_2$ at 410 °C and Bi_5O_7I at 500 °C by the hydrolysis method. In addition, calcination was used to produce $Bi_{24}O_{31}Br_{10}$ at different temperatures (300, 400, 500, 600, 700, and 800 °C) (Shang et al. 2014), where the obtained material was an excellent candidate for photoreduction of Cr(VI) under UV-vis irradiation. The strategy of using molecular precursors has not been extensively studied (Jin et al. 2017); in addition to the studies shown in Table 10.4, recently, hierarchical structures of $Bi_xO_yX_z$ were produced (Bai et al. 2017; Xiao et al. 2018; Wu et al. 2018) that possessed a narrow band gap and demonstrated good visible light harvesting. In addition, two $Bi_4O_5X_2$ (X = Br and I) photocatalysts were produced using glycerol and a hydrolysis route to achieve nanosheets with dominant {1 0 1} facets that showed high activity for photocatalytic H_2 production (Bai et al. 2016a, b, c, d). In addition, a previous study (Li et al. 2016a, b, c, d) is recognized for incorporating carbon into Bi_3O_4Cl, which enhanced its internal electric field by 126 times, resulting in a bulk charge separation efficiency of 80%.

The study of $Bi_xO_yX_z$ materials is currently highly relevant due to their high selectivity. $Bi_4O_5Br_2$ photocatalysts with different morphologies were reported (Xiao et al. 2018; Zheng et al. 2017), which clearly demonstrated excellent selectivity and conversion (>99%); the microspheres had a short reaction time (2.5 h) and the nanoflakes a longer time of 24 h to oxidize benzyl alcohol into benzaldehyde under blue LED irradiation (12 W). In addition, these materials showed higher stability and a more suitable band structure for photocatalysis compared with pure bismuth oxyhalides (Shang et al. 2014). However, $Bi_xO_yX_z$ materials show the same limitations as other semiconductors, including fast recombination of photoinduced carriers and a low photon absorption efficiency (Jin et al. 2017). In order to overcome these limitations, a $C_3N_4/Bi_4O_5I_2$ nanojunction photocatalyst with high photocatalytic activity for decomposition of RhB and BPA was prepared (Xia et al. 2016), while other $Bi_xO_yX_z$-based hybrid materials have also been reported (Zheng et al. 2018; Chang et al. 2017).

10.3.3 Hybrids

To achieve efficient photoactivity under UV-vis irradiation, some physicochemical properties should be optimized. For example, the semiconductors should have low charge recombination rates, low band gap energies, and more positive VB and more negative CB potentials. However, a single photocatalyst cannot achieve all these factors simultaneously. Hence, other methods to enhance the photocatalytic efficiency of bismuth oxyhalides semiconductor include developing co-catalyst and heterojunction photocatalytic systems (Natarajan et al. 2018).

Co-catalysts

Co-catalysts have played an important role in the evolution of photocatalyst systems. The discovery of new and more efficient co-catalysts has contributed to the understanding of the fundamentals of photochemical reactions, as well as to the technological advance of the field. Generally, co-catalysts can act as active sites for interfacial transfer of photogenerated electrons and promote charge separation, thereby enhancing the photocatalytic activity due to the rapid transfer of electrons. Well-known electron co-catalysts include noble metals (such as Pt, Au, and Ag), transition metal ions (such as Fe(III), Cu(II), and Cr(III)), transition metal oxides (such as NiO, RuO_2, CoO_X, and NiO_X), and transition metal sulfides (such as MoS_2 and CoS_2) (Wang et al. 2017). Co-catalyst materials have been classified into electron-deriving and hole-deriving types (Liqun Ye 2018). An example of a hole-deriving type is a heterogeneous nanostructured photocatalyst prepared by photodeposition (Ye et al. 2013). MnOx–BiOI co-catalysts exhibited higher photoactivity than BiOI for the degradation of aqueous RhB solutions under visible light irradiation. Therefore, facile electron transfer from bismuth oxyhalides to MnOx can greatly decrease electron–hole recombination, increase the lifetime of charge carriers, and finally enhance the photocatalytic efficiency. Related studies of

$Bi_xO_yX_z$ co-catalysts, $MoS_2/Bi_{12}O_{17}C_{12}$ monolayers, have been reported (Li et al. 2016a, b, c, d), showing electron-deriving behavior and good photocatalytic H_2 generation.

Heterojunction Photocatalytic Systems

Heterojunction photocatalytic systems are based on the junction of two or more semiconductors with energetic differences between their VB and CB energies. A classification of these materials was proposed (Du et al. 2017) considering the transfer mechanism of photogenerated charge carriers: heterojunctions and heterojunctions with the Z-scheme. These semiconductors can increase the photocatalytic efficiency via two mechanisms: (i) minimizing recombination of the electron–hole pair due to heterounion formation and (ii) sensitization to the material with the largest E_g by the other photocatalyst as the latter must be able to absorb in the UV-vis region (Xiangchao Meng and Zhang 2016). Heterojunctions can be classified as n–n, p–n, or p–p heterojunctions. To obtain a better understanding on the p–n and p–p heterojunctions and their enhanced visible light photocatalysis mechanism, ZnO and BiOBr were combined to form a heterojunction that effectively separated electron–hole pairs (Geng et al. 2017); this was attributed to the CB electrons from the n-type semiconductor (ZnO) being transferred to that of the p-type semiconductor (BiOBr), while holes moved in the opposite direction. Hence, electrons and holes were accumulated in the p–n type BiOBr/ZnO heterojunctions, reducing recombination and further increasing photocatalytic activity. Another study (Yosefi and Haghighi 2018) synthesized p–p heterojunctions of BiOI/NiOI; when the BiOI/NiO is exposed to visible light, only BiOI can be excited due to its narrow E_g (1.86 eV), the VB electrons in BiOI are easily excited due to the E_g ($\lambda >$ 400 nm). Thus, the CB of BiOI changes its potential to a position more negative than that of NiO, and the CB electrons of BiOI can easily migrate to the NiO CB via the interface, resulting in efficient separation of photogenerated carriers, enhanced efficiency, and increased photocatalytic activity. The electrons accumulated in the NiO CB can reduce O_2 to form the O_2^- reactive species. The O_2^- oxidants degrade acid orange to form CO_2 and H_2O. Both n–n and p–n heterojunctions can improve the separation efficiency of photogenerated electron–hole pairs and increase the lifetime of photogenerated carriers (Xiangchao Meng and Zhang 2016). To form n–n and p–n heterojunctions, many different compounds have been coupled with BiOX (X:Cl, Br, I), including BiOBr/Co–Ni (Ao et al. 2015), BiOI–zeolite (Zhao et al. 2014), Bi-OI/BiOCl (Sun et al. 2015), TiO/BiOBr (Wang et al. 2015a, b), and Ag-based compounds AgI/BiOI (Cheng et al. 2013) (Ning et al. 2017). Some examples are shown in Table 10.5.

Table 10.5 Summary of heterojunction synthesis methods and experimental conditions to test photocatalytic activity

Heterojunction	Type	Synthesis method	Operating conditions	References
p-BiOI/p-NiO	p–p	Solvothermal-precipitation $Ni(NO_3)_3*6H_2O$ NiO NaOH Bi $(NO_3)_3*5H_2O$ KI	Substrate: acid orange 7; 400 W halogen lamp (OSRAM, Germany)	Yosefi and Haghighi (2018)
BiOBr/ZnO	p–n	Hydrothermal $Zn(NO_3)_2*6H_2O$ KOH KBr $Bi(NO_3)_3$ $Bi(NO_3)_3*5H_2O$	Substrate: methyl blue; 300 W iodine-wolfram lamp	Geng et al. (2017)
BiOI@$Bi_{12}O_{17}Cl_2$	p–n	Hydrothermal $Bi_{12}O_{17}Cl_2$ In situ deposition Bi $(NO_3)_3*5H_2O$ KCl EG	Substrate: Rhodamine B, 2,4-dichlorophenol phenol, bisphenol A and tetracycline hydrochloride; 500 W Xe lamp, $\lambda > 420$ nm)	Huang et al. (2016)
BiOBr/CeO_2	p–n	Deposition–precipitation ethylene glycol Ce $(NO_3)_3 \cdot 6H_2O$ $C_6H_8O_7 \cdot H_2O$ Bi $(NO_3)_3 \cdot 5H_2O$ KBr	Substrate: RhB, methyl blue and phenol; 300 W Xe lamp (CEL-HXF300) $\lambda > 420$ nm	Wen et al. (2017)
g-C_3N_4/BiOBr/ Au-B g-C_3N_4/BiOBr/ Au-S	Z-scheme	Precipitation g-C_3N_4 BiOBr $Bi(NO_3)_3*5H_2O$ $AuCl_3$ KIO_3 $AgNO_3$	Substrate: RhB CO_2; 300 W Xe lamp ($\lambda = 380$ and 500 nm) (PLS-SXE300C)	Bai et al. (2016a, b, c, d)
BiOI/rGO/Bi_2S_3	Z-scheme	Electrostatic self-assembly Bi_2S_3 BiOI NH_4I EG Reduced graphene oxide (rGO)	Substrate: Cr(VI) and phenol; 500 W Xe lamp (400 nm cutoff filter)	Chen, A. et al. (2017a)

Heterojunctions with the Z-Scheme

The first-generation Z-scheme photocatalytic system was introduced by Bard in 1979 (Low 1979) and consists of an electron mediator, a photo-oxidizing agent, and a photo-reducing agent, which show a strong redox performance to simultaneously achieve reduction and oxidation reactions (Natarajan et al. 2018). In a Z-scheme system, oxidation and reduction reactions occur separately in two different photocatalysts to avoid recombination of photoinduced electron–hole pairs. Typically, the materials with a Z-scheme system have an enhanced range of light absorption and provide many selective reaction sites for oxidation and reduction processes (Zhou et al. 2014). However, the development of direct Z-scheme photocatalysts is still at an early stage. Table 10.5 shows a summary of various heterojunctions and synthesis processes.

Briefly, the photocatalytic activity of the photocatalysts is related to the composition of the CB and VB in the material as any modifications to the band structures of BiOX materials or their hybrids will have implications on the visible light-driven and photocatalytic activity. When doping with elements, the improved photocatalytic activity is mostly due to the enhanced separation of electric charges, efficiently achieved through network arrangements with the appropriate dopants, or to the redesign of the band gap (formation of a new energy level in the band gap); however, the dopant concentration has an upper limit. The synthesis routes most commonly employed to obtain doped bismuth oxyhalides or hybrids structures are the same as those used to obtain pure oxyhalides (solvothermal, hydrothermal, and coprecipitation). The modifications applied to BiOX (doping, Bi-rich materials, co-catalysts, and heterojunctions) have shown to be effective for increasing photoactivity. However, there are still many challenges; for example, the utilization of direct sunlight remains low and it is a major limitation for the development of practical sunlight-based applications. In the next section, we will discuss various applications for these bismuth-based materials.

10.4 Applications of BiOX-Based Materials

The use of solar radiation is considered as promising solution for solving various environmental problems. To achieve this, it is important to produce nanomaterials that allow efficient harvesting of solar irradiation. In this chapter, we will discuss the mechanisms by which BiOX-based materials are activated by UV and visible radiation, have high thermal stability, and their band gap can be easily tuned by changing the synthesis conditions (Jing et al. 2013; Li et al. 2015b; Lu et al. 2016; Xiong et al. 2016) or by synthesizing hybrid materials or doping the BiOX structure (Zhang and Zhang 2010; Cheng et al. 2014; Xiong et al. 2016). As the major photocatalyst, BiOX-based materials have several environmental applications which can be divided into three major areas: environmental remediation, energy production, and green chemistry (Fig. 10.5).

10.4.1 Environmental Remediation

Human activity has caused severe disturbance to the natural environment by contaminating the air, soil, and water to levels that are toxic for both humans and other

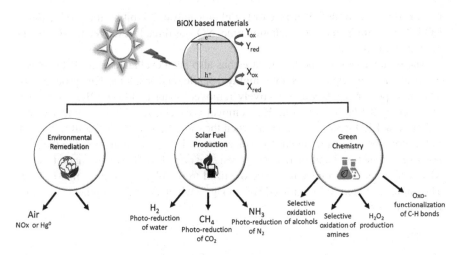

Fig. 10.5 Environmental applications of BiOX-based materials

species (Khin et al. 2012). Thus, different technologies have been developed to remediate the environment (Finnegan et al. 2018); among these, photocatalysis with photocatalysts such as BiOX is considered advantageous due to its simplicity of operation, low cost, and versatility. Remediation with BiOX-based materials has been performed in both aqueous and gas systems.

Air Remediation

In gas systems, air remediation by photocatalysis with BiOX has been studied. There are several reports of BiOX photocatalysts being used to remove nitrogen oxides (NO_x), which are common contaminants produced by fuel combustion in automobiles. For NOx removal, it is important that photoproduction of reactive oxygen species (e.g., $O_2^{\bullet-}$ and $\bullet OH$) occurs. In a comparative study of BiOX photoactivity to remove NO under visible light ($\lambda > 420$ nm) performed by Zhang et al. (2013a, b), it was observed that BiOBr showed the best photoremediation properties. Interestingly, BiOI showed poor NO removal (ten times lower than that of BiOBr), even though it has a band gap with the best visible light absorption; this was attributed to an increase in the VB potential and the narrow band gap. To improve NO removal, Liao et al. (2018) synthesized BiOBr nanoplates with oxygen vacancies that facilitated an increase in the adsorption and activation of O_2 to produce ROS. The addition of a photosensitizer as RhB by adsorption over hierarchical BiOCl spheres with OVs was performed by Li et al. (2014a, b, c), which enabled the BiOCl photocatalyst to remove NO by ROS produced under visible light. The BiOI hollow microspheres synthetized by Dong et al. (2015) showed particular behavior; the initial nonselective NO removal under visible light became selective to NO_2 production. According to the authors, the initial NO_3^- produced by the oxidation of NO remained adsorbed on the surface of the photocatalyst, inhibiting the production of ROS. Thus, with increasing irradiation time, NO was directly oxidized to NO_2 by the photogenerated holes.

The removal of other polluting gases has also been evaluated using BiOX-based materials. For example, the stable, highly volatile, and insoluble Hg^0 produced by coal-fired power plants was effectively removed using a $BiOI/BiOIO_3$ composite (Zhou et al. 2017a, b); near total Hg^0 removal was achieved (98.53%) under visible light (24 W LED lamp, $\lambda = 420$ nm) when the molar ratio of $BiOI:BiOIO_3$ was 3:1. In this composite, the photogenerated electrons in BiOI were transferred to the more positive conduction band of $BiOIO_3$. Then, the adsorbed O_2 was reduced to $O_2^{\bullet-}$. Meanwhile, the H_2O molecules were oxidized to •OH by the hole formed in BiOI. The photogenerated ROS oxidized Hg^0 to Hg^{2+}, which can be solubilized to remove it from the air.

Water Remediation

Water is vital for all types of life. However, human activity generates waste streams that are constantly discharged into all water sources in high quantities. Specifically, the dumping of chemicals, such as dyes from the textile industry, pesticides and fertilizers from the agricultural industry, heavy metals from mining and industrial activities, and a wide range of pharmaceutical compounds (Hernández-Ramírez and Medina-Ramírez 2015).

Dyes

As they are easily detected, the degradation of dyes is one of the most commonly used probes of BiOX photocatalytic activity (Choi et al. 2015; Yoon et al. 2015; Zhang et al. 2015b; Tian et al. 2016; Xu et al. 2017a; Alansi et al. 2018; Li et al. 2018a; Peng et al. 2018; Wang et al. 2018a; Yosefi and Haghighi 2018). However, the photocatalytic activity of this material depends on the photocatalyst morphology, type of dye, and irradiation conditions. For example, using the same dye (RhB), nanosheets prepared by An et al. (2008) showed a reactivity of BiOCl > BiOBr and BiOI showing no activity. However, years later, hierarchical flower-like BiOX synthesized by Chen et al. (2013) showed a different order of reactivity: BiOBr > BiOI > BiOCl. This difference in the BiOCl photoactivity seems to be due to the good coupling between the adsorbed RhB LUMO with the conduction band of the BiOCl when the structure is in the form of nanosheets. Hence, the dye acted as photosensitizer and was oxidized when the photo-excited electrons were transferred to the conduction band of BiOCl.

Pharmaceuticals

The removal of pharmaceuticals in water is extremely important as they represent a risk to human health and other species as they are pharmacologically active, highly persistent in aqueous solution, and can sometimes bioaccumulate (Rivera-Utrilla et al. 2013). Several pharmaceuticals have been successfully removed with BiOX-based materials. The photooxidation of carbamazepine (Gao et al. 2015, 2018; Meribout et al. 2016; Chen et al. 2017b), tetracycline (Di et al. 2016), ciprofloxacin (Di et al. 2016; Su and Wu 2018), sulfadiazine (Li et al. 2017c), and ibuprofen (Arthur et al. 2018) was achieved using simulated sunlight. Specifically, the mineralization of ibuprofen with BiOBr microspheres (Tian et al. 2014) reached nearly 60% or 12 mg·L^{-1}. Moreover, Li et al. (Li et al. 2016a, b, c, d) found a selective interaction between BiOBr and ibuprofen, which increased the adsorption of ibuprofen due to the formation of a surface complex between carboxylic groups in the ibuprofen and the Bi in the photocatalyst. Complete mineralization of ampicillin and oxytetracycline was reported in a system irradiated with real sunlight using BiOCl/Bi$_2$O$_3$ composite photocatalysts (Priya et al. 2016) which improve the photooxidation ability of nanoplate composites by supporting the photocatalyst on graphene sand or chitosan. A synergic effect in the photodegradation was observed as the antibiotic was adsorbed on the support and simultaneously photooxidized by the •OH formed during photocatalysis.

Heavy Metals

Among the heavy metals currently detected in water, removal of Cr(VI) with BiOX has been widely studied as it can cause cancer and other diseases; levels of this contaminant in natural water supplies are increasing due to industrial discharges (Owlad et al. 2009). Considering the reports of Cr(VI) removal, it is important to distinguish works focused on removal by adsorption (Li et al. 2012; Wang et al. 2014) and photoreduction (Fan et al. 2016). All BiOX materials prepared by Li et al. (2013) showed Cr(VI) removal ability, the efficiency of which depended on the morphology (related to the BET surface area and isoelectric point). Flower-like nanostructures of BiOBr with the highest isoelectric point (2.6) showed the highest Cr(VI) removal and best removal capacity (63.5 mg·g^{-1}). The photoreduction of Cr(VI) can be successfully achieved by the photogenerated holes in BiOX-based photocatalysts (Wang et al. 2016). However, a decrease in the photocatalyst activity at neutral or basic pH was observed (Han et al. 2015; Xu, H. et al. 2017b). This change in the photoactivity was related to the ability of the Cr(VI) to be adsorbed on the photocatalyst surface.

Table 10.6 Photocatalysts used for water disinfection and the major reactive species responsible for bacterial inactivation, as proposed by the authors

Photocatalyst	Bacteria	Light source	Main reactive species	References
Ag/BiOI	*E. coli* 8099	150 W Xe lamp with filter (λ>420 nm)	•OH	Zhu et al. (2012)
CuI-BiOI/Cu	E. coli K-12	300 W Xe lamp with filter (λ>400 nm)	h^+	Zhang, Y. et al. (2018b)
Fe_3O_4@BiOI@AgI	E. coli ATCC15597	300 W Xe lamp with filter (λ>400 nm)	e^- and $O_2^{\bullet-}$	Liang et al. (2018)
Ag@AgI/Bi-BiOI	E. coli K-12	300 W Xe lamp with filter (λ>420 nm)	h^+, e^- and $O_2^{\bullet-}$	Ning et al. (2017)

Water Disinfection

Water pollution can be caused by organic and inorganic substances and by pathogenic microorganisms. The removal of *E. coli* (disinfection) with BiOX has been mainly studied using hybrid materials as better yields of bacterial inactivation were achieved by modifying the photocatalyst structure (Wu et al. 2016). Unlike the removal of other substances, there is not an exclusive reactive species responsible for the disinfection (Table 10.6), which could be due to the different hybrid microstructures, and the complexity of the substrate being degraded.

10.4.2 Energy Generation by Solar Fuel Production

H_2 Production

Water splitting to form O_2 and H_2 using photocatalysts is a kind of artificial photosynthesis. The water molecule is oxidized to produce O_2, while simultaneously being reduced to produce H_2 that can be used as fuel (Bard and Fox 1995). The production of H_2 fuels by photocatalysis using sunlight is an interesting challenge and a sustainable solution that could solve current energy problems (Yu et al. 2013a, b). To achieve water splitting by one-step photoexcitation, the photocatalyst must satisfy two conditions: (1) to obtain H_2 from the photoreduction, the CB energy needs to be more negative than the reduction potential of H_2O ($E°H^+/H_2 = 0.0$ V at pH $= 0$); and (2) the VB energy must be more positive than the oxidation potential of H_2O ($E°$ $H_2O/O_2 = +1.23$ V at pH $= 0$) to photooxidize H_2O (Natarajan et al. 2015). Thus, water splitting thermodynamically requires photons with wavelengths less than 1000 nm (Gratzel 1981), corresponding to a photocatalyst band gap of 1.23 eV (Fig. 10.6). However, the real required band gap is 3.21 eV (much higher than

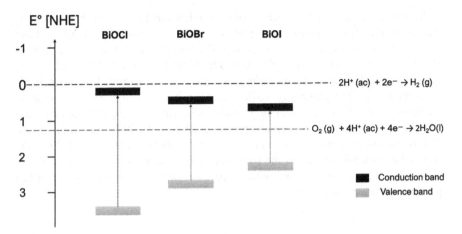

Fig. 10.6 Relative band edge positions of BiOX photocatalysts compared to the water splitting oxidation and reduction reactions

1.23 eV), due to the high overpotentials related to the redox reaction between water molecules and the charge carriers in the photocatalyst (Wang et al. 2018b).

In general, BiOX energy bands show electrons in the CB with low ability to produce H_2, but holes in the CB that can effectively oxidize H_2O and produce O_2 (Meng and Zhang 2016). However, it is possible to find several reports of photoreduction of H_2O using modified BiOX photocatalysts, including heterojunctions of p–n photocatalysts, such as BiOBr/α-Fe_2O_3 (Si et al. 2017) and BiOBr with Cu_2S quantum dots (Wang et al. 2015a, b); under 300 W Xe lamp irradiation with UV filters, these composites showed a photocatalytic activity that reached 16.08 $\mu mol \cdot h^{-1} \cdot g^{-1}$ H_2 and 239 $\mu mol \cdot h^{-1} \cdot g^{-1}$ H_2, respectively. In both systems, silver nanoparticles were required to facilitate the reduction reaction that produces H_2, and sacrificial reagents (S^{2-} and SO_3^{2-}) to scavenge the holes and avoid charge carrier recombination. Ternary structures, such as BiOI-CdS/MoS_2 composites (Zhou, C. et al. 2017a), show good photocatalytic activity. These types of structures provide more reactive sites for H_2 evolution and enhance charge transfer and, hence, charge recombination.

In order to achieve direct photoreduction of water, the photocatalytic activity of pure BiOX was enhanced by increasing the internal electric field (by reducing the thickness) and adding defects; for example, black ultrathin BiOCl nanosheets with oxygen vacancies (Ye et al. 2015). In the same year, Zhang et al. (2015a) synthetized BiOCl ultrathin nanosheets with surface/subsurface vacancies. The Bi and O defects resulted in band gap narrowing and an upshift of both the CB and VB; specifically, the O vacancies catalyzed the H_2 evolution reaction (Li et al. 2018a, b). The O vacancies promote the adsorption of water molecules, reducing the O–H bond energy barrier and reducing the sum of the recombination and desorption energies of H_2.

Recently, Lee et al. (2018) successfully synthesized hierarchical BiOX flower-like microspheres that achieved complete water splitting under 350 W Xenon lamp radiation with a cutoff filter for $\lambda < 400$ nm. The best photocatalytic activity (highest H_2 production) was observed at pH 7 with a BiOI catalyst dosage of 0.20 g·L^{-1}. The high photocatalytic activity of BiOI was related to its narrow E_g with suitable absorption edge for harvesting visible light, in addition to low charge-transfer recombination. The photoproduction of H_2 using BiOX-based materials can also be achieved by photoelectrochemical processes (Fan et al. 2017) or by coupling the H_2O photoreduction with the oxidation of organic pollutants (Zhang et al. 2013a, b) where the contaminants act as the sacrificial electron donors. The hybridization of these types of processes enable reduced cost, increased efficiency, and simplified processes (Park et al. 2009).

Photoreduction of CO_2

The transformation of CO_2 into carbon fuel sources, such as CO, CH_4, or CH_3OH, could be a solution for the energy crisis and the environmental problems related to the greenhouse effect. The photoreduction of the molecules involves a multielectron process, as is illustrated in Eqs. 10.1, 10.2, 10.3, 10.4 and 10.5 to the CO_2 reduction (Kubacka et al. 2012), so the photocatalyst must provide high numbers of catalytic sites and generate enough electron–hole pairs with long lifetimes in order to successfully photoreduce CO_2 (Li, K. et al. 2014b).

$$CO_2 + 2e^- + 2H^+ \rightarrow CO + H_2O \qquad E^\circ = -0.77 \text{ V} \qquad (10.1)$$

$$CO_2 + 6e^- + 6H^+ \rightarrow CH_3OH \qquad E^\circ = -0.62 \text{ V} \qquad (10.2)$$

$$CO_2 + 8e^- + 8H^+ \rightarrow CH_4 \qquad E^\circ = -0.48 \text{ V} \qquad (10.3)$$

$$CO_2 + 2e^- + 2H^+ \rightarrow HCOOH \qquad E^\circ = -0.85 \text{ V} \qquad (10.4)$$

$$CO_2 + 4e^- + 2H^+ \rightarrow HCOH \qquad E^\circ = -0.72 \text{ V} \qquad (10.5)$$

The photoreduction of CO_2 by BiOX with different halides, morphologies, and modifications showed significant difference in the photocatalytic activity (evaluated via the production of CO and/or CH_4). Although most studies of CO_2 photoreduction used Xenon lamps (300 W), UV cutoff filters were not always used. Thus, it was difficult to compare the photocatalytic activity of the different BiOX samples. Moreover, diverse heterogeneous systems (solid/liquid and solid/gas) were used. However, the main goal is to tailor the CB minimum, rather than the reduction potential of CO_2. One of the first reports of BiOI flower-like photocatalysts with ability of reduce CO_2 to CH_4 under simulated solar light (Xenon lamp) was published by Zhang et al. in 2014 (Zhang et al. 2014). In their gas/solid system, the CH_4 production was ~800 µmol·g^{-1} after 4 h. An ultrathin nanosheet of BiOCl

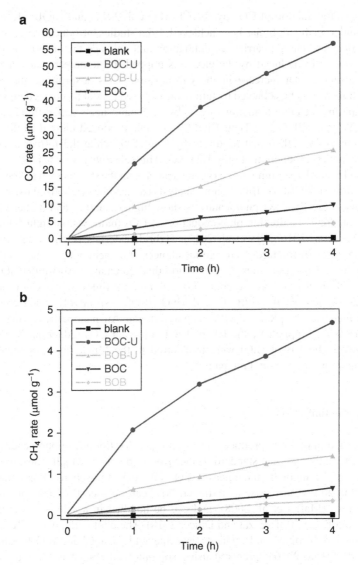

Fig. 10.7 Yields of CO and CH$_4$ achieved by photoreduction of CO$_2$ with ultrathin nanosheet of BiOCl (BOC) and BiOBr (BOB). (Reproduced from Ding et al. (2017) with permission from Elsevier)

and BiOBr synthetized by Ding et al. (Ding et al. 2017), with a high electron–hole separation efficiency, simultaneously produced CO and CH$_4$. The best yields in water were observed for ultrathin BiOCl (more than 55 µmol·g^{-1} of CO and 4.5 µmol·g^{-1} of CH$_4$) as this photocatalyst has a CB at −1.20 V, which is higher than the potential of ultrathin BiOBr (−0.60 V) (Fig. 10.7).

The photoreduction of CO_2 by BiOCl (Ma et al. 2017) and BiOBr (Kong et al. 2016) under visible light has been achieved. Both studies introduced oxygen vacancies to decrease charge carrier recombination and enhance the adsorption of CO_2, which was easily reduced by the electrons trapped in the defect states created by these vacancies. An increase in the CB position, CO_2 adsorption, and efficient charge transfer can be achieved by tuning the exposed crystal facets, as demonstrated by comparing the photoreduction of CO_2 by BiOI nanosheets with different exposed facets (Ye et al. 2016a, b). Their BiOI-001 sample produced more than three times the amount of CO than that produced by BiOI-100, while the former showed an increase in CH_4 production of only 1.2 times. This selectivity for CO production was attributed to the large number of oxygen vacancies on the 001 facet, which increased the stability of BiOI-001. BiOX photocatalysts can also be synthetized using Bi-rich materials; these $Bi_xO_yX_z$ photocatalysts show higher stability and higher CB minimum potentials, facilitating photoreduction of CO_2 under visible light (Ding et al. 2016; Ye et al. 2016b; Jin et al. 2017). Hybrid materials, such as co-catalyst or Z-scheme heterostructures, are good choices for achieving better yields of photoreduced CO_2. For example, the hybrid dual-co-catalyst $Au/BiOI/MnO_x$ material (Bai et al. 2016a) showed about 7.0, 2.8, and 5.9 times higher CO production than that of pure BiOI, Au/BiOI, and $MnO_x/BiOI$, respectively. Each co-catalyst helped increase the photocatalytic activity. In the Au nanoparticles, the excited electrons were supplied by the CB of BiOI, allowing CO_2 reduction. Meanwhile, the holes in the VB of BiOI were transferred to the MnO_x layer, preventing the recombination of electron–hole pairs.

Photoreduction of N_2

Photoreduction of N_2 to produce NH_3 or nitrogen fixation using photocatalysts is a green alternative to the standard Haber-Bosch process, which consumes large amounts of fossil fuels and releases CO_2 into the atmosphere (Kandemir et al. 2013). In addition to its use as fertilizer, NH_3 can be used for stationary fuel cell applications (Maffei et al. 2008), and can be handled in a safer manner than H_2 (Reiter and Kong 2011; Ezzat and Dincer 2018). The photoreduction of N_2 requires the activation of triple bonds, which was achieved by BiOCl and BiOBr nanosheets (Li et al. 2015a, 2016a) prepared using solvothermal synthesis to create oxygen vacancies. The photocatalytic activity of BiOCl was facet-dependent, showing different mechanisms for N_2 reduction due to changes in the substrate interaction with the oxygen vacancies (Fig. 10.8). Under solar light (500 W), the best rate of NH_3 production was 1.19 $\mu mol \cdot h^{-1}$ for BiOCl-001. The photoreduction of N_2 by BiOBr-001 nanosheets showed a low-energy barrier that enabled a rate of 104.2 $mmol \cdot h^{-1} \cdot g^{-1}$ (without use of co-catalyst or scavengers) under visible light (300 W Xenon lamp with 420 nm cutoff filter). In the case of NH_3 production using BiOX photocatalysts, the oxygen vacancies mimic the role of FeMo-cofactor in nitrogenase, promoting N_2 adsorption and the activation of the N_2 bond, indicated by an increase in the bond length (Li et al. 2017b). A similar facet dependence was

Fig. 10.8 Adsorption of N_2 in BiOCl depending on the facet and reduction mechanism. (Modified from Li et al. (2016a, b, c, d))

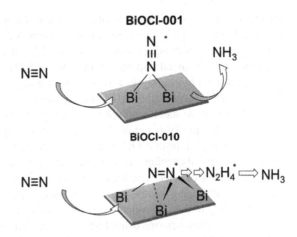

BiOCl-001

BiOCl-010

observed by Bai et al. (2016a) who synthetized Bi-rich nanosheets of BiOX that showed the best N_2 reduction using Bi_5O_7I-001 with a rate of 195.2 $\mu mol \cdot L^{-1}$ using a 300 W Xenon lamp and CH_3OH as a sacrificial reagent.

10.4.3   Green Chemistry: Obtaining Value-Added Products

As photocatalytic activity of BiOX-based materials is easily tuned to increase yields or promote selectivity, photocatalysis using such materials represents an alternative green process to obtain chemical products in an environmentally friendly manner. The production of solar fuels by photoreduction, detailed in the previous section, is an example of a green process. Selective oxidation with BiOX has been reported for many years; however, a green chemistry focus has been applied only in the last few years. In 1989, BiOX-based catalysts were employed to oxidize CH_4 at high temperatures (Thomas et al. 1989). From all oxidized CH_4, 71.1% was selective C2 products, obtaining 4.3 times more ethene than ethane. In the same year, Williams et al. (1989) compared the selectivity in the CH_4 oxidation at 700°C by BiOCl with the selectivity observed with isostructural photocatalyst where bismuth ions were replaced by samarium (Sm) or lanthanum (La). Despite being less thermally stable, BiOCl showed the best C2 selectivity (19.4%) with a ratio of $C_2H_4/C_2H_6 = 56.1$. Although neither of these studies considered the photocatalytic properties of BiOCl, they were the first reports of selectivity associated with this catalyst.

Selective Oxidation of Alcohols

Due to their importance in organic synthesis (Davis et al. 2013), the selective photocatalytic oxidation of alcohol to aldehyde under mild conditions has been widely studied. For BiOX photocatalysis, the oxidation of benzyl alcohol (BA) to benzaldehyde (BAD) is the most common controlled reaction. The strategies have been focused on increasing the conversion of BA, for which BiOBr and BiOCl have been modified by metal doping. An increase of 8.1% in the conversion of BA was observed when photocatalysis was performed with H_2O_2 and BiOBr modified with Fe(III), mainly because the Fe(III) ions promoted charge separation (Yuan et al. 2017). When noble metal nanoparticles (Au and Pd) were deposited on BiOCl-001 nanosheets rich in OV (Li et al. 2018a), the improved charge separation due to the vacancies in the ultrathin BiOCl nanosheet resulted in selective reaction of BA as the metal facilitates adsorption and activation of the alcohol. The conversion of BA was 99% with >99% of selectivity for BAD after 8 h of reaction with Pd-BiOCl ultrathin nanosheets and using visible light. Additionally, Pd-BiOCl shows high efficiency and selectivity in the controlled oxidation of other aromatic alcohols. Better results have been recently reported using Pt-BiOCl nanocomposites (Liu et al. 2018), which achieved 99.6% conversion at almost 100% selectivity to BAD. The conversion of other alcohols was also effectively increased after 5 h of reaction.

Changes in the Bi:O:X ratios can produce photoactive materials with VB potentials to control the oxidation. Thus, Bi-rich photocatalysts have shown a high selectivity (>99%) under visible light at room temperature (Xiao et al. 2013; Liu et al. 2016). However, only $Bi_4O_5Br_2$ nanoflakes (Zheng et al. 2017) and $Bi_{24}O_{31}Br_{10}$ flower-like particles (Xiao et al. 2018) reached conversions of BA to BAD of >99% under blue LED irradiation (~450 nm) after 8 and 2.5 h, respectively. The selective photocatalytic conversion of aliphatic alcohols was studied by Xie et al. (Xie et al. 2018) with a full BiOX series under 300 W Xenon lamp radiation and an acetonitrile solution. BiOCl showed the best photoactivity, reaching a conversion of >90% with 1-pentanol and 1-hexanol. Conversions using BiOBr were <30%, while BiOI did not show oxidative ability. All oxidations of the alcohols showed selectivity values >99%.

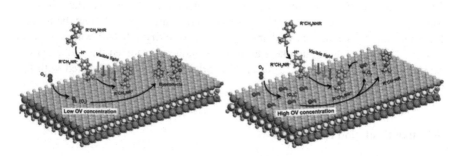

Fig. 10.9 Amine to imine oxidation by BiOCl depending on oxygen vacancies and the crystal facet. (Reproduced from Mao et al. (2018) with permission from Elsevier)

Oxidation of Amines to Imines

In a similar process, the selective oxidation of amines to imines has been reported under visible light with BiOCl-001 with OVs (Mao et al. 2018) (Fig. 10.9). The preferential exposed facet and high amounts of OVs promoted the conversion and selectivity of amine oxidation as the "side-on" adsorption of O_2 was facilitated. This difference in the adsorption resulted in the production of H_2O_2 by two-electron reduction, rather than $O_2^{\bullet-}$ by single-electron reduction.

H_2O_2 Production

A green process to produce H_2O_2 was evaluated in an aqueous solution (Su et al. 2018) using BiOCl nanoplates and HCOOH under 500 W Xenon radiation (Fig. 10.10). The H_2O_2 was produced by the reaction of 2•OH. These radicals were indirectly photogenerated by HCOOH, which acted as a charge carrier. The H_2O_2 evolution was 685 μmol·h^{-1}, which increased to 1020 μmol·h^{-1} when O_2 was bubbled in the system. This was attributed to further oxidation of CO_2 being effectively avoided by the presence of O_2.

Selective C-H Bond Activation

The oxo-functionalization of C–H bonds in cyclohexane was studied by Henriquez et al. (2017) using BiOX photocatalysts under irradiation from a 400 W metal halide lamp to simulate sunlight. After 3 h, the cyclohexanol yields were linearly related to the photocatalyst band gap ($r = -0.94$). BiOI showed the best selectivity to cyclohexanol, which was 4.3 times higher than the cyclohexanone production.

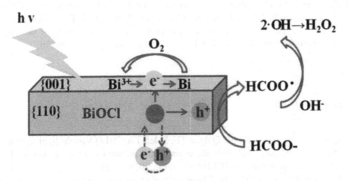

Fig. 10.10 Scheme of H_2O_2 photoproduction by BiOCl nanoplates in the presence of HCOOH. (Reproduced from Su et al. (2018) with permission from Elsevier)

10.5 Challenges in BiOX Research

BiOX-based materials have been intensively studied over the last decades, mainly due to their activation by visible light, which is an advantage for the growing trend of photocatalysis using sunlight, particularly compared with common materials such as TiO_2. The main challenges in relation to photocatalytic applications of bismuth oxyhalides are the optimization of the photocatalytic process at the laboratory level before scaling up this technology to accomplish *green chemistry* principles. Future approaches may be focused on the orientation of BiOX crystal growth to tune the surface properties of the BiOX materials. Thus, the polarity, the amount of surface defects in the form of oxygen vacancies, can be modified. Consequently, the adsorption–desorption process over the BiOX could be modulated to obtain desired oxidation products.

Despite the body of work published regarding synthesis and environmental applications of BiOX, there have not been systematic studies comparing the photocatalytic properties and reactivity of these materials. Hence, to select the best photocatalyst for a specific application, researchers should use standardized experimental conditions to facilitate efforts related to applications at larger scales. To the best of our knowledge, most experimental procedures were performed under different conditions, making it very difficult to compare the results. Moreover, experiments with natural sunlight at ambient conditions, where the radiation and temperature vary, are necessary to explore real applications using solar radiation.

Ensuring that the BiOX materials can be recuperated after the photocatalytic process and minimizing bismuth leaching are other challenges that need to be addressed in order to take this technology from academia to industry. Some potential strategies have been developed, such as magnetic-hybrid photocatalysts. However, this modification affects the photoreactivity, restricting possible applications. Thus, immobilizing the photocatalyst on a support should be seriously considered by researchers.

Acknowledgements All the authors acknowledge to FONDECYT n°1160100. V. Melin acknowledges the support given by FONDECYT postdoctoral n°3180202. G. Pérez-González thanks the fellowships given by Faculty of Chemistry and postgraduate department of the University of Concepcion. L. González thanks CONICYT – doctoral fellowship n°21170226.

References

Alansi AM, Al-qunaibit M, Alade IO, Qahtan TF, Saleh TA (2018) Visible-light responsive BiOBr nanoparticles loaded on reduced graphene oxide for photocatalytic degradation of dye. J Mol Liq 253:297–304. https://doi.org/10.1016/j.molliq.2018.01.034

An H, Du Y, Wang T, Wang C, Hao W, Zhang J (2008) Photocatalytic properties of BiOX (X = Cl , Br , and I). Rare Metals 27(3):243–250. https://doi.org/10.1016/S1001-0521(08)60123-0

Anastas P, Eghbali N (2009) Green chemistry: principles and practice. Chem Soc Rev 39 (1):301–312. https://doi.org/10.1039/b918763b

Andreozzi R, Marotta R (2004) Removal of benzoic acid in aqueous solution by Fe (III) homogeneous photocatalysis. Water Res 38(5):1225–1236. https://doi.org/10.1016/j.watres.2003.11.020

Ao Y, Wang D, Wang P, Wang C, Hou J, Quian J (2015) A BiOBr/Co-Ni layered double hydroxide nanocomposite with excellent adsorption and photocatalytic properties. RSC Adv 5 (67):54613–54621. https://doi.org/10.1039/c5ra05473g

Armand M, Endres F, MacFarlane D, Ohno H, Scrosati B (2009) Ionic-liquid materials for the electrochemical challenges of the future. Nat Mater 8:621–629. https://doi.org/10.1038/nmat2448

Arthur RB, Bonin JL, Ardill LP, Rourk EJ, Patterson HH, Stemmler EA (2018) Photocatalytic degradation of ibuprofen over BiOCl nanosheets with identification of intermediates. J Hazard Mater 358:1–9. https://doi.org/10.1016/j.jhazmat.2018.06.018

Bai Y, Ye L, Wang L, Shi X, Wang P, Bai W (2016a) A dual-cocatalyst-loaded Au/BiOI/MnO$_x$ system for enhanced photocatalytic greenhouse gas conversion into solar fuels. Environ Sci Nano 4:902–909. https://doi.org/10.1039/C6EN00139D

Bai Y, Chen T, Wang P, Wang L, Ye L (2016b) Bismuth-rich Bi4O5X2(X = Br, and I) nanosheets with dominant {1 0 1} facets exposure for photocatalytic H2evolution. Chem Eng J 304:454–460. https://doi.org/10.1016/j.cej.2016.06.100

Bai Y, Ye L, Chen T, Wang L, Shi X, Zhang X, Chen D (2016c) Facet-dependent photocatalytic N$_2$ fixation of bismuth-rich Bi$_5$O$_7$I nanosheets. ACS Appl Mater Interfaces 8:27661–27668. https://doi.org/10.1021/acsami.6b08129

Bai Y, Chen T, Wang P, Wang L, Ye L, Shi X, Bai W (2016d) Size-dependent role of gold in g-C3N4/BiOBr/Au system for photocatalytic CO2reduction and dye degradation. Sol Energy Mater Sol Cells 157:406–414. https://doi.org/10.1016/j.solmat.2016.07.001

Bai Y, Ye L, Chen T, Wang P, Wang L, Shi X, Keung Wong P (2017) Synthesis of hierarchical bismuth-rich Bi4O5BrxI2-xsolid solutions for enhanced photocatalytic activities of CO2conversion and Cr(VI) reduction under visible light. Appl Catal B 203:633–640. https://doi.org/10.1016/j.apcatb.2016.10.066

Bannister FA, Hey MH (1935) The crystal structure of the bismuth oxyhalides. Mineral Mag 24 (149):49–58. https://doi.org/10.1180/minmag.1935.024.149.01

Bard AJ, Fox MA (1995) Artificial photosynthesis: solar splitting of water to hydrogen and oxygen. Acc Chem Res 28(3):141–145. https://doi.org/10.1021/ar00051a007

Bastow TJ, Whitfield HJ (1977) Nuclear ouadrupole resonance of 209Bi in bismuth oxyhalides. J Magn Reson 468:461–468. https://doi.org/10.1016/0022-2364(77)90097-X

Bent SF (2008) Semiconductor surface chemistry. In: Nilsson A (ed) Chemical bonding at surfaces and interfaces, 1st edn. Elsevier, Amsterdam, pp 323–395

Blesa MA, Navntoft C, Dawidowski L (2005) Capitulo 7. Modelado de la radiación solar UV para aplicaciones en tratamiento de aguas, in Solar Safe Water: Tecnologías solares para el tratamiento y la descontaminación de las aguas, 1st edn. OCRE, La Plata, Buenos Aires, pp 99–118

Bletskan DI, Kopinets IF, Rubish ID, Turyanitsa II, Shtilika MV (1972) Photoconductivity and photoluminescence in BiOCl crystals. Russ Phys J 16(5):646–648. https://doi.org/10.1007/BF00898801

Bohórquez-Ballen J, Pérez JF (2007) Radiación ultravioleta. Dig CTSVO 9:97–104

Braslavsky SE, Braun AM, Cassano AE, Emeline AV, Litter MI, Palmisano L, Parmon VN, Serpone N (2011) Glossary of terms used in photocatalysis and radiation catalysis (IUPAC recommendations 2011). IUPAC 83(4):931–1014. https://doi.org/10.1351/gold-book

Catalá M, Dominguez-Morueco N, Migens A, Molina R, Martinez F, Valcarcel Y, Mastroianni N, Lopez de Alda M, Barcelo D, Segura Y (2015) Elimination of drugs of abuse and their toxicity from natural waters by photo-Fenton treatment. Sci Total Environ 520:198–205. https://doi.org/10.1016/j.scitotenv.2015.03.042

Chang F, Wang X, Luo J, Wang J, Xie Y, Deng B, Hu X (2017) Ag/Bi12O17Cl2composite: a case study of visible-light-driven plasmonic photocatalyst. J Mol Catal 427:45–53. https://doi.org/10.1016/j.molcata.2016.11.028

Chen D, Huan F, Cheng Y, Caruso R (2009) Mesoporous anatase TiO2 beads with high surface areas and controllable pore sizes: a superior candidate for high-performance dye-sensitized solar cells. Adv Mater 21:2206–2210. https://doi.org/10.1002/adma.200802603

Chen L, Huang R, Xiong M, Yuan Q, He J, Jia J, Yao M, Luo S, Au S, Yin S (2013) Room-temperature synthesis of flower-like BiOX (X=Cl, Br, I) hierarchical structures and their visible-light photocatalytic activity. Inorg Chem 52(19):11118–11125. https://doi.org/10.1021/ic401349j

Chen G, Chen G, Wang Y, Wang Q, Zhang Z (2015) Facile synthesis of N-doped BiOCl photocatalyst by an ethylenediamine-assisted hydrothermal method. Dig J Nanomater. https://doi.org/10.1155/2015/316057

Chen A, Bian Z, Xu J, Xin X, Wang H (2017a) Simultaneous removal of Cr(VI) and phenol contaminants using Z-scheme bismuth oxyiodide/reduced graphene oxide/bismuth sulfide system under visible-light irradiation. Chemosphere 188:659–666. https://doi.org/10.1016/j.chemosphere.2017.09.002

Chen H, Wang X, Bi W, Wu Y, Dong W (2017b) Photodegradation of carbamazepine with BiOCl/Fe3O4catalyst under simulated solar light irradiation. J Colloid Interface Sci 502:89–99. https://doi.org/10.1016/j.jcis.2017.04.031

Chen Z, Zeng J, Di J, Zhao D, Ji M, Xia J, Li H (2017c) Facile microwave-assisted ionic liquid synthesis of sphere-like BiOBr hollow and porous nanostructures with enhanced photocatalytic performance. GEE 2:124–133. https://doi.org/10.1016/j.gee.2017.01.005

Cheng H, Wang W, Huang B, Wang Z, Zhan J, Quin X, Zhang X, Dai Y (2013) Tailoring AgI nanoparticles for the assembly of AgI/BiOI hierarchical hybrids with size-dependent photocatalytic activities. J Mater Chem A 1(24):7131–7136. https://doi.org/10.1039/c3ta10849j

Cheng H, Huang B, Dai Y (2014) Engineering BiOX (X = Cl, Br, I) nanostructures for highly efficient photocatalytic applications. Nanoscale 4:2009. https://doi.org/10.1039/c3nr05529a

Cheremisinoff NP (ed) (2002) Handbook of water and wastewater treatment technologies, 1st edn. Elsevier, New Delhi

Choi YI, Kim Y, Cho DW, Kan J, Leung KT, Sohn Y (2015) Recyclable magnetic CoFe2O4/BiOX (X = Cl, Br and I) microflowers for photocatalytic treatment of water contaminated with methyl orange, rhodamine B, methylene blue, and a mixed dye. RSC Adv 5(97):79624–79634. https://doi.org/10.1039/c5ra17616f

Ciamician G (1912) The photochemistry of the future. Science 36(926):385–394. https://doi.org/10.1126/science.36.926.385

Ciesla P, Kocot P, Mytych P (2004) Homogeneous photocatalysis by transition metal complexes in the environment. J Mol Catal A Chem 224:17–33. https://doi.org/10.1016/j.molcata.2004.08.043

Cui P, Wang J, Wang Z, Chen J, Xing X, Wang L, Yu R (2016) Bismuth oxychloride hollow microspheres with high visible light photocatalytic activity. Nano Res 9(3):593–601. https://doi.org/10.1007/s12274-015-0939-z

Davididou K, Chatzisymeon E, Perez-Estrada L, Oller I, Malato S (2018) Photo-Fenton treatment of saccharin in a solar pilot compound parabolic collector: use of olive mill wastewater as iron chelating agent, preliminary results. Dig J Hazard Mater https://doi.org/10.1016/j.jhazmat.2018.03.016

Davis SE, Ide MS, Davis RJ (2013) Selective oxidation of alcohols and aldehydes over supported metal nanoparticles. Green Chem 15(1):17–45. https://doi.org/10.1039/c2gc36441g

Di J, Xia J, Ge Y, Xu L, He M, Zhang Q, Li H (2014) Reactable ionic liquid-assisted rapid synthesis of BiOI hollow microspheres at room temperature with enhanced photocatalytic activity. J Mater Chem A 2(38):15864–15874. https://doi.org/10.1039/c4ta02400a

Di J, Jiexiang X, Ji M, Wang B, Yin S, Zhang Q, Zhigang C, Li H (2016) Advanced photocatalytic performance of graphene-like BN modified BiOBr flower-like materials for the removal of pollutants and mechanism insight. Appl Catal B Environ 183:254–262. https://doi.org/10.1016/j.apcatb.2015.10.036

Di J, Jiexiang X, Li H, Guo S, Dai S (2017) Bismuth oxyhalide layered materials for energy and environmental applications. Nano Energy 41:172–192. https://doi.org/10.1016/j.nanoen.2017.09.008

Ding C, Ye L, Zhao Q, Zhong Z, Liu K, Xie H, Bao K, Zhang X, Huang Z (2016) Synthesis of BixOyIzfrom molecular precursor and selective photoreduction of CO2 into CO. J CO2 Util 14:135–142. https://doi.org/10.1016/j.jcou.2016.04.012

Ding C, Ma Z, Han C, Liu X, Jia Z, Xie H, Bao K (2017) Large-scale preparation of BiOX (X=Cl, Br) ultrathin nanosheets for efficient photocatalytic CO2 conversion. J Taiwan Inst Chem Eng 78:395–400. https://doi.org/10.1016/j.jtice.2017.06.044

Dong G, Ho W, Zhang L (2015) Photocatalytic NO removal on BiOI surface: the change from nonselective oxidation to selective oxidation. Appl Catal B Environ 168–169:490–496. https://doi.org/10.1016/j.apcatb.2015.01.014

Du H, Liu Y, Shen C, Xu A (2017) Nanoheterostructured photocatalysts for improving photocatalytic hydrogen production. Chin J Catal 38(8):1295–1306. https://doi.org/10.1016/S1872-2067(17)62866-3

Ekar SU, Shekhar G, Khollam YB, et al (2016) Green synthesis and dye-sensitized solar cell application of rutile and anatase TiO2 nanorods. J Solid State Electrochem 1–6. https://doi.org/10.1007/s10008-016-3376-3

Elmorsi TM, Riyad YM, Mahamed ZH, Abd El Barry HM (2010) Decolorization of Mordant red 73 azo dye in water using H2O2/UV and photo-Fenton treatment. J Hazard Mater 174:352–358. https://doi.org/10.1016/j.jhazmat.2009.09.057

Etacheri V, Di Valentin C, Schneider J, Bahnemann D, Pillai S (2015) Visible-light activation of TiO2 photocatalysts: advances in theory and experiments. J Photochem Photobiol C 25:1–29. https://doi.org/10.1016/j.jphotochemrev.2015.08.003

Ezzat MF, Dincer I (2018) Development and assessment of a new hybrid vehicle with ammonia and hydrogen. Appl Energy 219:226–239. https://doi.org/10.1016/j.apenergy.2018.03.012

Fan Z, Zhao Y, Zhai W, Qiu L, Li H, Hoffmann MR (2016) Facet-dependent performance of BiOBr for photocatalytic reduction of Cr(VI). RSC Adv 6(3):2028–2031. https://doi.org/10.1039/c5ra18768k

Fan W, Li C, Bai H, Zhao Y, Luo B, Li Y, Ge Y, Shi W, Li H (2017) An in situ photoelectroreduction approach to fabricate Bi/BiOCl heterostructure photocathodes: understanding the role of Bi metal for solar water splitting. J Mater Chem 5(10):4894–4903. https://doi.org/10.1039/C6TA11059B

Finnegan C, Ryan D, Enright A, Garcia-Cabellos G (2018) A review of strategies for the detection and remediation of organotin pollution. Crit Rev Environ Sci Technol 48(1):77–118. https://doi.org/10.1080/10643389.2018.1443669

Fu H, Pan CS, Zhang L, Zhu Y (2007) Synthesis, characterization and photocatalytic properties of nanosized Bi2WO6, PbWO4 and ZnWO4 catalysts. Mater Res Bull 42:696–706. https://doi.org/10.1016/j.materresbull.2006.07.017

Fujishima A, Rao TN, Tryk DA (2000) Titanium dioxide photocatalysis. J Photochem Photobiol C 1:1–21. https://doi.org/10.1016/S1389-5567(00)00002-2

Ganose AM, Cuff M, Butler KT, Walsh A, Scanlon DO (2016) Interplay of orbital and relativistic effects in bismuth oxyhalides: BiOF, BiOCl, BiOBr, and BiOI. Chem Mater 28(7):1980–1984. https://doi.org/10.1021/acs.chemmater.6b00349

Gao X, Zhang X, Peng S, Yue B, Fan C (2015) Photocatalytic degradation of carbamazepine using hierarchical BiOCl microspheres: some key operating parameters, degradation intermediates and reaction pathway. Chem Eng J 273:156–165. https://doi.org/10.1016/j.cej.2015.03.063

Gao X, Peng W, Tang G, Guo Q, Luo Y (2018) Highly efficient and visible-light-driven BiOCl for photocatalytic degradation of carbamazepine. J Alloys Compd 757:455–465. https://doi.org/10.1016/j.jallcom.2018.05.081

Geng Y, Li N, Ma J, Sun Z (2017) Preparation, characterization and photocatalytic properties of BiOBr/ZnO composites. J Energy Chem 26(3):416–421. https://doi.org/10.1016/j.jechem.2017.01.002

Gnanaprakasam A, Sivakumar VM, Thirumarimurugan M (2017) Facile one-step synthesis of Nd-doped BiOCl nanoparticles with the excellent photocatalytic behavior. Asia Pac J Chem Eng 12(5):723–731. https://doi.org/10.1002/apj.2112

Gratzel M (1981) Artificial photosynthesis: water cleavage into hydrogen and oxygen by visible light. Acc Chem Res 14(12):376–384. https://doi.org/10.1021/ar00072a003

Gupta A, Thakur IS (2017) Treatment of organic recalcitrant contaminants in wastewater waste-water. In: Farooq R, Ahmad Z (eds) Biological wastewater treatment and resource recovery, 1st edn. Intechopen, Rijeka, pp 3–16

Hajem B, Hamzaoui H, Adel M (2007) Chemical interaction between industrial acid effluents and the hydrous medium. Desalination 206-154:162. https://doi.org/10.1016/j.desal.2006.03.565

Han J, Zhu G, Hojamberdiev M, Peng J, Zhang X, Liu Y, Ge B, Liu P (2015) Rapid adsorption and photocatalytic activity for Rhodamine B and Cr(VI) by ultrathin BiOI nanosheets with highly exposed {001} facets. New J Chem 39(3):1874–1882. https://doi.org/10.1039/c4nj01765j

He C, Gu M (2006) Preparation, characterization and photocatalytic properties of Bi12SiO20 powders. Scr Mater 55:481–484. https://doi.org/10.1016/j.scriptamat.2006.04.039

He R, Cao S, Zhou P, Yu J (2014a) Recent advances in visible light Bi-based photocatalysts. Chin J Catal 35(7):989–1007. https://doi.org/10.1016/S1872-2067(14)60075-9

He X, Zhang GA, de la Cruz A, O'Shea K, Dionysiou D (2014b) Degradation mechanism of cyanobacterial toxin cylindrospermopsin by hydroxyl radicals in homogeneous UV/H2O2 process. Environ Sci Technol 48:4495–4504. https://doi.org/10.1021/es403732s

He M, Weibing L, Jiexiang X, Li X, Jun D, Hui X, Sheng Y, Huaming L, Mengua L (2015) The enhanced visible light photocatalytic activity of yttrium-doped BiOBr synthesized via a reactable ionic liquid. Appl Surf Sci 331:170–178. https://doi.org/10.1016/j.apsusc.2014.12.141

He R, Cao S, Yu J, Yang Y (2016) Microwave-assisted solvothermal synthesis of Bi4O5I2hierarchical architectures with high photocatalytic performance. Catal Today 264:221–228. https://doi.org/10.1016/j.cattod.2015.07.029

He R, Xu D, Cheng B, Yu J, Ho W (2018) Nanoscale horizons review on nanoscale Bi-based photocatalysts. Nanoscale Horiz https://doi.org/10.1039/c8nh00062j

Henríquez A, Mansilla H, Martinez-de la Cruz AM, Freer J, Contreras D (2017) Selective oxofunctionalization of cyclohexane over titanium dioxide–based and bismuth oxyhalide (BiOX, X=Cl−, Br−, I−) photocatalysts by visible light irradiation. Appl Catal B Environ 206:252–262. https://doi.org/10.1016/j.apcatb.2017.01.022

Hernández-Ramírez A, Medina-Ramírez I (eds) (2015) Photocatalytic semiconductors: synthesis, characterization, and environmental applications. Photocatalytic semiconductors: synthesis, characterization, and environmental applications. Springer, Cham

Hu J, Weng S, Zheng Z, Pei Z, Huang M, Liu P (2014) Solvents mediated-synthesis of BiOI photocatalysts with tunable morphologies and their visible-light driven photocatalytic perfor-mances in removing of arsenic from water. J Hazard Mater 264:293–302. https://doi.org/10.1016/j.jhazmat.2013.11.027

Huang HW, Xiao K, He Y, Zhang TR, Doug F, Du X, Zhang YH (2016) In situ assembly of BiOI@Bi12O17Cl2p-n junction: charge induced unique front-lateral surfaces coupling heterostructure with high exposure of BiOI {001} active facets for robust and nonselective photocatalysis. Appl Catal B Environ 199:75–86. https://doi.org/10.1016/j.apcatb.2016.06.020

Huang H, Xiao K, Zhang T, Dong F, Zhang Y (2017) Rational design on 3D hierarchical bismuth oxyiodides via in situ self-template phase transformation and phase-junction construction for optimizing photocatalysis against diverse contaminants. Appl Catal B Environ 203:879–888. https://doi.org/10.1016/j.apcatb.2016.10.082

Huizhong A, Yi D, Tianmin W, Cong W, Weichang H, Junying Z (2008) Photocatalytic properties of BiOX (X =Cl, Br, and I). Rare Metals 27(3):243–250. https://doi.org/10.1016/S1001-0521 (08)60123-0

Husaini SN, Zaidi JH, Malik F, Arif M (2008) Application of nuclear track membrane for the reduction of pollutants in the industrial effluent. Radiat Meas 43:S607–S611. https://doi.org/10.1016/j.radmeas.2008.03.070

Jamett FJ, Melendez MF, Mera AC, Valdes H (2017) BiOBr microspheres for photocatalytic degradation of an anionic dye. Solid State Sci 65. https://doi.org/10.1016/j.solidstatesciences.2017.01.001

Jiang J, Zhao K, Xiao X, Zhang L (2012) Synthesis and facet-dependent photoreactivity of BiOCl single- crystalline nanosheets. J Am Chem Soc 110:8–11. https://doi.org/10.1021/ja210484t

Jiang Z, Liu Y, Jing T, Huang B, Wang Z, Zhang X, Quin X, Dai Y (2015) One-pot solvothermal synthesis of S doped BiOCl for solar water oxidation. RSC Adv 5(58):47261–47264. https://doi.org/10.1039/c5ra07776a

Jin X, Ye L, Wang H, Su Y, Xie H, Zhong Z, Zhang H (2015) Bismuth-rich strategy induced photocatalytic molecular oxygen activation properties of bismuth oxyhalogen: the case of Bi24O31Cl10. Appl Catal B Environ 165:668–675. https://doi.org/10.1016/j.apcatb.2014.10.075

Jin X, Ye L, Xie H, Chen G (2017) Bismuth-rich bismuth oxyhalides for environmental and energy photocatalysis. Coord Chem Rev 349:84–101. https://doi.org/10.1016/j.ccr.2017.08.010

Jing L, Zhou W, Tian G, Fu H (2013) Surface tuning for oxide-based nanomaterials as efficient photocatalysts. Chem Soc Rev 42(24):9509. https://doi.org/10.1039/c3cs60176e

Kandemir T, Schuster M, Senyshyn A, Behrens M, Schlöl R (2013) The Haber-Bosch process revisited: on the real structure and stability of "ammonia iron" under working conditions. Angew Chem Int Ed 52(48):12723–12726. https://doi.org/10.1002/anie.201305812

Kayan B, Gözmen B, Demirel M, Gizir AM (2010) Degradation of acid red 97 dye in aqueous medium using wet oxidation and electro-Fenton techniques. J Hazard Mater 177:95–102. https://doi.org/10.1016/j.jhazmat.2009.11.076

Khan SU, Al-Shary M, Ingler WB (2002) Efficient photochemical water splitting by a chemically modified n-TiO2. Science 297:2243–2245. https://doi.org/10.1126/science.1075035

Khin MM, Nair AS, Babu VJ, Murugan R, Ramakrishna S (2012) A review on nanomaterials for environmental remediation. Energy Environ Sci 5(8):8075–8109. https://doi.org/10.1039/c2ee21818f

Kong XY, Cathie Lee WP, Ong WJ, Chai SP, Mohamed AR (2016) Oxygen-deficient BiOBr as a highly stable photocatalyst for efficient CO2 reduction into renewable carbon-neutral fuels. ChemCatChem 8(19):3074–3081. https://doi.org/10.1002/cctc.201600782

Kubacka A, Fernández-García M, Colón G (2012) Advanced nanoarchitectures for solar photocatalytic applications. Chem Rev 112(3):1555–1614. https://doi.org/10.1021/cr100454n

Kudo A, Omori K, Kato H (1999) A novel aqueous process for preparation of crystal form-controlled and highly crystalline BiVO4 powder from layered vanadates at room temperature and its photocatalytic and photophysical properties. J Am Chem Soc 14:11459–11467

Lee GJ, Zheng YC, Wu JJ (2018) Fabrication of hierarchical bismuth oxyhalides (BiOX, X = Cl, Br, I) materials and application of photocatalytic hydrogen production from water splitting. Catal Today 307:197–204. https://doi.org/10.1016/j.cattod.2017.04.044

Lei L, Dai Q, Zhou M, Zhang X (2007) Decolorization of cationic red X-GRL by wet air oxidation: performance optimization and degradation mechanism. Chemosphere 68(6):1135–1142. https://doi.org/10.1016/j.chemosphere.2007.01.075

Li Y, Gang C, Hongjie Z, Zonghua L (2009) Electronic structure and photocatalytic water splitting of lanthanum-doped Bi2AlNbO7. Mater Res Bull 44:741–746. https://doi.org/10.1016/j.materresbull.2008.09.020

Li R, Chen W, Ma C (2010) Platinum-nanoparticle-loaded bismuth oxide : an efficient plasmonic photocatalyst active under visible light. Green Chem 12:212–215. https://doi.org/10.1039/b917233e

Li G, Qin F, Yang H, Lu Z, Sun H, Chen R (2012) Facile microwave synthesis of 3D flowerlike BiOBr nanostructures and their excellent CrVI removal capacity. Eur J Inorg Chem 15:2508–2513. https://doi.org/10.1002/ejic.201101427

Li G, Qin F, Wang R, Xiao S, Sun H, Chen R (2013) BiOX (X=Cl, Br, I) nanostructures: mannitol-mediated microwave synthesis, visible light photocatalytic performance, and Cr(VI) removal capacity. J Colloid Interface Sci 409:43–51. https://doi.org/10.1016/j.jcis.2013.07.068

Li G, Jiang B, Xiao S, Lian Z, Zhang D, Yu JC, Li H (2014a) An efficient dye-sensitized BiOCl photocatalyst for air and water purification under visible light irradiation. Environ Sci Processes Impacts 16(8):1975–1980. https://doi.org/10.1039/C4EM00196F

Li K, An X, Park KH, Khraisheh M, Tang J (2014b) A critical review of CO2 photoconversion: catalysts and reactors. Catal Today 224:3–12. https://doi.org/10.1016/j.cattod.2013.12.006

Li KL, Lee WW, Lu CS, Dai YM, Chou SY, Chen HL, Lin HP, Chen CC (2014c) Synthesis of BiOBr, Bi3O4Br, and Bi12O17Br2 by controlled hydrothermal method and their photocatalytic properties. J Taiwan Inst Chem E 45(5):2688–2697. https://doi.org/10.1016/j.jtice.2014.04.001

Li H, Shang J, Ai Z, Zhang L (2015a) Efficient visible light nitrogen fixation with BiOBr nanosheets of oxygen vacancies on the exposed {001} facets. J Am Chem Soc 137 (19):6393–6399. https://doi.org/10.1021/jacs.5b03105

Li R, Gao X, Fan C, Zhang X, Wang Y, Wang Y (2015b) A facile approach for the tunable fabrication of BiOBr photocatalysts with high activity and stability. Appl Surf Sci 355:1075–1082. https://doi.org/10.1016/j.apsusc.2015.07.216

Li WT, Huang WZ, Zhou H, Yin H, Zheng YF, Song XC (2015c) Synthesis of Zn2+ doped BiOCl hierarchical nanostructures and their exceptional visible light photocatalytic properties. J Alloys Compd 638:148–154. https://doi.org/10.1016/j.jallcom.2015.03.103

Li H, Shang J, Shi J, Zhao K, Zhang L (2016a) Facet-dependent solar ammonia synthesis of BiOCl nanosheets via a proton-assisted electron transfer pathway. Nanoscale 8(4):1986–1993. https://doi.org/10.1039/c5nr07380d

Li J, Cai L, SHng J, Yu Y, Zhang L (2016b) Giant enhancement of internal electric field boosting bulk charge separation for photocatalysis. Adv Mater 28(21):4059–4064. https://doi.org/10.1002/adma.201600301

Li J, Zhan G, Yu Y, Zhang L (2016c) Superior visible light hydrogen evolution of Janus bilayer junctions via atomic-level charge flow steering. Nat Commun 7:1–9. https://doi.org/10.1038/ncomms11480

Li J, Sun S, Quian C, He L, Kenneth C, Zhang T, Chen Z, Ye M (2016d) The role of adsorption in photocatalytic degradation of ibuprofen under visible light irradiation by BiOBr microspheres. Chem Eng J 297:139–147. https://doi.org/10.1016/j.cej.2016.03.145

Li H, Yang Z, Zhang J, Huang Y, Ji H, Tong X (2017a) Indium doped BiOI nanosheets: preparation, characterization and photocatalytic degradation activity. Appl Surf Sci 423:1188–1197. https://doi.org/10.1016/j.apsusc.2017.06.301

Li J, Li H, Zhan G, Zhang L (2017b) Solar water splitting and nitrogen fixation with layered bismuth oxyhalides. Acc Chem Res 50(1):112–121. https://doi.org/10.1021/acs.accounts.6b00523

Li Q, Guan Z, Wu D, Zhao X, Bao S, Tian B, Zhang J (2017c) Z-scheme BiOCl-Au-CdS heterostructure with enhanced sunlight-driven photocatalytic activity in degrading water dyes and antibiotics. ACS Sustain Chem Eng 5(8):6958–6968. https://doi.org/10.1021/acssuschemeng.7b01157

Li B, Shao L, Wang R, Dong X, Zhao F, Gao P, Li Z (2018a) Interfacial synergism of Pd-decorated BiOCl ultrathin nanosheets for the selective oxidation of aromatic alcohols. J Mater Chem A 6 (15):6344–6355. https://doi.org/10.1039/c8ta00449h

Li H, Li J, Ai Z, Jia F, Zhang L (2018b) Oxygen vacancy-mediated photocatalysis of BiOCl: reactivity, selectivity, and perspectives. Angew Chem Int Ed 57(1):122–138. https://doi.org/10.1002/anie.201705628

Liang J, Liu F, Li M, Liu W, Tong M (2018) Facile synthesis of magnetic Fe3O4@BiOI@AgI for water decontamination with visible light irradiation: different mechanisms for different organic pollutants degradation and bacterial disinfection. Water Res 137:120–129. https://doi.org/10.1016/j.watres.2018.03.027

Liao J, Chen L, Sun M, Lei B, Zeng X, Sun Y, Dong F (2018) Improving visible-light-driven photocatalytic NO oxidation over BiOBr nanoplates through tunable oxygen vacancies. Chin J Catal 39(4):779–789. https://doi.org/10.1016/S1872-2067(18)63056-6

Liqun Ye (2018) BiOX (X = Cl, Br, and I) photocatalysts. In Cao W (edn) Semiconductor photocatalysis-meterials, mechanisms and applications. Intech open. https://doi.org/10.5772/62626

Liu QC, Ma DK, Hu YY, Zeng YW, Huang SM (2013a) Various bismuth oxyiodide hierarchical architectures: alcohothermal- controlled synthesis, photocatalytic activities, and adsorption capabilities for phosphate in water. ACS Appl Mater Interfaces 5(22):11927–11934. https://doi.org/10.1021/am4036702

Liu X, Zhang Q, Yip JN, Xiong Q, Sum TC (2013b) Wavelength tunable single nanowire lasers based on surface plasmon polariton enhanced Burstein-Moss effect. Nano Lett 13 (11):5336–5343. https://doi.org/10.1021/nl402836x

Liu Z, Wu B, Yulong Z, Niu J, Zhu Y (2014) Solvothermal synthesis and photocatalytic activity of Al-doped BiOBr microspheres. Ceram Int 40(4):5597–5603. https://doi.org/10.1016/j.ceramint.2013.10.152

Liu X, Su Y, Zhao Q, Du C, Liu Z (2016) Constructing Bi24O31Cl10/BiOCl heterojunction via a simple thermal annealing route for achieving enhanced photocatalytic activity and selectivity. Sci Rep 6:1–13. https://doi.org/10.1038/srep28689

Liu J, Yuan Q, Zhao H, Zou S (2018) Efficient room-temperature selective oxidation of benzyl alcohol into benzaldehyde over Pt/BiOCl nanocomposite. Catal Lett 148(4):1093–1099. https://doi.org/10.1007/s10562-018-2337-0

Low J (1979) Photoelectrochemistry and heterogeneous photo-catalysis at semiconductors. Small Methods 10:59–75. https://doi.org/10.1016/0047-2670(79)80037-4

Lu L, Zhou M, Yin L, Zhou G, Jian T, Wan X, Shi H (2016) Tuning the physicochemical property of BiOBr via pH adjustment: towards an efficient photocatalyst for degradation of bisphenol A. J Mol Catal A Chem 423:379–385. https://doi.org/10.1016/j.molcata.2016.07.017

Luan J, Zou Z, Lu M, Zheng S, Chen Y (2004) Growth, structural and photophysical properties of Bi2GaTaO7. J Cryst Growth 273:241–247. https://doi.org/10.1016/j.jcrysgro.2004.08.127

Luo SQ, Tang C, Huang Z, Liu C, Chen J, Fang M (2016) Effect of different Bi/Ti molar ratios on visible-light photocatalytic activity of BiOI/TiO2 heterostructured nanofibers. Ceram Int 42 (14):15780–15786. https://doi.org/10.1016/j.ceramint.2016.07.043

Lv Y, Yao W, Zong R, Zhu Y (2016) Fabrication of wide–range–visible photocatalyst Bi2WO (6−x) nanoplates via surface oxygen vacancies. Sci Rep 6:1–9. https://doi.org/10.1038/srep19347

Ma Z, Li P, Ye L, Zhou Y, Su F, Ding C, Xie H, Bai Y, Wong P (2017) Oxygen vacancies induced exciton dissociation of flexible BiOCl nanosheets for effective photocatalytic CO2 conversion. J Mater Chem A 5(47):24995–25004. https://doi.org/10.1039/c7ta08766g

Maeda K (2013) Z – scheme water splitting using two different semiconductor photocatalysts. ACS Catal 3:1486–1503. https://doi.org/10.1021/cs4002089

Maffei N, Pelletier L, McFarlan A (2008) A high performance direct ammonia fuel cell using a mixed ionic and electronic conducting anode. J Power Sources 175(1):221–225. https://doi.org/10.1016/j.jpowsour.2007.09.040

Mao C, Cheng H, Tian H, Li H, Xiao WJ, Xu H, Zhao J, Zhang L (2018) Visible light driven selective oxidation of amines to imines with BiOCl: does oxygen vacancy concentration matter? Appl Catal B 228:87–96. https://doi.org/10.1016/j.apcatb.2018.01.018

Matamoros V, Rodríguez Y, Bayona JM (2017) Mitigation of emerging contaminants by full-scale horizontal flow constructed wetlands fed with secondary treated wastewater. Ecol Eng 99:222–227. https://doi.org/10.1016/j.ecoleng.2016.11.054

McDonald KJ, Choi KS (2012) A new electrochemical synthesis route for a BiOI electrode and its conversion to a highly efficient porous BiVO4 photoanode for solar water oxidation. Energy Environ Sci 5(9):8553–8557. https://doi.org/10.1039/c2ee22608a

Meng X, Zhang Z (2016) Bismuth-based photocatalytic semiconductors: introduction, challenges and possible approaches. J Mol Catal A Chem 423:533–549. https://doi.org/10.1016/j.molcata.2016.07.030

Mera AC, Rodriguez CA, Melendez MF, Valdes H (2017) Synthesis and characterization of BiOI microspheres under standardized conditions. J Mater Sci 52(2):944–954. https://doi.org/10.1007/s10853-016-0390-x

Meribout R, Zuo Y, Khodja A, Piram A, Lebarillier S, Cheng J, Wang C, Wing-Wah-Chung P (2016) Photocatalytic degradation of antiepileptic drug carbamazepine with bismuth oxychlorides (BiOCl and BiOCl/AgCl composite) in water: efficiency evaluation and elucidation degradation pathways. J Photochem Photobiol A 328:105–113. https://doi.org/10.1016/j.jphotochem.2016.04.024

Miklos DB et al (2018) UV/H2O2 process stability and pilot-scale validation for trace organic chemical removal from wastewater treatment plant effluents. Water Res 136:169–179. https://doi.org/10.1016/j.watres.2018.02.044

Muktha B, Darriet J, Madras G, Guru Row TN (2006) Crystal structures and photocatalysis of the triclinic polymorphs of BiNbO4 and BiTaO4. J Solid State Chem 179:3919–3925. https://doi.org/10.1016/j.jssc.2006.08.032

Nakata K, Fujishima A (2012) TiO 2 photocatalysis: design and applications. J Photochem Photobiol C 13:169–189. https://doi.org/10.1016/j.jphotochemrev.2012.06.001

Natarajan K, Kalithasan N, Kureshy R, Bajaj H, Jo W, Tayade RJ (2015) Photocatalytic H2 production using semiconductor nanomaterials via water splitting – an overview. Adv Mater Res 1116:130–156. https://doi.org/10.4028/www.scientific.net/AMR.1116.130

Natarajan TS, Thampi KR, Tayade RJ (2018) Visible light driven redox-mediator-free dual semiconductor photocatalytic systems for pollutant degradation and the ambiguity in applying Z-scheme concept. Appl Catal B 227:296–311. https://doi.org/10.1016/j.apcatb.2018.01.015

Ning S, Lin H, Tong Y, Zhang X, Lin Q, Zhang Y, Long J, Wang X (2017) Dual couples Bi metal depositing and Ag@AgI islanding on BiOI 3D architectures for synergistic bactericidal mechanism of E. coli under visible light. Appl Catal B 204:1–10. https://doi.org/10.1016/j.apcatb.2016.11.006

Owlad M, Aroua MK, Wan Daud WA, Baroutian S (2009) Removal of hexavalent chromium-contaminated water and wastewater: a review. Water Air Soil Pollut 200:59–77. https://doi.org/10.1007/s11270-008-9893-7

Pan ML, Zhang H, Gao G, Liu L, Chen W (2015) Facet-dependent catalytic activity of nanosheet-assembled bismuth oxyiodide microspheres in degradation of bisphenol A. Environ Sci Technol 49(10):6240–6248. https://doi.org/10.1021/acs.est.5b00626

Pare B, Sarwan B, Jonnalagadda SB (2011) Photocatalytic mineralization study of malachite green on the surface of Mn-doped BiOCl activated by visible light under ambient condition. Appl Surf Sci 258(1):247–253. https://doi.org/10.1016/j.apsusc.2011.08.040

Park H, Vecitis CD, Hoffmann MR (2009) Electrochemical water splitting coupled with organic compound oxidation: the role of active chlorine species. J Phys Chem C 113(18):7935–7945. https://doi.org/10.1021/jp810331w

Parmon VN (1997) Photocatalysis as a phenomenon: aspects of terminology. Catal Today 39:137–144. https://doi.org/10.1016/S0920-5861(97)00095-3

Peng JH, Zhao YJ, Hassan QU, Li HY, Liu YB, Ma SH, Mao DL, Li HQ, Mng LC, Hojamberdiev M (2018) Rapid microwave-assisted solvothermal synthesis and visible-light-induced photocatalytic activity of Er3+–doped BiOI nanosheets. Adv Power Technol 29(5):1158–1166. https://doi.org/10.1016/j.apt.2018.02.007

Pfeiffer M, Leo K, Zhou X, Huang JS, Hofmann M, Werner A, Blocohwitz-Nimoth J (2003) Doped organic semiconductors: physics and application in light emitting diodes. Org Electron 4:89–103. https://doi.org/10.1016/j.orgel.2003.08.004

Pouran SR, Raman AA, Wan Daud WM (2015) Chemistry review on the main advances in photo-Fenton oxidation system for recalcitrant wastewaters. Ind Eng Chem Res 21:53–69. https://doi.org/10.1016/j.jiec.2014.05.005

Poznyak SK, Kulak AI (1990a) Electrochemical formation of bismuth oxyhalide films in neutral halide solutions. J Electroanal Chem 278:227–247. https://doi.org/10.1016/0022-0728(90)85136-S

Poznyak SK, Kulak AI (1990b) Photoelectrochemical properties oxyhalide films. Electrochim Acta 35(11):1941–1947. https://doi.org/10.1016/0013-4686(90)87103-9

Prasse C (2015) Spoilt for choice: a critical review on the chemical and biological assessment of current wastewater treatment technologies. Water Res 87:237–270. https://doi.org/10.1016/j.watres.2015.09.023

Priya N, Modak JM, Raichur AM (2009) LbL fabricated poly (Styrene Sulfonate)/TiO2 applications. ACS Appl Mater Interfaces 1(11):2–11. https://doi.org/10.1021/am900566n

Priya B, Raizada P, Singh N, Thakut P, Singh P (2016) Adsorptional photocatalytic mineralization of oxytetracycline and ampicillin antibiotics using Bi2O3/BiOCl supported on graphene sand composite and chitosan. J Colloid Interface Sci 479:271–283. https://doi.org/10.1016/j.jcis.2016.06.067

Qin XY, Cheng H, Wang W, Huang B, Zhang X, Dai Y (2013) Three dimensional BiOX (X=Cl, Br and I) hierarchical architectures: facile ionic liquid-assisted solvothermal synthesis and photocatalysis towards organic dye degradation. Mater Lett 100:285–288. https://doi.org/10.1016/j.matlet.2013.03.045

Rehman S, Ullah R, Butt AM, Gohar ND (2009) Strategies of making TiO2 and ZnO visible light active. J Hazard Mater 170:560–569. https://doi.org/10.1016/j.jhazmat.2009.05.064

Reiter AJ, Kong SC (2011) Combustion and emissions characteristics of compression-ignition engine using dual ammonia-diesel fuel. Fuel 90(1):87–97. https://doi.org/10.1016/j.fuel.2010.07.055

Rivera-Utrilla J, Sanchez-Polo M, Ferro-Garcia MA, Prados-Joya G, Ocampo-Perez R (2013) Pharmaceuticals as emerging contaminants and their removal from water. Rev Chemosphere 93(7):1268–1287. https://doi.org/10.1016/j.chemosphere.2013.07.059

Saravanan R, Gracia F, Stephen A (2017) Basic principles, mechanism, and challenges of photocatalysis. In: Mansoob M (ed) Nanocomposites of visible light-induced photocatalysis. Springer, Cham. https://doi.org/10.1007/978-3-319-62446-4

Schneider J, Matsuoka M, Takeuchi M, Zhang J, Horiuchi Y, Anpo M, Bahnemann DW (2014) Understanding TiO 2 Photocatalysis: mechanisms and materials. Chem Rev 114:9919–9986. https://doi.org/10.1021/cr5001892

Schubert EF (ed) (2005) Doping in III-V semiconductors. Cambridge University Press, Cambridge

Serpone N (2006) Is the band gap of pristine TiO2 narrowed by anion- and cation-doping of titanium dioxide in second-generation photocatalysts? J Phys Chem B 110(48):24287–24293. https://doi.org/10.1021/jp065659r

Shang J, Hao W, Lv X, Wang T, Wang X, Di Y, Dou S, Xie T, Wang D, Wang J (2014) Bismuth oxybromide with reasonable photocatalytic reduction activity under visible light. ACS Catal 4 (3):954–961. https://doi.org/10.1021/cs401025u

Sheldon RA (2016) Green chemistry and resource efficiency: towards a green economy. Green Chem 18:3180–3183. https://doi.org/10.1039/c6gc90040b

Shevlin S, Woodley S (2010) Electronic and optical properties of doped and undoped (TiO2) (n) nanoparticles. J Phys Chem C 114(41):17333–17343. https://doi.org/10.1021/jp104372j

Si HY, Mao CJ, Xie YA, Sun XG, Zhao JJ, Zhou N, Wang JQ, Feng WJ, Li YT (2017) P-N depleted bulk BiOBr/α-Fe2O3heterojunctions applied for unbiased solar water splitting. Dalton Trans 46(1):200–206. https://doi.org/10.1039/c6dt03683j

Škodič L, Vajnhandl S, Valh JV, Zeljko T, Voncina B, Lobnik A (2017) Comparative study of reactive dyes oxidation by H2O2/UV, H2O2/UV/Fe(II) and H2O2/UV/Fe° processes. Ozone Sci Eng 9512:14–23. https://doi.org/10.1080/01919512.2016.1229173

Su X, Wu D (2018) Facile construction of the phase junction of BiOBr and Bi4O5Br2nanoplates for ciprofloxacin photodegradation. Mater Sci Semicond Process 80:123–130. https://doi.org/10.1016/j.mssp.2018.02.034

Su Y, Zhang L, Wang W, Shao D (2018) Internal electric field assisted photocatalytic generation of hydrogen peroxide over BiOCl with HCOOH. ACS Sustain Chem Eng 6(7):8704–8710. https://doi.org/10.1021/acssuschemeng.8b01023

Sun L, Xiang L, Zhao X, Jia CJ, Yang J, Jin Z, Cheng X, Fan W (2015) Enhanced visible-light photocatalytic activity of BiOI/BiOCl heterojunctions: key role of crystal facet combination. ACS Catal 5(6):3540–3551. https://doi.org/10.1021/cs501631n

Tang J, Zou Z, Ye J (2004) Photocatalytic decomposition of organic contaminants by Bi2WO6 under visible light irradiation. Catal Lett 92:53–56. https://doi.org/10.1023/B:CATL.0000011086.20412.aa

Thomas JM, Ueda W, Williams J, Harris K (1989) New families of catalysts for the selective oxidation of methane. Faraday Discuss Chem Soc 87:33–45. https://doi.org/10.1039/DC9898700033

Tian H, Fan Y, Zhao Y, Liu L (2014) Elimination of ibuprofen and its relative photo-induced toxicity by mesoporous BiOBr under simulated solar light irradiation. RSC Adv 4(25):13061–13070. https://doi.org/10.1039/c3ra47304j

Tian F, Zhao H, Dai Z, Cheng G, Chen R (2016) Mediation of valence band maximum of BiOI by Cl incorporation for improved oxidation power in photocatalysis. Ind Eng Chem Res 55(17):4969–4978. https://doi.org/10.1021/acs.iecr.6b00847

Umebayashi T, Yamaki T, Itoh H, Asai K (2002) Analysis of electronic structures of 3d transition metal-doped TiO 2 based on band calculations. J Phys Chem Solids 63(10):1909–1920. https://doi.org/10.1016/S0022-3697(02)00177-4

Van Mieghem P (1992) Theory of bands tails in heavily doped semiconductors. Rev Mod Phys 64:755. https://doi.org/10.1103/RevModPhys.64.755

Walsh A, Da Silva JLF, Wei SH (2008) Origins of band-gap renormalization in degenerately doped semiconductors. Phys Rev B 78(7):1–5. https://doi.org/10.1103/PhysRevB.78.075211

Wang J, Zou Z, Ye J (2005) Surface modification and photocatalytic activity of distorted pyrochlore-type Bi2M(M=In, Ga and Fe)TaO7 photocatalysts. J Phys Chem Solids 66:349–355. https://doi.org/10.1016/j.jpcs.2004.07.014

Wang J, Yu Y, Zhang L (2013) Highly efficient photocatalytic removal of sodium pentachlorophenate with Bi3O4Br under visible light. Appl Catal B 136–137:112–121. https://doi.org/10.1016/j.apcatb.2013.02.009

Wang X, Liu W, Tian J, Zhao Z, Hao P, Kang X, Sang Y, Liu H (2014) Cr(vi), Pb(ii), Cd (ii) adsorption properties of nanostructured BiOBr microspheres and their application in a continuous filtering removal device for heavy metal ions. J Mat Chem A 2(8):2599–2608. https://doi.org/10.1039/c3ta14519k

Wang B, An W, Liu L, Chen W, Liang Y, Cui W (2015a) Novel Cu2S quantum dots coupled flower-like BiOBr for efficient photocatalytic hydrogen production under visible light. RSC Adv 5(5):3224–3231. https://doi.org/10.1039/c4ra12172d

Wang XJ, Yang WY, Li FT, Zhao J, Liu RH, Liu SJ, Li B (2015b) Construction of amorphous TiO/BiOBr heterojunctions via facets coupling for enhanced photocatalytic activity. J Hazard Mater 292:126–136. https://doi.org/10.1016/j.jhazmat.2015.03.030

Wang Q, Shi X, Liu E, Crittenden JC, Ma X, Zhang Y, Cong Y (2016) Facile synthesis of AgI/BiOI-Bi2O3multi-heterojunctions with high visible light activity for Cr(VI) reduction. J Hazard Mater 317:8–16. https://doi.org/10.1016/j.jhazmat.2016.05.044

Wang P, Lu Y, Wang X, Yu H (2017) Co-modification of amorphous-Ti(IV) hole cocatalyst and Ni (OH)2electron cocatalyst for enhanced photocatalytic H2-production performance of TiO2. Appl Surf Sci 391:259–266. https://doi.org/10.1016/j.apsusc.2016.06.108

Wang D, Hou P, Yang P, Cheng X (2018a) BiOBr@SiO2flower-like nanospheres chemically-bonded on cement-based materials for photocatalysis. Appl Surf Sci 430:539–548. https://doi.org/10.1016/j.apsusc.2017.07.202

Wang Y, Suzuki H, Xie J, Tomita O, Martin DJ, Higashi M, Kong D, Abe R, Tang J (2018b) Mimicking natural photosynthesis: solar to renewable H2 fuel synthesis by Z-Scheme water splitting systems. Chem Rev 118:5201–5241. https://doi.org/10.1021/acs.chemrev.7b00286

Wen XJ, Zhang C, Niu CG, Zhang L, Zeng GM, Zhang XG (2017) Highly enhanced visible light photocatalytic activity of CeO2 through fabricating a novel p–n junction BiOBr/CeO2. Catal Commun 90:51–55. https://doi.org/10.1016/j.catcom.2016.11.018

Williams J, Jones R, Thomas JM, Kent J (1989) A comparison of the catalytic performance of the layered oxychlorides of bismuth, lanthanum and samarium in the conversion of methane to ethylene. Catal Lett 3(3):247–255. https://doi.org/10.1007/BF00766400

Wu T, Liu G, Zhao J (1998) Photoassisted degradation of dye pollutants. V. Self-photosensitized oxidative transformation of rhodamine B under visible light irradiation in aqueous TiO2 dispersions. J Phys Chem B 102(30):5845–5851. https://doi.org/10.1021/jp980922c

Wu D, Wang B, Wang W, An T, Li G, Ng TW, Yip HY, Xiong C, Lee HK, Wong PK (2015) Visible-light-driven BiOBr nanosheets for highly facet-dependent photocatalytic inactivation of *Escherichia coli*. J Mater Chem A 3(29):15148–15155. https://doi.org/10.1039/c5ta02757h

Wu D, Yue S, Wang W, An T, Li G, Yip HY, Zhao H, Wong PK (2016) Boron doped BiOBr nanosheets with enhanced photocatalytic inactivation of *Escherichia coli*. Appl Catal B 192:35–45. https://doi.org/10.1016/j.apcatb.2016.03.046

Wu X, Ng YH, Wang L, Du Y, Dou SX, Amal R, Scott J (2017) Improving the photo-oxidative capability of BiOBr: via crystal facet engineering. J Mater Chem A 5(17):8117–8124. https://doi.org/10.1039/c6ta10964k

Wu G, Zhao Y, Li Y, Ma H, Zhao J (2018) pH-dependent synthesis of iodine-deficient bismuth oxyiodide microstructures: visible-light photocatalytic activity. J Colloid Interface Sci 510:228–236. https://doi.org/10.1016/j.jcis.2017.09.053

Xia J, Xu L, Zhang J, Yin S, Li H, Xu H, Di J (2013) Improved visible light photocatalytic properties of Fe/BiOCl microspheres synthesized via self-doped reactable ionic liquids. Cryst Eng Comm 15(46):10132–10141. https://doi.org/10.1039/c3ce41555d

Xia J, Ji M, Di J, Wang B, Yin B, Zhang Q, He M, Li H (2016) Construction of ultrathin C3N4/Bi4O5I2layered nanojunctions via ionic liquid with enhanced photocatalytic performance and mechanism insight. Appl Catal B 191:235–245. https://doi.org/10.1016/j.apcatb.2016.02.058

Xiangchao M, Zhang Z (2016) Bismuth-based photocatalytic semiconductors: introduction , challenges and possible approaches. J Mol Catal A 423:533–549. https://doi.org/10.1016/j.molcata.2016.07.030

Xiao X, Zhang W (2010) Facile synthesis of nanostructured BiOI microspheres with high visible light-induced photocatalytic activity. J Mater Chem 28:5866–5870. https://doi.org/10.1039/c0jm00333f

Xiao X, Jiang J, Zhang L (2013) Selective oxidation of benzyl alcohol into benzaldehyde over semiconductors under visible light: the case of Bi12O17Cl2nanobelts. Appl Catal B Environ 142–143:487–493. Elsevier B.V. https://doi.org/10.1016/j.apcatb.2013.05.047

Xiao X, Liu C, Zuo X, Liu J, Nan J (2016) Microwave synthesis of hierarchical BiOCl microspheres as a green adsorbent for the pH-dependent adsorption of methylene blue. J Nanosci Nanotechnol 16(12):12517–12525. https://doi.org/10.1166/jnn.2016.12969

Xiao X, Zheng C, Lu M, Zhang L, Liu F, Zuo X, Nan J (2018) Deficient Bi24O31Br10as a highly efficient photocatalyst for selective oxidation of benzyl alcohol into benzaldehyde under blue LED irradiation. Appl Catal B 228:142–151. https://doi.org/10.1016/j.apcatb.2018.01.076

Xie F, Mao X, Fan C, Wang C (2014) Facile preparation of Sn-doped BiOCl photocatalyst with enhanced photocatalytic activity for benzoic acid and rhodamine B degradation. Mater Sci Semicond Process 27(1):380–389. https://doi.org/10.1016/j.mssp.2014.07.020

Xie F, Zhang Y, He X, Li H, Qui X, Zhou W, Hou S, Tang Z (2018) First achieving highly selective oxidation of aliphatic alcohols to aldehydes over. J Mater Chem A. https://doi.org/10.1039/c8ta03680b

Xiong X, Ding L, Qang Q, Li Y, Jiang Q, Hu J (2016) Synthesis and photocatalytic activity of BiOBr nanosheets with tunable exposed (0 1 0) facets. Appl Catal B 188:283–291. https://doi.org/10.1016/j.apcatb.2016.02.018

Xu HY, Han X, Tan Q, Wu KJ, Qi SY (2017a) Crystal-chemistry insight into the photocatalytic activity of BiOCl x Br1−x nanoplate solid solutions. Front Mater Sci 11(2):120–129. https://doi.org/10.1007/s11706-017-0379-7

Xu H, Zhang W, Mingmei D, Xiaohong G (2017b) Microwave-assisted synthesis of flower-like BN/BiOCl composites for photocatalytic Cr(VI) reduction upon visible-light irradiation. Mat Des 114:129–138. https://doi.org/10.1016/j.matdes.2016.10.057

Xu MM, Zhao YL, Yan QS (2017c) Degradation of aniline by bismuth oxyiodide (BiOI) under visible light irradiation. JESAM 20(1):18–25

Yao WF, Xu XH, Zhou JT, Yang XN, Zhang Y, Shang SX, Wang H, Huang B (2004) Photocatalytic property of sillenite $Bi24AlO39$ crystals. J Mol Catal A 212(3):323–328. https://doi.org/10.1016/j.molcata.2003.11.012

Ye L, Liu X, Zhao Q, Xie H, Zan L (2013) Dramatic visible light photocatalytic activity of MnOx-BiOI heterogeneous photocatalysts and the selectivity of the cocatalyst. J Mat Chem A 1(31):8978–8983. https://doi.org/10.1039/c3ta11441d

Ye L, Su Y, Jin X, Xie H, Zhang C (2014) Recent advances in BiOX (X=Cl, Br and I) photocatalysts: synthesis, modification, facet effects and mechanisms. Envirom Sci Nano. (1):90–112. https://doi.org/10.1039/c3en00098b

Ye L, Jin X, Leng Y, Su Y, Xie H, Liu C (2015) Synthesis of black ultrathin BiOCl nanosheets for efficient photocatalytic H2 production under visible light irradiation. J Power Sources 293:409–415. https://doi.org/10.1016/j.jpowsour.2015.05.101

Ye L, Jin X, Ji X, Liu C, Su Y, Xie H, Liu C (2016a) Facet-dependent photocatalytic reduction of CO2 on BiOI nanosheets. J Chem Eng 291:39–46. https://doi.org/10.1016/j.cej.2016.01.032

Ye L, Jin X, Liu C, Ding C, Xie H, Chu KH, Wong PK (2016b) Thickness-ultrathin and bismuth-rich strategies for BiOBr to enhance photoreduction of CO2 into solar fuels. Appl Catal B 187:281–290. https://doi.org/10.1016/j.apcatb.2016.01.044

Yin B, Fang Z, Luo B, Zhang G, Shi W (2017) Facile preparation of $Bi24O31Cl10$ nanosheets for visible-light-driven photocatalytic degradation of tetracycline hydrochloride. Catal Lett 147(8):2167–2172. https://doi.org/10.1007/s10562-017-2115-4

Yoon TP, Ischay MA, Du J (2010) Visible light photocatalysis as a greener approach to photochemical synthesis. Nat Chem 2(7):527–532. https://doi.org/10.1038/nchem.687

Yoon HJ et al (2015) Graphene, charcoal, ZnO, and ZnS/BiOX (X = Cl, Br, and I) hybrid microspheres for photocatalytic simulated real mixed dye treatments. J Ind Eng Chem. The Korean Society of Industrial and Engineering Chemistry 32:137–152. https://doi.org/10.1016/j.jiec.2015.08.010

Yosefi L, Haghighi M (2018) Applied catalysis B : environmental fabrication of nanostructured flowerlike p-BiOI/p-NiO heterostructure and its efficient photocatalytic performance in water treatment under visible-light irradiation. Appl Catal B 220:367–378. https://doi.org/10.1016/j.apcatb.2017.08.028

Yu J, Wei B, Zhu L, Gao H, Sun W, Xu L (2013a) Flowerlike C-doped BiOCl nanostructures: facile wet chemical fabrication and enhanced UV photocatalytic properties. Appl Surf Sci 284:497–502. https://doi.org/10.1016/j.apsusc.2013.07.124

Yu Y, Chen G, Wang G, Lv Z (2013b) Visible-light-driven ZnIn2S4/CdIn2S4composite photocatalyst with enhanced performance for photocatalytic H2evolution. Int J Hydrog Energy 38(3):1278–1285. https://doi.org/10.1016/j.ijhydene.2012.11.020

Yu N, Chen Y, Zhang W, Wen M, Zhang L, Chen Z (2016) Preparation of Yb3+/Er3+-co-doped BiOCl sheets as efficient visible-light-driven photocatalysts. Mater Lett 179:154–157. https://doi.org/10.1016/j.matlet.2016.05.071

Yuan M, Tian F, Li G, Zhao H, Liu Y, Chen R (2017) Fe(III)-modified BiOBr hierarchitectures for improved photocatalytic benzyl alcohol oxidation and organic pollutants degradation. Ind Eng Chem Res 20:5935–5943. https://doi.org/10.1021/acs.iecr.7b00905

Zhang X, Zhang L (2010) Electronic and band structure tuning of ternary semiconductor photocatalysts by self doping: the case of BiOI. J Phys Chem C 114(42):18198–18206. https://doi.org/10.1021/jp105118m

Zhang L, Wang W, Yang J, Chen Z, Zhang W, Zhou L, Liu S (2006) Sonochemical synthesis of nanocrystallite Bi2O3 as a visible-light-driven photocatalyst. Appl Catal A 308:105–110. https://doi.org/10.1016/j.apcata.2006.04.016

Zhang H, Liu L, Zhou Z (2012) Towards better photocatalysts: first-principles studies of the alloying effects on the photocatalytic activities of bismuth oxyhalides under visible light. Phys Chem Chem Phys 14(3):1286–1292. https://doi.org/10.1039/c1cp23516h

Zhang L, Wang W, Sun S, Sun Y, Gao E, Zu J (2013a) Water splitting from dye wastewater: a case study of BiOCl/copper(II) phthalocyanine composite photocatalyst. Appl Catal B 132–133:315–320. https://doi.org/10.1016/j.apcatb.2012.12.003

Zhang W, Zhang Q, Dong F (2013b) Visible-light photocatalytic removal of NO in air over BiOX (X = Cl, Br, I) single-crystal nanoplates prepared at room temperature. Ind Eng Chem Res 52 (20):6740–6746. https://doi.org/10.1021/ie400615f

Zhang G, Su A, Qu J, Xu Y (2014) Synthesis of BiOI flowerlike hierarchical structures toward photocatalytic reduction of CO2to CH4. Mater Res Bull 55(3):43–47. https://doi.org/10.1016/j.materresbull.2014.04.012

Zhang L, Han Z, Wang W, Li X, Su Y, Jiang D, Lei X, Sun S (2015a) Solar-light-driven pure water splitting with ultrathin BiOCl nanosheets. Chem Eur J 21(50):18089–18094. https://doi.org/10.1002/chem.201503778

Zhang Y, Sun X, Yang G, Zhu Y, Si HY, Zhang J, Li Y (2015b) Preparation and characterization of bifunctional BiOClxIysolid solutions with excellent adsorption and photocatalytic abilities for removal of organic dyes. Mater Sci Semicond Process 41:193–199. https://doi.org/10.1016/j.mssp.2015.08.040

Zhang L, Ma Z, Xu H, Xie R, Zhong Y, Sui X, Wang B, Mao Z (2018a) Preparation of upconversion Yb3+doped microspherical BiOI with promoted photocatalytic performance. Solid State Sci 75:45–52. https://doi.org/10.1016/j.solidstatesciences.2017.11.008

Zhang Y, Lin C, Lin Q, Jin Y, Wang Y, Zhang Z, Lin H, Long J, Wang X (2018b) CuI-BiOI/Cu film for enhanced photo-induced charge separation and visible-light antibacterial activity. Appl Catal B 235:238–245. https://doi.org/10.1016/j.apcatb.2018.05.001

Zhao L, Liu Z, Zhang X, Cui T, Han J, Guo K, Wang B, Li Y, Hong T, Liu J, Liu Z (2014) Three-dimensional flower-like hybrid BiOI-zeolite composites with highly efficient adsorption and visible light photocatalytic activity. RSC Adv 4(85):45540–45547. https://doi.org/10.1039/c4ra07049f

Zhao M, Dong L, Zhang Q, Dong H (2016) Novel plate-stratiform nanostructured Bi12O17Cl2 with visible-light photocatalytic performance. Powder Diffract 31(01):2–7. https://doi.org/10.1017/S0885715615000901

Zheng C, He G, Xiao X, Lu M, Zhong H (2017) Selective photocatalytic oxidation of benzyl alcohol into benzaldehyde with high selectivity and conversion ratio over Bi4O5Br2 nanoflakes under blue LED irradiation. Appl Catal B 205:201–210. https://doi.org/10.1016/j.apcatb.2016.12.026

Zheng J, Chang F, Jiao M, Xu Q, Deng B, Hu X (2018) A visible-light-driven heterojuncted composite WO3/Bi12O17Cl2: synthesis, characterization, and improved photocatalytic performance. J Colloid Interface Sci 510:20–31. https://doi.org/10.1016/j.jcis.2017.07.119

Zhou L, Wang W, Xu H, Sun S, Shang M (2009) Bi2O3 hierarchical nanostructures: controllable synthesis, growth mechanism, and their application in photocatalysis. Chem Eur J 15 (7):1776–1782. https://doi.org/10.1002/chem.200801234

Zhou P, Yu J, Jaroniec M (2014) All-solid-state Z-scheme photocatalytic systems. Adv Mater 26 (29):4920–4935. https://doi.org/10.1002/adma.201400288

Zhou C, Yang H, Wang GG, Chen J, Wang R, Jian C (2017a) BiOI-promoted nano-on-micro BiOI-MoS2/CdS system for high-performance on photocatalytic H2evolution under visible light irradiation. Int J Hydrog Energy 2(47):28337–28348. https://doi.org/10.1016/j.ijhydene.2017. 09.098

Zhou R, Wu J, Zhang J, Tian H, Liang P, Zeng T, Lu P, Ren J, Huang T, Zhou X (2017b) Photocatalytic oxidation of gas-phase Hg0on the exposed reactive facets of BiOI/ BiOIO3heterostructures. Appl Catal B 204:465–474. https://doi.org/10.1016/j.apcatb.2016.11. 013

Zhu L, Che H, Huang Y, Chen Z, Xia D, Su M, Xiong Y, Li S, Shu D (2012) Enhanced photocatalytic disinfection of E. coli 8099 using Ag/BiOI composite under visible light irradiation. Sep Purif Technol 91:59–66. https://doi.org/10.1016/j.seppur.2011.10.026

Zou Z, Ye J, Arakawa H (2001a) Optical and structural properties of the BiTa(1-x)NbxO4 (0<x<1) compounds. Solid State Commun 119:471–475. https://doi.org/10.1016/S0038-1098(01) 00250-2

Zou Z, Ye J, Arakawa H (2001b) Photocatalytic and photophysical properties of a novel series of solid potocatalyst, Bi2MNbO7 (M=Al(III), Ga(III) and In(III)). Chem Phys Lett 333:57–62. https://doi.org/10.1016/S0009-2614(00)01348-8

Zou Z, Ye J, Arakawa H (2001c) Substitution effects of In(III) by Al(III) and Ga(III) on the photocatalytic and structural properties of the Bi2InNbO7. Photocatalyst Chem Mat 13 (5):1765–1769. https://doi.org/10.1021/cm000687m

Index

© Springer Nature Switzerland AG 2020
M. Naushad et al. (eds.), *Green Photocatalysts*, Environmental Chemistry
for a Sustainable World 34, https://doi.org/10.1007/978-3-030-15608-4